Methods in Enzymology

Volume 413
AMYLOID, PRIONS, AND OTHER PROTEIN AGGREGATES
PART C

METHODS IN ENZYMOLOGY

EDITORS-IN-CHIEF

John N. Abelson   Melvin I. Simon

DIVISION OF BIOLOGY
CALIFORNIA INSTITUTE OF TECHNOLOGY
PASADENA, CALIFORNIA

FOUNDING EDITORS

Sidney P. Colowick and Nathan O. Kaplan

Methods in Enzymology

Volume 413

Amyloid, Prions, and Other Protein Aggregates Part C

EDITED BY

Indu Kheterpal

PENNINGTON BIOMEDICAL
RESEARCH CENTER
LOUISIANA STATE UNIVERSITY SYSTEM
BATON ROUGE, LOUISIANA

Ronald Wetzel

UNIVERSITY OF TENNESSEE
GRADUATE SCHOOL OF MEDICINE
KNOXVILLE, TENNESSEE

AMSTERDAM • BOSTON • HEIDELBERG • LONDON
NEW YORK • OXFORD • PARIS • SAN DIEGO
SAN FRANCISCO • SINGAPORE • SYDNEY • TOKYO
Academic Press is an imprint of Elsevier

ELSEVIER

Academic Press is an imprint of Elsevier
525 B Street, Suite 1900, San Diego, California 92101-4495, USA
84 Theobald's Road, London WC1X 8RR, UK

For information on all Elsevier Academic Press publications
visit our Web site at www.books.elsevier.com

ISBN-13: 978-0-12-182818-9
ISBN-10: 0-12-182818-2

PRINTED IN THE UNITED STATES OF AMERICA
06 07 08 09 9 8 7 6 5 4 3 2 1

Table of Contents

Contributors to Volume 413

Article numbers are in parentheses following the names of contributors.
Affiliations listed are current.

TADATO BAN (5), *Osaka University, Institute for Protein Research, Suita, Osaka, Japan*

VALERIE BERTHELIER (16), *University of Tennessee Medical Center–Graduate School of Medicine, Knoxville, Tennessee*

ANUSRI M. BHATTACHARYYA (3), *University of Washington, Bellevue, Washington*

GAL BITAN (12), *UCLA, Department of Neurology, David Geffen School of Medicine, Los Angeles, California*

STEPHEN P. BOTTOMLEY (1), *Monash University, Department of Biochemistry and Molecular Biology, Clayton, Victoria, Australia*

JAMES R. BURKE (14), *Duke University Medical Center, Department of Medicine (Neurology) and Deane Laboratory, Durham, North Carolina*

LISA D. CABRITA (1), *Monash University, Department of Biochemistry and Molecular Biology, Clayton, Victoria, Australia*

JOHN F. CARPENTER (13), *University of Colorado HSC, Department of Pharmaceutical Sciences, Denver, Colorado*

SONGMING CHEN (3), *University of Tennessee Medical Center, Graduate School of Medicine, Knoxville, Tennesse*

FABRIZIO CHITI (4), *Università di Firenze, Dipartimento di Scienze Biochimiche, Firenze, Italy*

MICHELLE K. M. CHOW (1), *Monash University, Department of Biochemistry and Molecular Biology, Clayton, Victoria, Australia*

KELSEY D. COOK (8), *University of Tennessee, Department of Chemistry, Knoxville, Tennessee*

ANDREW M. ELLISDON (1), *Monash University, Department of Biochemistry and Molecular Biology, Clayton, Victoria, Australia*

CHARLES G. GLABE (17), *UC Irvine, Department of Molecular Biology and Biochemistry, Irvine, California*

DAVID J. GORDON (15), *The University of Chicago, Department of Pathology, Chicago, Illinois*

YUJI GOTO (5), *Osaka University, Institute for Protein Research, Suita, Osaka 565-0871, Japan*

GEOFFREY J. HOWLETT (11), *University of Melbourne, Biochemistry and Molecular Biology, Melbourne, Victoria, Australia*

RAKEZ KAYED (17), *UC Irvine, Department of Molecular Biology and Biochemistry, Irvine, California*

DANIEL J. KENAN (14), *Duke University Medical Center, Department of Medicine (Neurology) and Deane Laboratory, Durham, North Carolina*

INDU KHETERPAL (8), *Pennington Biomedical Research Center, Louisiana State University System, Baton Rouge, Louisiand*

YONG-SUNG KIM (13), *Ajou University, Department of Molecular Science and Technology, Yeongtong-gu, Suwon, Korea*

RALF LANGEN (7), *University of Southern California, Department of Biochemistry and Molecular Biology, Zilkha Neurogenetic Institute, Keck School of Medicine, Los Angeles, California*

GIORDANA MARCON (4), *Università di Firenze, Dipartimento di Scienze Biochimiche, Firenze, Italy*

MARTIN MARGITTAI (7), *University of Southern California, Department of Biochemistry and Molecular Biology, Los Angeles, California*

STEPHEN C. MEREDITH (15), *The University of Chicago, Department of Pathology, Chicago, Illinois*

YEE-FOONG MOK (11), *University of Melbourne, Biochemistry and Molecular Biology, Melbourne, Victoria, Australia*

ALEXIS NAZABAL (9), *Swiss Federal Institute of Technology, ETH, Department of Chemistry and Applied Biosciences, Zurich, Switzerland*

BRIAN O'NUALLAIN (3), *University of Tennessee Medical Center, Graduate School of Medicine, Knoxville, Tennessee*

GEORGIA PLAKOUTSI (4), *Università di Firenze, Dipartimento di Scienze Biochimiche, Firenze, Italy*

THEODORE W. RANDOLPH (13), *University of Colorado, Department of Chemical and Bilogical Engineering, Boulder, Colorado*

JEAN-MARIE SCHMITTER (9), *Swiss Federal Institute of Technology, ETH, Department of Chemistry and Applied Biosciences, Zurich, Switzerland*

KIMBERLY L. SCIARRETTA (15), *The University of Chicago, Department of Pathology, Chicago, Illinois*

MATTHEW B. SEEFELDT (13), *BaroFold Inc., Lafayette, Colorado*

SHANKARAMMA SHIVAPRASAD (10), *University of Tennessee Medical Center, Graduate School of Medicine, Knoxville, Tennessee*

WARREN J. STRITTMATTER (14), *Duke University Medical Center, Department of Medicine (Neurology) and Deane Laboratory, Durham, North Carolina*

DAVID B. TEPLOW (2), *David Geffen School of Medicine at UCLA, Department of Neurology, Los Angeles, California*

ASHWANI K. THAKUR (3), *University of Tennessee Medical Center, Graduate School of Medicine, Knoxville, Tennessee*

GEETHA THIAGARAJAN (3), *University of Tennessee Medical Center, Graduate School of Medicine, Knoxville, Tennessee*

ROBERT TYCKO (6), *National Institutes of Health, Laboratory of Chemical Physics NIDDK, Bethesda, Maryland*

RONALD WETZEL (3, 8, 10, 16), *University of Tennessee Medical Center, Graduate School of Medicine, Knoxville, Tennessee*

ANGELA D. WILLIAMS (3), *University of Tennessee Medical Center, Graduate School of Medicine, Knoxville, Tennessee*

Preface

After one of us suffered through a series of unsuccessful attempts to obtain funding for protein aggregation projects in the early 1990s, a biophysicist colleague pointed out that a major problem with such applications was that there simply weren't many good tools available for obtaining useful information about protein aggregate structures and how they are formed. While this feedback was hard to hear, accept, and constructively respond to, it was largely accurate. The state-of-the-art methods commonly available and in use for following aggregation kinetics in those years were turbidity assays, or perhaps, with luck, ThT fluorescence. What was considered to be high-end structural information on aggregates, meanwhile, came from electron microscopy, FTIR, and X-ray fiber diffraction. The biophysicist colleague was too kind to point out another major problem with these applications: very few scientists really cared about protein aggregation.

Times have changed, and by 2006 misfolding and aggregation studies have moved to center stage in the protein science and biomedical funding theatres. Protein aggregates are found in many of the common neurodegenerative diseases in addition to a well-established group of peripheral amyloidoses and other deposition diseases. At the same time, some protein aggregates are known to play useful roles in a variety of life forms, possibly even in humans. The fossilized footprints of protein aggregation in molecular and cellular evolution are revealed by the existence of a growing list of cellular systems involved in preventing, managing, and/or limiting the consequences of aggregation. It is now recognized that disordered polypeptides, whether exiting the ribosome or a membrane channel or emerging from denaturant into native buffer, often undergo kinetic partitioning between a native state maturation pathway and one or more alternative aggregation pathways. Protein aggregation also plays important roles in biotechnology, and our accumulated experience in working with amyloid fibrils and other ordered aggregates may yet prepare the way for exploiting these structures as nanomaterials.

Not so very long ago, the conventional wisdom in the protein folding community was that all proteins know how to fold properly, and if one has the misfortune to encounter aggregation in a folding experiment, it is not because there is anything interesting or fundamental going on in the experiment, it is only because one don't know the "proper conditions" for avoiding aggregation. In the past decade, as aggregation studies have become not only acceptable but almost glamorous, this wisdom has been turned on its head, such that now if one's protein can't be induced to form amyloid fibrils, it must be because one hasn't found the correct protocol. Aggregation is now thought of as a central feature of the polypeptide molecules that do Nature's heavy lifting, and, as such, protein aggregation has a central place in the biophysics of life processes. As a major constraint in the "other genetic code" that controls how and whether the linear sequence of amino acids encoded in a stretch of DNA expresses its intrinsic functionality, aggregation is also important as a mode by which biological information transfer can ultimately be corrupted, and as such plays a continuing role in molecular evolution.

Not only has the funding environment improved over the past 15 years, so have the tools. Thanks to the foresight, commitment and ingenuity of many research groups working on amyloid and other protein aggregation, the quantity, quality and breadth of the technology available to study protein aggregation has greatly expanded. This volume (Part C, Volume 413) and its companion volume (Part B, Volume 412), which along with Volume 309 constitute what amounts to a *Methods in Enzymology* amyloid trilogy, celebrate this achievement of the field by highlighting a variety of new methods now being applied to protein aggregation and aggregates in a variety of settings and from a variety of points of view. Part C focuses on methods for characterizing the structures and mechanisms of assembly of misfolded aggregates *in vitro*.

Amyloidogenic globular proteins present a number of challenges. On the one hand, some of these proteins are so close to the brink of amyloid formation that it can be very difficult to prepare these molecules in a non-aggregated state so that the native state as well as its aggregation can be studied. In Chapter 1, Bottomley and co-workers provide a description of how this is done with a member of the particularly difficult polyglutamine proteins. On the other hand, amyloidogenic globular proteins also present the challenge of how to sort out the energetics of unfolding of a globular protein from its off-pathway assembly into amyloid. In Chapter 4, Chiti and colleagues describe methods for meeting this challenge in order to better focus on the amyloid assembly process. Relatively unstructured amyloidogenic peptides present their own challenges, and Chapters 2 and 3 describe different approaches for generating aggregate-free solutions of monomeric peptides for structural and assembly studies. Chapter 3 also discusses the use of a sedimentation assay for quantifying amyloid formation kinetics and critical concentrations. Ban and Goto in Chapter 5 describe an elegant and exciting method for directly following the growth of a single amyloid fibril by fluorescence microscopy.

Chapters 6–10 describe methods for analyzing the structures of amyloid fibrils and other aggregates. Chapter 6 deals with solid state NMR, Chapter 7 with spin labeling, and Chapters 8 and 9 with hydrogen deuterium exchange methods. Chapter 10 describes several ways of exploiting cysteine mutants of amyloidogenic proteins to dissect amyloid structure and structural dynamics.

Chapters 11 and 12 discuss methods for approaching the oligomeric structure of aggregates, using sedimentation velocity analysis (Chapter 11) and photo-crosslinking (Chapter 12). Chapter 11 also includes a useful discussion of the requirements for centrifugal sedimentation of different molecular weights of particles. In Chapter 13, Carpenter and colleagues describe the response of protein aggregates to high pressures.

Chapters 14–16 discuss methods for discovering compounds capable of modulating amyloid fibril growth. In Chapter 14, Burke and colleagues describe a phage display approach to identifying amino acid sequences capable of binding to protein aggregates, some of which prove to be effective inhibitors. In Chapter 15, Meredith and colleagues discuss the strategies for designing peptide–based inhibitors of amyloid formation. Chapter 16 describes a microtiter plate–based screening assay for discovering small molecule modulators of amyloid elongation.

Finally, Chapter 17 describes the creation and use of antibodies specific for oligomeric assemblies of amyloidogenic peptides. The antibodies are useful diagnostic agents and may also have therapeutic potential.

For the editors, preparing these volumes was a labor of love. The labor, along with a love of the subject matter, was shared by many colleagues who participated in a variety of ways. We are grateful to all of the authors, who responded to our urgings to make their methods both available and accessible, and to the anonymous reviewers, who helped the authors to meet that challenge by reading these chapters critically and carefully and providing frank feedback. We are also especially grateful for the administrative and organizational skills of Pam Trentham at the University of Tennessee and to our editor at Elsevier, Cindy Minor, who both did their best to keep us on track. We gladly acknowledge our sources of support that helped make these volumes possible: the National Institutes of Health grants (RW), the National Science Foundation (IK) and the Louisiana Board of Regents (IK).

INDU KHETERPAL
RONALD WETZEL

METHODS IN ENZYMOLOGY

[1] Purification of Polyglutamine Proteins[1]

By MICHELLE K. M. CHOW, ANDREW M. ELLISDON,
LISA D. CABRITA, and STEPHEN P. BOTTOMLEY

Abstract

The misfolding and formation of fibrillar-like aggregates by polyglutamine proteins is believed to be a key factor in the development of the neurodegenerative polyglutamine diseases; however, relatively little is known about structural and conformational aspects of polyglutamine-induced misfolding and aggregation. This is largely attributable to the fact that polyglutamine proteins have proved difficult to purify in quantities suitable for biochemical and biophysical analyses, thus limiting the extent to which the proteins can be conformationally characterized. Recent advances, however, have seen the development of a number of protocols enabling the expression and purification of these proteins in more significant quantities. In this report, we describe a purification protocol for ataxin-3, which, in its polyglutamine-expanded form, causes Machado-Joseph disease. Purification of different length ataxin-3 variants, including one of pathological length, is facilitated by an N-terminal hexa-histidine tag, which enables binding to a nickel-chelated agarose resin. A key issue that arose during purification was the undesirable proteolysis of ataxin-3 by a trace contaminant protease. We solved this problem by the addition of a benzamidine-binding step during purification, which greatly reduced the level of proteases present. We found that the inclusion of this step had a significant positive impact on the quality of the purified protein product. We also inactivated trace amounts of proteases during experiments by the addition of specific protease inhibitors. Finally, we also describe initial structural and functional analyses that confirm the integrity of the purified protein.

Introduction

Polyglutamine diseases have attracted growing interest in recent years as protein misfolding disorders. Central to the pathogenesis of these diseases is the expansion of a polymorphic length glutamine tract in the disease-causing protein (Margolis and Ross, 2001; Trottier *et al.*, 1995).

---

[1] Michelle K. M. Chow and Andrew M. Ellisdon contributed equally to this work.

METHODS IN ENZYMOLOGY, VOL. 413   0076-6879/06 $35.00
          DOI: 10.1016/S0076-6879(06)13001-3

The formation of neuronal nuclear inclusions (NIs) containing polygluta-mine-expanded protein in disease states (Davies *et al.*, 1998; DiFiglia *et al.*, 1997; Malinchik *et al.*, 1998; Paulson *et al.*, 1997) and the ability of polyglutamine-containing peptides and proteins to form fibrillar aggregates both *in vivo* and *in vitro* (Chen *et al.*, 2002a; Chow *et al.*, 2004a; Diaz-Hernandez *et al.*, 2004; DiFiglia *et al.*, 1997; Huang *et al.*, 1998; Scherzinger *et al.*, 1997; Tanaka *et al.*, 2001) strongly support a role for protein misfold-ing in the pathogenesis of polyglutamine disorders. Although many general principles of protein misfolding and fibrillogenesis have been delineated for various proteins (Chow *et al.*, 2004b; Ellisdon and Bottomley, 2004; Horwich, 2002; Stefani and Dobson, 2003), the exact mechanisms of such events as pertains to polyglutamine proteins remain largely unclear.

Until relatively recently, *in vitro* analysis of polyglutamine proteins has been relatively scarce, with the bulk of the research being performed at a cellular level. Over the past few years, however, there has been increasing interest in the conformational behavior and characteristics of these pro-teins, and a number of studies have suggested various different mechanisms by which polyglutamine-induced fibrillogenesis may occur (Bennett *et al.*, 2002; Chen *et al.*, 2002b; Chow *et al.*, 2004d; Poirier *et al.*, 2002; Thakur and Wetzel, 2002). Although numerous model proteins and peptides have been analyzed (Chen *et al.*, 2002a,b; Popiel *et al.*, 2004; Tanaka *et al.*, 2003, 2001; Thakur and Wetzel, 2002), relatively few of the naturally occurring disease-associated proteins have been expressed and purified *in vitro*. A major reason for the lack of data specifically addressing disease-associated polyglutamine proteins is that the purification of recombinant proteins, especially those of pathological length, has often proved to be technically challenging. Table I displays some examples of polyglutamine proteins that have been purified in various contexts. As with many proteins, purification has often been aided by the co-expression of fusion tags, which facilitate solubilization and affinity purification. It is interesting to note that only two of the nine identified polyglutamine proteins (huntingtin and ataxin-3) have been successfully purified with polyglutamine lengths in the patholog-ical range, and even then, huntingtin has only been purified when attached to a solubilizing fusion partner.

In our laboratory, we have focused our attention on ataxin-3, which, in its expanded form (>45 glutamine repeats [Padiath *et al.*, 2005]), causes spinocerebellar ataxia type 3, also known as Machado-Joseph disease (Kawaguchi *et al.*, 1994; Paulson *et al.*, 1997). Our interest lies in under-standing the conformational characteristics of the protein and how this relates to misfolding and aggregation. In order to understand these pro-cesses further, purified protein is required for biophysical and biochemical analyses. In this chapter, we describe a protocol that we have developed for the purification of ataxin-3 expressed in *Escherichia coli* cells. This protocol is applicable to proteins of both pathological and nonpathological length, and the purified protein is free of any fusion protein partners, although it contains an N-terminal hexa-histidine (His_6) tag. We also describe a method for genetically manipulating the polyglutamine tract to produce variants of different lengths and consider alternative methods of purifying ataxin-3 as reported by other researchers, often involving the expression of the protein with fusion partner proteins. Finally, we describe the biophysical and biochemical characterization of the native protein, which enables verification of the integrity of the purified products.

Ataxin-3 Constructs Used

The primary structures of the ataxin-3 constructs purified are depicted in Fig. 1. The gene used codes for the more common ataxin-3 isoform, containing two ubiquitin-interacting motifs (UIMs) (Kawaguchi *et al.*, 1994). We used two different constructs, both incorporating an N-terminal His_6 tag, to facilitate purification. The first construct contains a four-residue linker between the His_6 tag and the start of the ataxin-3 sequence (pQE-His_6-Atax3). We later found the solubility and purification yields of protein were increased using a T7 promotor system based on a modification of the pET21b vector (Novagen); therefore, we subcloned the ataxin-3 gene into this vector to create a different N-terminally His_6-tagged protein (pET-His_6-Atax3), which could be cleaved by the tobacco etch virus (TEV) protease to remove the His_6 tag if required.

Manipulation of the Polyglutamine Tract Length Using Cassette Mutagenesis

In order to study the role of polyglutamine length on ataxin-3, variants containing different length tracts were purified and analyzed. In addition to atax3(Q28) and atax3(Q50), which contain 28 and 50 glutamine residues, respectively, and were cloned from cDNAs into *E. coli* expression vectors (Chow *et al.*, 2004a,d), a third shorter variant (Q15) was produced by manipulation of the number of glutamine repeats. Using mutagenesis techniques to incorporate additional glutamine residues is often difficult because of the repeating nature of the CAG triplets in the sequence. We used a cassette mutagenesis approach; however, great care had to be taken to avoid introducing additional amino acids when creating appropriate restriction sites. Although such sites may be removed with multiple rounds of mutagenesis, this can be a time-consuming process. With this issue in mind, we successfully designed a construct with restrictions sites that allow

TABLE I

EXAMPLES OF HUMAN POLYGLUTAMINE PROTEINS THAT HAVE BEEN RECOMBINANTLY EXPRESSED AND PURIFIED

Protein	Domain/construct	Purification tags (polyQ length)	Expression system	References
Ataxin-1	AXH domain	His$_6$-tagged	E. coli	Chen et al., 2004[a]
		His$_6$-GST fusion (cleaved)	E. coli	de Chiara et al., 2003[b]
Ataxin-3	Josephin domain	GST-fusion (cleaved)	E. coli	Masino et al., 2004
		His$_6$-tagged	E. coli	Chow et al., 2004c
	Whole protein	MBP-fusion (intact) Q27, Q78	E. coli	Bevivino and Loll, 2001 Burnett et al., 2003
		GST-fusion (intact) QHQ, Q28, Q50, Q64, Q84	E. coli	Chai et al., 2004
		GST-fusion (cleaved)		
		Q18	E. coli	Masino et al., 2003, 2004.
		Q26, Q36	E. coli	Marchal et al., 2003 Shehi et al., 2003
		His$_6$-tagged		
		Q15, Q28, Q50	E. coli	Chow et al., 2004a,d
		Q22	E. coli	Chai et al., 2004
		Q26	E. coli	Masino et al., 2003
Huntingtin	Exon-1	GST fusion (intact)	E. coli	Scherzinger et al., 1997, 1999[c]
		Q20–Q51		Wanker et al., 1999[d]
		MBP fusion (intact) Q16, Q44	E. coli	Poirier et al., 2002
TBP	Core domain	His$_6$-tagged	E. coli	Coleman et al., 1995[e] Jackson-Fisher et al., 1999[f]
		GST-fusion (cleaved)	E. coli	Zhao and Herr, 2002[g]
	Whole protein	No tag Q38	E. coli	Peterson et al., 1990[h] Kao et al., 1990[i] Thompson et al., 2004[j] Pugh, 1995[k]
		His$_6$-tagged Q38	E. coli	Jackson-Fisher et al., 1999
		GST-fusion (cleaved) Q38	E. coli	Zhao and Herr, 2002
Androgen receptor	Whole protein	GST fusion (cleaved) Q23	Insect cells	Roehrborn et al., 1992[l]
		His$_6$-tagged polyQ not stated	Insect cells	Zhu et al., 2001[m] Liao et al., 1999[n] Liao and Wilson, 2001[o]

[a] Chen, Y. W., Allen, M. D., Veprintsev, D. B., Lowe, J., and Bycroft, M. (2004). The structure of the AXH domain of spinocerebellar ataxin-1. *J. Biol. Chem.* **279**, 3758–3765.

[b] de Chiara, C., Giannini, C., Adinolfi, S., de Boer, J., Guida, S., Ramos, A., Jodice, C., Kioussis, D., and Pastore, A. (2003). The AXH module: An independently folded domain common to ataxin-1 and HBP1. *FEBS Lett.* **551**, 107–112.

[c] Scherzinger, E., Sittler, A., Schweiger, K., Heiser, V., Lurz, R., Hasenbank, R., Bates, G. P., Lehrach, H., and Wanker, E. E. (1999). Self-assembly of polyglutamine-containing huntingtin fragments into amyloid-like fibrils: Implications for Huntington's disease pathology. *Proc. Natl. Acad. Sci. USA* **96**, 4604–4609.

for the manipulation of polyglutamine length without the incorporation of additional amino acids, as described below.

Of the commercially available restriction enzymes, it was found that EcoN1 was a suitable candidate; thus, this restriction site was introduced at either side of the polyglutamine tract using atax-3(Q50)/pQE30 as the template (Fig. 2A). The EcoN1 sites were introduced using the QuikChange method (Stratagene) to create atax3(Q50)/pQE30(EcoN1) (Fig. 2B, C). Introduction of the EcoN1 sites initially involved the substitution of existing amino acids (Fig. 2C); however, on successful ligation of the cassette into the vector, these residues were replaced and the sites were removed (Fig. 2D). Cassettes containing varying glutamine repeats were created using pairs of

[d] Wanker, E. E., Scherzinger, E., Heiser, V., Sittler, A., Eickhoff, H., and Lehrach, H. (1999). Membrane filter assay for detection of amyloid-like polyglutamine-containing protein aggregates. *Methods Enzymol.* **309,** 375–386.

[e] Coleman, R. A., Taggart, A. K., Benjamin, L. R., and Pugh, B. F. (1995). Dimerization of the TATA binding protein. *J. Biol. Chem.* **270,** 13842–13849.

[f] Jackson-Fisher, A. J., Chitikila, C., Mitra, M., and Pugh, B. F. (1999). A role for TBP dimerization in preventing unregulated gene expression. *Mol. Cell* **3,** 717–727.

[g] Zhao, X., and Herr, W. (2002). A regulated two-step mechanism of TBP binding to DNA: A solvent-exposed surface of TBP inhibits TATA box recognition. *Cell* **108,** 615–627.

[h] Peterson, M. G., Tanese, N., Pugh, B. F., and Tjian, R. (1990). Functional domains and upstream activation properties of cloned human TATA binding protein. *Science* **248,** 1625–1630.

[i] Kao, C. C., Lieberman, P. M., Schmidt, M. C., Zhou, Q., Pei, R., and Berk, A. J. (1990). Cloning of a transcriptionally active human TATA binding factor. *Science* **248,** 1646–1650.

[j] Thompson, N. E., Foley, K. M., and Burgess, R. R. (2004). Antigen-binding properties of monoclonal antibodies reactive with human TATA-binding protein and use in immunoaffinity chromatography. *Protein Expr. Purif.* **36,** 186–197.

[k] Pugh, B. F. (1995). Purification of the human TATA-binding protein, TBP. *Methods Mol. Biol.* **37,** 359–367.

[l] Roehrborn, C. G., Zoppi, S., Gruber, J. A., Wilson, C. M., and McPhaul, M. J. (1992). Expression and characterization of full-length and partial human androgen receptor fusion proteins. Implications for the production and applications of soluble steroid receptors in Escherichia coli. *Mol. Cell. Endocrinol.* **84,** 1–14.

[m] Zhu, Z., Bulgakov, O. V., Scott, S. S., and Dalton, J. T. (2001). Recombinant expression and purification of human androgen receptor in a baculovirus system. *Biochem. Biophys. Res. Commun.* **284,** 828–835.

[n] Liao, M., Zhou, Z., and Wilson, E. M. (1999). Redox-dependent DNA binding of the purified androgen receptor: Evidence for disulfide-linked androgen receptor dimers. *Biochemistry* **38,** 9718–9727.

[o] Liao, M., and Wilson, E. M. (2001). Production and purification of histidine-tagged dihydrotestosterone-bound full-length human androgen receptor. *Methods Mol. Biol.* **176,** 67–79.

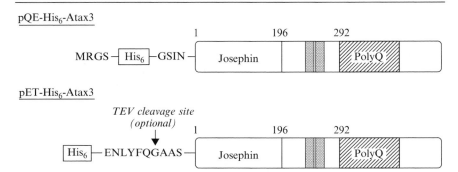

FIG. 1. Schematic representation of ataxin-3 constructs. Purification is facilitated by the presence of the hexa-histidine (His$_6$) tag. In the second construct, the His$_6$-glutathione S-transferase fusion partner is removed by tobacco etch virus (TEV) cleavage.

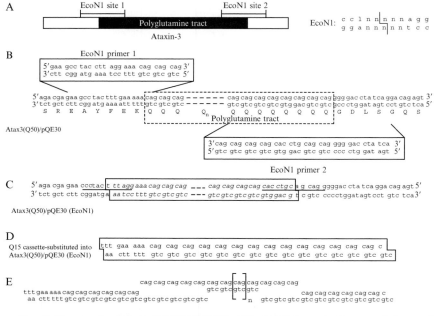

FIG. 2. Construction of atax(Q50)/pQE30(EcoN1). (A) Schematic diagram of the atax3 (Q50)/pQE30(EcoN1) construct. The restriction enzyme recognition site for EcoN1 is shown at the right. (B) Regions where the EcoN1 sites were engineered around the polyglutamine tract and the corresponding oligonucleotides were used for mutagenesis (EcoN1 primers 1 and 2). The sites of coding substitutions are highlighted in bold. (C) Polyglutamine region of the atax3(Q50)/pQE30(EcoN1) construct with the EcoN1 sites underlined and the introduced substitutions shown in bold. (D) Example of a cassette with 15 glutamine repeats that was introduced into the ataxin-3(Q50)/pQE30 (EcoN1) construct. (E) Multiple polyglutamine cassettes were used to extend the polyglutamine tract further.

annealed oligonucleotides that were ligated into the vector using standard molecular biology techniques (Fig. 2D). The cassettes were limited to a single pair of oligonucleotides in this instance; however, it is possible that extended polyglutamine tracts can be created by ligating several cassettes in succession (Fig. 2E).

Expression and Purification of Ataxin-3

Buffers for Expression and Purification

The procedures for expression and purification of the ataxin-3 variants use phosphate-based buffers, as listed below. The primary distinction between lysis, wash, and elution buffers is the concentration of imidazole present.

Lysis buffer: 50 mM NaH$_2$PO$_4$, 300 mM NaCl, 10% (v/v) glycerol, 0.1% (v/v) Triton X-100, 10 mM imidazole, 1 mM β-mercaptoethanol, pH 8.0

Wash buffer: 50 mM NaH$_2$PO$_4$, 300 mM NaCl, 10% (v/v) glycerol, 0.1% (v/v) Triton X-100, 20 mM imidazole, 1 mM β-mercaptoethanol, pH 8.0

Elution buffer: 50 mM NaH$_2$PO$_4$, 300 mM NaCl, 10% (v/v) glycerol, 0.1% (v/v) Triton X-100, 500 mM imidazole, 1 mM β-mercaptoethanol, pH 8.0

Phosphate-buffered saline/glycerol (PBS/G): 137 mM NaCl, 2.68 mM KCl, 10.1 mM NaH$_2$PO$_4$, 1.76 KH$_2$PO$_4$, 10% glycerol, 1 mM β-mercaptoethanol, pH 7.4

Expression of Ataxin-3 Variants

Expression of all ataxin-3 variants is induced by isopropyl-β-D-thiogalactopyranoside (IPTG), as described below. Generally, 8 L of cells is grown up.

1. Inoculate 100 ml 2TY medium (5 g NaCl, 10 g yeast extract, and 16 g tryptone per liter) with a single colony of *E. coli* cells containing the relevant plasmid. For pQE-His$_6$-Atax3, both JM101 and SG13009 (pREP4) cells are suitable for expression, whereas pET-His$_6$-Atax3 is best expressed in C41(DE3) cells, which are particularly suitable for the overexpression of toxic proteins (Miroux and Walker, 1996). Add ampicillin (0.1 mg/ml) and/or kanamycin (0.025 mg/ml) as necessary. Incubate overnight with shaking at 37°.

2. Inoculate 1 L 2TY medium containing required antibiotics with 10 ml overnight culture. Incubate at 37° with shaking until A_{600} ≈0.6–0.7.
3. Lower incubation to temperature to 28°. Add IPTG to a final concentration of 0.5 mM, and incubate with shaking for a further 3.5 h.
4. Harvest cells by centrifugation at 6000g for 15 min at 4°. Resuspend cells in 40 ml lysis buffer, snap freeze, and store at −20° until purification.

Purification of His₆-Tagged Ataxin-3 Variants

The method outlined below describes the purification of ataxin-3 variants expressed with an N-terminal His$_6$ tag. This protocol has been used successfully for variants containing 15, 28, and 50 residues in the polyglutamine tract. Typical yields range between 5 and 10 mg of protein for 8 L of cells when purifying pQE-His$_6$-Atax3 and between 10 and 20 mg of protein for 8 L of cells for pET-His$_6$-Atax3.

1. Add lysozyme (final concentration, 1.4 mM) and the proteinase inhibitor 4-(2-aminoethyl) benzenesulphonyl fluoride (AEBSF, purchased as Pefabloc from Roche) (final concentration, 0.5 mM) to the resuspended cells. Incubate on ice for 30 min, and then lyse cells by sonication and remove insoluble protein and cellular debris by centrifugation (53,000g, 60 min at 4°).
2. Filter the supernatant (0.45 μm) and mix for 2 h at 4° with 2 ml p-aminobenzamidine-agarose (Sigma) previously equilibrated with lysis buffer.
3. Elute unbound protein with lysis buffer and mix with 5 ml Ni-NTA resin previously equilibrated with lysis buffer and containing 0.5 mg/ml AEBSF for at least 2 h at 4°.
4. Elute protein not bound to the Ni-NTA resin, and then wash resin with 5–10 column volumes of wash buffer.
5. Elute protein from Ni-NTA in 5-ml fractions with elution buffer. Suitable fractions for further purification may then be identified by sodium dodecyl sulfate-polyacrylamide gel electrophoresis (SDS-PAGE) analysis. Add AEBSF (final concentration, 0.5 mg/ml) to fractions containing ataxin-3.
6. Purify the selected fractions further by size exclusion chromatography using a Hi-Load 16/60 Superdex 200 column (Amersham Biosciences) (Fig. 3A). The equilibration and elution buffer is PBS/G buffer. Up to 5 ml of protein may be loaded onto the Superdex 200 column in any single run. Three-milliliter fractions are collected

and analyzed by SDS-PAGE analysis. Select and pool fractions of suitably pure protein, which may be concentrated up to 4 mg/ml and are stored at $-80°$.

Purification of Ataxin-3 Using Fusion Partner Proteins

In addition to the purification of the His_6-tagged ataxin-3 variant as described here, ataxin-3 has been produced by a number of researchers using fusion protein constructs (Bevivino and Loll, 2001; Li *et al.*, 2002; Marchal *et al.*, 2003; Masino *et al.*, 2003, 2004; Shehi *et al.*, 2003). As a common tool in protein purification, fusion proteins are often generated with one or two motives: (1) to increase the soluble expression of the target

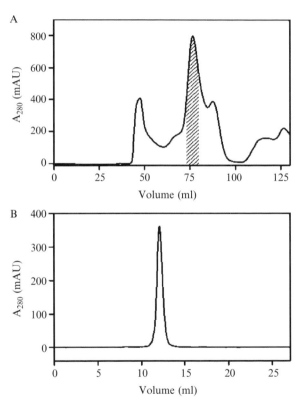

FIG. 3. Size exclusion chromatography of ataxin-3. (A) Typical elution profile of hexa-histidine (His6)-tagged atax3(Q15) purification using a Superdex 200 column. The shaded area indicates the volume at which fractions were collected. Similar profiles were observed during the purification of atax3(Q28) and atax3(Q50). (B) Typical elution profile of purified atax3(Q15) by Superose 12 analysis. The void volume of the column is 7.4 ml.

protein and/or (2) to facilitate purification of the protein by affinity chromatography. In the case of ataxin-3, the most commonly reported fusion construct comprises a glutathione S-transferase (GST) fusion partner linked N-terminally to ataxin-3 (Chai *et al.*, 2004; Marchal *et al.*, 2003; Masino *et al.*, 2003, 2004; Shehi *et al.*, 2003). In each of these systems, the protein is expressed in *E. coli* BL21 cells and undergoes an initial purification step involving binding of the fusion protein to glutathione sepharose affinity resin. Removal of the GST tag may be achieved by cleavage by a protease corresponding to the specific recognition site included in the construct, often PreScission Protease (Marchal *et al.*, 2003; Shehi *et al.*, 2003) or TEV protease (Masino *et al.*, 2004). Often, the cleavage of the fusion tag is performed while still attached to the glutathione sepharose resin so as to release ataxin-3, which is then separated from the protease by another purification step exploiting the affinity tag of the protease itself (Masino *et al.*, 2004; Shehi *et al.*, 2003). To date, all these procedures involving the use and subsequent removal of solubility tags have been reported for nonpathological length protein only; as such, the yields of longer constructs by these methods remain unknown.

On the other hand, retention of the GST fusion partner to the protein allows for expression and purification of pathological length proteins up to 84 glutamine residues (Chai *et al.*, 2004). Similarly, ataxin-3 has also been produced using a maltose-binding protein (MBP) fusion partner construct (Bevivino and Loll, 2001; Burnett *et al.*, 2003; Li *et al.*, 2002); however, once again, the purified protein is not cleaved from the fusion partner but, instead, remains as a fusion construct of the two proteins.

Quality Control and Analysis of Purified Ataxin-3

Prevention of Aggregation During Purification

A major issue when working with polyglutamine proteins in their purified form is their inherent instability and propensity to aggregate. This is hardly surprising, because it is this very conformational misbehavior that makes these proteins the focus of such intense interest. Indeed, it has been observed that polyglutamine peptides alone are extremely difficult to solubilize (Chen and Wetzel, 2001), and this is also consistent with the fact that polyglutamine proteins are so often expressed and purified as fusion proteins (Table I).

As such, it was not particularly unexpected that the expression of ataxin-3 results in the majority of the protein being expressed as insoluble inclusion bodies, with the soluble fraction barely detectable in comparison (Fig. 4). A number of parameters were adjusted to optimize the solubility

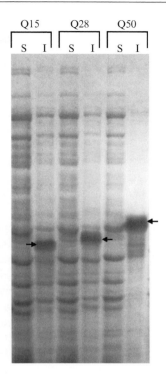

Fig. 4. Sodium dodecyl sulfate-polyacrylamide gel electrophoresis (SDS-PAGE) of ataxin-3 expression. Atax3(Q15), atax3(Q28), and atax(Q50) were expressed as described. Whole cells were lysed and centrifuged, and samples taken from the soluble (S) and insoluble (I) fractions were run on a 10% SDS-PAGE gel. Although for all variants, almost all the protein is present in the insoluble pellet (indicated by arrows), sufficient quantities of protein could be purified from the soluble fraction.

of the expressed protein; specifically, a reduction in expression temperature from 37° to 28° and limiting the incubation time to 3.5 h. Furthermore, use of the pET-His$_6$-Atax3 construct in the CD41 cell line also increased the quantity of soluble protein produced. Although a large proportion of the protein still remained insoluble on initial expression, the overall final yield of soluble purified protein was nevertheless significantly improved with the incorporation of these adjustments in the expression process. Interestingly, no significant difference was observed in the insoluble-to-soluble ratio of expressed protein between different length polyglutamine variants.

During the purification process, great care was also taken to minimize aggregation of the protein. Several trials with different buffer conditions revealed that the inclusion of Triton X-100 and glycerol in the purification buffers significantly increased the stability and yield of monomeric protein. Furthermore, the presence of β-mercaptoethanol is imperative in preventing the formation of inter- and intramolecular disulfide bonds, which increase the rate and extent of aggregation of variants of all lengths. The incorporation of a gel filtration chromatography step as the very last stage of the purification procedure is also important; this is vital to ensuring that the protein obtained in the final stage is monomeric. Furthermore, following purification and storage at $-80°$, on being thawed for use, the protein is also passed down a Superose 12 (Pharmacia) analytical gel filtration column in order to verify the conformational integrity of the protein (Fig. 3B).

Undesirable Proteolysis of the Purified Protein

A significant problem encountered during the development of the purification protocols was the proteolysis of ataxin-3. It has been documented that this protein is quite prone to cleavage (Masino et al., 2003), and we have also observed this in the course of our work. Although the protein was purified to $\approx90\%$ purity, it seemed that a minor contaminant, present in trace amounts, was cleaving the protein over a period of days during experiments. Given that ataxin-3 has protease activity (Burnett et al., 2003; Chow et al., 2004c; Scheel et al., 2003), it is also conceivable that an autocatalytic reaction may be responsible for the degradation. The activity of such de-ubiquitinating enzymes is generally substrate-specific to ubiquitin-linked proteins, and, furthermore, mutation of the active Cys14 residue to serine in atax3(Q28) did not prevent proteolysis of the purified protein (data not shown); thus, autocatalysis seems highly unlikely. We also observed no difference in the extent to which the different polyglutamine length variants were cleaved.

In order to address this issue of proteolysis, a range of protease inhibitors were screened for their efficacy in preventing ataxin-3 cleavage (Table II). Ataxin-3 was incubated at $37°$ in the presence of the various inhibitors over 3 days. Of all the inhibitors tested, as judged by SDS-PAGE analysis, AEBSF and PMSF were the most effective, suggesting that the culprit protein may be a serine protease, further ruling out the possibility of autocatalysis by ataxin-3. It was found that incubation of ataxin-3 in the presence of either AEBSF, 2 mM PMSF, or 2 mM PMSF and 5 mM ethylenediaminetetraacetic acid (EDTA) effectively reduced the level of proteolysis. A further complication arose when AEBSF was found to accelerate aggregation of the protein on prolonged incubation over 3 days (data not shown), however. Based on these results, AEBSF is used only during the early stages of purification in order to inactivate serine proteases released

TABLE II
PROTEASE INHIBITORS TESTED AGAINST ATAXIN-3 PROTEOLYSIS

Inhibitor	Inhibitor concentration	Effectiveness
1,10 phenantholine	10 mM	++
AEBSF	0.5 mg/ml	+++
Antipain	5 μM	+
Aprotinin	1 μg/ml	+
Benzamidine HCl	5 mM	+
Bestatin	10 μM	−
Caproic acid	100 mM	−
Chymostatin	3 μg/ml	++
E-64	10 μM	−
EDTA	5 mM	+
EDTA/PMSF	5 mM/2 mM	+++
Leupeptin	100 μM	−
Pepstatin	1 μg/ml	−
PMSF	2 mM	+++
Ppack	1 μM	−
Soybean trypsin inhibitor	50 μg/ml	−

No effect, −; minimal effect, +; some effect, ++; very effective, +++.

on cell lysis. Once the protein is purified, PMSF alone or PMSF and EDTA may be included in assays at concentrations of 2 and 5 mM, respectively, over periods of days in order to prevent proteolysis of ataxin-3 by trace contaminants.

The purification protocol also incorporates a benzamidine resin purification step to remove contaminating proteases from the cell lysate prior to binding to the nickel resin. It was found that the inclusion of the benzamidine-binding step greatly improved the quality and purity of ataxin-3 produced and, importantly, significantly reduced the extent to which the protein was proteolyzed on extended incubation during experimental assays (data not shown).

Conformational Analysis of Purified Proteins

The protocols described above may be utilized to produce ataxin-3 that is at least 90% pure. To confirm the purity and integrity of the final product, the protein is analyzed by size exclusion chromatography using a Superose 12 HR 10/30 column (Amersham Biosciences) (Fig. 3B). A 100-μl sample of protein at 1 mg/ml is sufficient to produce a gel filtration profile suitable for analysis. This is an important quality control step, because ataxin-3 can be conformationally labile and a wholly monomeric species cannot be guaranteed merely by selection of fractions from the Superdex 200 purification step and observation of SDS-PAGE gels.

The Superose 12 elution profile of ataxin-3 should produce a single peak corresponding to the monomeric protein (Fig. 3B). Initial calculation of the molecular weight according to the elution profiles suggests that all three variants have a much larger molecular weight than their theoretical values. Analyses by mass spectrometry and sedimentation velocity confirmed that the protein is of the correct molecular weight and monomeric (data not shown), however. Furthermore, the size exclusion chromatography results may be explained by the observations that ataxin-3 has an elongated shape with an 11:1 length-to-width ratio (Bevivino and Loll, 2001) and the C-terminal portion of the protein, which includes the polyglutamine tract, is highly flexible and largely unstructured (Masino *et al.*, 2003). Thus, the extended conformation of the protein would result in a nonconventional migration through the gel filtration matrix, and the anomalous gel filtration elution profile of ataxin-3 is consistent with other documented characteristics of the protein (Bevivino and Loll, 2001; Masino *et al.*, 2003).

Ataxin-3 variants containing 15, 28, and 50 glutamine residues were also analyzed by far-ultraviolet (UV) circular dichroism (CD) spectroscopy (Fig. 5). The spectra obtained were concordant with data reported by ourselves and others (Bevivino and Loll, 2001; Chow *et al.*, 2004d; Masino *et al.*, 2003; Shehi *et al.*, 2003), indicating that the protein contains a significant amount of α-helical content. Furthermore, the spectra of all three variants were very similar to each other, suggesting that there were no major differences in secondary structure, an observation that was confirmed by spectral deconvolution of the far-UV spectra (Fig. 5, tables).

Functional Analysis of Purified Proteins

Ataxin-3 is proposed to be a de-ubiquitinating protease that is capable of binding tetraubiquitin constructs (Burnett *et al.*, 2003; Chai *et al.*, 2004; Donaldson *et al.*, 2003; Doss-Pepe *et al.*, 2003; Scheel *et al.*, 2003), although the link between these two functional capacities, if any, is yet to be understood and delineated fully. At the time of writing, ataxin-3 has been reported to display ubiquitin protease activity with respect to two different substrates: the fluorescent substrate ubiquitin-AMC (Burnett *et al.*, 2003) and a GST-ubiquitin fusion construct (Chow *et al.*, 2004c).

The functional assay for ataxin-3 ubiquitin protease activity described here involves SDS-PAGE analysis of the interaction between ataxin-3 and the linear substrate GST-Ub52. This substrate comprises a GST purification tag protein N-terminally attached to Ub52, which is composed of

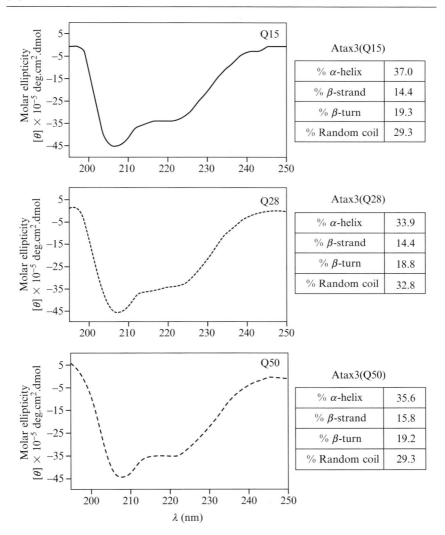

Atax3(Q15)

% α-helix	37.0
% β-strand	14.4
% β-turn	19.3
% Random coil	29.3

Atax3(Q28)

% α-helix	33.9
% β-strand	14.4
% β-turn	18.8
% Random coil	32.8

Atax3(Q50)

% α-helix	35.6
% β-strand	15.8
% β-turn	19.2
% Random coil	29.3

FIG. 5. Far-ultraviolet (UV) CD emission spectra of purified ataxin-3 variants. Spectra at pH 7.4 and 25° are shown for atax3(Q15), atax3(Q28), and atax3(Q50) as indicated. Protein concentrations used are 0.1 mg/ml. The respective tables show spectral deconvolution analyses of each variant, calculated using the CONTINLL algorithm (Provencher and Glockner, 1981; van Stokkum *et al.*, 1990) as provided by the DICHROWEB online facility (Lobley and Wallace, 2001; Lobley *et al.*, 2002).

ubiquitin expressed as a fusion protein with a 52-amino acid ribosomal protein L40 at the C-terminal end (Baker and Board, 1991; Everett *et al.*, 1997). Following incubation of ataxin-3 variants with GST-Ub52 at 37°

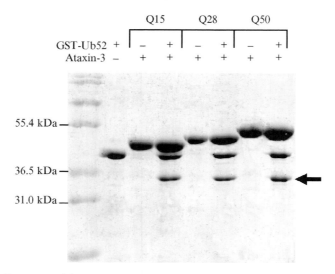

FIG. 6. Protease activity assay of ataxin-3 variants. Variants were incubated at 37° for 60 min in the presence and absence of the linear substrate glutathione S-transferase (GST)-Ub52 as marked. The position of the cleaved substrate is marked by an arrow.

for 1 h, SDS-PAGE analysis reveals a band shift of the substrate consistent with the release of the 6-kDa L40 protein from the C-terminus of the ubiquitin molecule (Fig. 6). The reaction buffer used is 100 mM Tris, 1 mM dithiothreitol (DTT), pH 8.0. The enzyme-to-substrate molar ratio is 1:5. It is interesting to note that although a distinctive gel-shifted cleavage product is clearly observed, ataxin-3 appears to be an inefficient protease for this substrate, with much of the GST-Ub52 fusion protein remaining intact. It is also pertinent to note that as judged by this analysis, elongation of the polyglutamine tract does not appear to affect the ability of ataxin-3 to cleave the ubiquitin substrate, because all three variants cleaved the substrate to the same extent (Fig. 6).

Conclusion

In the development of protocols for the purification and analysis of ataxin-3 variants using different His$_6$-tagged constructs, our experience suggested that one of the fundamental issues when working with this protein is its ready susceptibility to proteolysis. As such, much of our work in refining the conditions of purification and assays has focused on preventing the unwanted cleavage of ataxin-3, and this problem has been addressed by the inclusion of a benzamidine-affinity resin step during purification to remove proteases and the use of specific serine protease inhibitors in experimental assays. The protein produced is 90% pure, monomeric, and active as a ubiquitin-specific protease, even in the expanded pathogenic form. The analysis of purified protein is essential to establish the native fold of the starting material for experimental assays. Thus, these protocols lay the groundwork for further studies into the conformational and functional behavior of ataxin-3 and, furthermore, the effects of polyglutamine expansion on these properties.

References

Baker, R. T., and Board, P. G. (1991). The human ubiquitin-52 amino acid fusion protein gene shares several structural features with mammalian ribosomal protein genes. *Nucleic Acids Res.* **19,** 1035–1040.

Bennett, M. J., Huey-Tubman, K. E., Herr, A. B., West, A. P., Jr., Ross, S. A., and Bjorkman, P. J. (2002). Inaugural article: A linear lattice model for polyglutamine in CAG-expansion diseases. *Proc. Natl. Acad. Sci. USA* **99,** 11634–11639.

Bevivino, A. E., and Loll, P. J. (2001). An expanded glutamine repeat destabilizes native ataxin-3 structure and mediates formation of parallel beta-fibrils. *Proc. Natl. Acad. Sci. USA* **98,** 11955–11960.

Burnett, B., Li, F., and Pittman, R. N. (2003). The polyglutamine neurodegenerative protein ataxin-3 binds polyubiquitylated proteins and has ubiquitin protease activity. *Hum. Mol. Genet.* **12,** 3195–3205.

Chai, Y., Berke, S. S., Cohen, R. E., and Paulson, H. L. (2004). Poly-ubiquitin binding by the polyglutamine disease protein ataxin-3 links its normal function to protein surveillance pathways. *J. Biol. Chem.* **279,** 3605–3611.

Chen, S., Berthelier, V., Hamilton, J. B., O'Nuallain, B., and Wetzel, R. (2002a). Amyloid-like features of polyglutamine aggregates and their assembly kinetics. *Biochemistry* **41,** 7391–7399.

Chen, S., Ferrone, F. A., and Wetzel, R. (2002b). Huntington's disease age-of-onset linked to polyglutamine aggregation nucleation. *Proc. Natl. Acad. Sci. USA* **99,** 11884–11889.

Chen, S., and Wetzel, R. (2001). Solubilization and disaggregation of polyglutamine peptides. *Protein Sci.* **10,** 887–891.

Chow, M. K., Ellisdon, A. M., Cabrita, L. D., and Bottomley, S. P. (2004a). Polyglutamine expansion in ataxin-3 does not affect protein stability: Implications for misfolding and disease. *J. Biol. Chem.* **279,** 47643–47651.

Chow, M. K., Lomas, D. A., and Bottomley, S. P. (2004b). Promiscuous beta-strand interactions and the conformational diseases. *Curr. Med. Chem.* **11,** 491–499.

Chow, M. K., Mackay, J. P., Whisstock, J. C., Scanlon, M. J., and Bottomley, S. P. (2004c). Structural and functional analysis of the Josephin domain of the polyglutamine protein ataxin-3. *Biochem. Biophys. Res. Commun.* **322,** 387–394.

Chow, M. K., Paulson, H. L., and Bottomley, S. P. (2004d). Destabilization of a non-pathological variant of ataxin-3 results in fibrillogenesis via a partially folded intermediate: A model for misfolding in polyglutamine disease. *J. Mol. Biol.* **335,** 333–341.

Davies, S. W., Beardsall, K., Turmaine, M., DiFiglia, M., Aronin, N., and Bates, G. P. (1998). Are neuronal intranuclear inclusions the common neuropathology of triplet-repeat disorders with polyglutamine-repeat expansions? *Lancet* **351,** 131–133.

Diaz-Hernandez, M., Moreno-Herrero, F., Gomez-Ramos, P., Moran, M. A., Ferrer, I., Baro, J., Avila, J., Hernandez, F., and Lucas, J. J. (2004). Biochemical, ultrastructural, and reversibility studies on huntingtin filaments isolated from mouse and human brain. *J. Neurosci.* **24,** 9361–9371.

DiFiglia, M., Sapp, E., Chase, K. O., Davies, S. W., Bates, G. P., Vonsattel, J. P., and Aronin, N. (1997). Aggregation of huntingtin in neuronal intranuclear inclusions and dystrophic neurites in brain. *Science* **277,** 1990–1993.

Donaldson, K. M., Li, W., Ching, K. A., Batalov, S., Tsai, C. C., and Joazeiro, C. A. (2003). Ubiquitin-mediated sequestration of normal cellular proteins into polyglutamine aggregates. *Proc. Natl. Acad. Sci. USA* **100,** 8892–8897.

Doss-Pepe, E. W., Stenroos, E. S., Johnson, W. G., and Madura, K. (2003). Ataxin-3 interactions with rad23 and valosin-containing protein and its associations with ubiquitin chains and the proteasome are consistent with a role in ubiquitin-mediated proteolysis. *Mol. Cell. Biol.* **23,** 6469–6483.

Ellisdon, A. M., and Bottomley, S. P. (2004). The role of protein misfolding in the pathogenesis of human diseases. *IUBMB Life* **56,** 119–123.

Everett, R. D., Meredith, M., Orr, A., Cross, A., Kathoria, M., and Parkinson, J. (1997). A novel ubiquitin-specific protease is dynamically associated with the PML nuclear domain and binds to a herpesvirus regulatory protein. *EMBO J.* **16,** 1519–1530.

Horwich, A. (2002). Protein aggregation in disease: A role for folding intermediates forming specific multimeric interactions. *J. Clin. Invest.* **110,** 1221–1232.

Huang, C. C., Faber, P. W., Persichetti, F., Mittal, V., Vonsattel, J. P., MacDonald, M. E., and Gusella, J. F. (1998). Amyloid formation by mutant huntingtin: Threshold, progressivity and recruitment of normal polyglutamine proteins. *Somat. Cell. Mol. Genet.* **24,** 217–233.

Kawaguchi, Y., Okamoto, T., Taniwaki, M., Aizawa, M., Inoue, M., Katayama, S., Kawakami, S., Nakamura, S., Nishimura, M., Akiguchi, I., Kimmura, J., Narumiya, S., and Kakizuka, A. (1994). CAG expansions in a novel gene for Machado-Joseph disease at chromosome 14q32.1. *Nat. Genet.* **8,** 221–228.

Li, F., Macfarlan, T., Pittman, R. N., and Chakravarti, D. (2002). Ataxin-3 is a histone-binding protein with two independent transcriptional corepressor activities. *J. Biol. Chem.* **277,** 45004–45012.

Lobley, A., and Wallace, B. A. (2001). DICHROWEB: A Website for the analysis of protein secondary structure from circular dichroism spectra. *Biophys. J.* **80,** 373a.

Lobley, A., Whitmore, L., and Wallace, B. A. (2002). DICHROWEB: An interactive website for the analysis of protein secondary structure from circular dichroism spectra. *Bioinformatics* **18,** 211–212.

Malinchik, S. B., Inouye, H., Szumowski, K. E., and Kirschner, D. A. (1998). Structural analysis of Alzheimer's beta(1-40) amyloid: Protofilament assembly of tubular fibrils. *Biophys. J.* **74,** 537–545.

Marchal, S., Shehi, E., Harricane, M. C., Fusi, P., Heitz, F., Tortora, P., and Lange, R. (2003). Structural instability and fibrillar aggregation of non-expanded human ataxin-3 revealed under high pressure and temperature. *J. Biol. Chem.* **278,** 31554–31563.

Margolis, R. L., and Ross, C. A. (2001). Expansion explosion: New clues to the pathogenesis of repeat expansion neurodegenerative diseases. *Trends Mol. Med.* **7,** 479–482.

Masino, L., Musi, V., Menon, R. P., Fusi, P., Kelly, G., Frenkiel, T. A., Trottier, Y., and Pastore, A. (2003). Domain architecture of the polyglutamine protein ataxin-3: A globular domain followed by a flexible tail. *FEBS Lett.* **549,** 21–25.

Masino, L., Nicastro, G., Menon, R. P., Dal Piaz, F., Calder, L., and Pastore, A. (2004). Characterization of the structure and the amyloidogenic properties of the Josephin domain of the polyglutamine-containing protein ataxin-3. *J. Mol. Biol.* **344,** 1021–1035.

Miroux, B., and Walker, J. E. (1996). Over-production of proteins in Escherichia coli: Mutant hosts that allow synthesis of some membrane proteins and globular proteins at high levels. *J. Mol. Biol.* **260,** 289–298.

Padiath, Q. S., Srivastava, A. K., Roy, S., Jain, S., and Brahmachari, S. K. (2005). Identification of a novel 45 repeat unstable allele associated with a disease phenotype at the MJD1/SCA3 locus. *Am. J. Med. Genet. B Neuropsychiatr. Genet.* **133,** 124–126.

Paulson, H. L., Perez, M. K., Trottier, Y., Trojanowski, J. Q., Subramony, S. H., Das, S. S., Vig, J. L., Mandel, J. L., Fischbeck, K. H., and Pittman, R. N. (1997). Intranuclear inclusions of expanded polyglutamine protein in spinocerebellar ataxia type 3. *Neuron* **19,** 333–344.

Poirier, M. A., Li, H., Macosko, J., Cai, S., Amzel, M., and Ross, C. A. (2002). Huntingtin spheroids and protofibrils as precursors in polyglutamine fibrillization. *J. Biol. Chem.* **277,** 41032–41037.

Popiel, H. A., Nagai, Y., Onodera, O., Inui, T., Fujikake, N., Urade, Y., Strittmatter, W. J., Burke, A., Ichikawa, A., and Toda, T. (2004). Disruption of the toxic conformation of the expanded polyglutamine stretch leads to suppression of aggregate formation and cytotoxicity. *Biochem. Biophys. Res. Commun.* **317,** 1200–1206.

Provencher, S. W., and Glockner, J. (1981). Estimation of globular protein secondary structure from circular dichroism. *Biochemistry* **20,** 33–37.

Scheel, H., Tomiuk, S., and Hofmann, K. (2003). Elucidation of ataxin-3 and ataxin-7 function by integrative bioinformatics. *Hum. Mol. Genet.* **12,** 2845–2852.

Scherzinger, E., Lurz, R., Turmaine, M., Mangiarini, L., Hollenbach, B., Hasenbank, R., Bates, S. W., Davies, S. W., Lehrach, H., and Wanker, E. E. (1997). Huntingtin-encoded polyglutamine expansions form amyloid-like protein aggregates *in vitro* and *in vivo*. *Cell* **90,** 549–558.

Shehi, E., Fusi, P., Secundo, F., Pozzuolo, S., Bairati, A., and Tortora, P. (2003). Temperature-dependent, irreversible formation of amyloid fibrils by a soluble human ataxin-3 carrying a moderately expanded polyglutamine stretch (Q36). *Biochemistry* **42,** 14626–14632.

Stefani, M., and Dobson, C. M. (2003). Protein aggregation and aggregate toxicity: New insights into protein folding, misfolding diseases and biological evolution. *J. Mol. Med.* **81,** 678–699.

Tanaka, M., Machida, Y., Nishikawa, Y., Akagi, T., Hashikawa, T., Fujisawa, T., and Nukina, N. (2003). Expansion of polyglutamine induces the formation of quasi-aggregate in the early stage of protein fibrillization. *J. Biol. Chem.* **278,** 34717–34724.

Tanaka, M., Morishima, I., Akagi, T., Hashikawa, T., and Nukina, N. (2001). Intra- and intermolecular beta-pleated sheet formation in glutamine-repeat inserted myoglobin as a model for polyglutamine diseases. *J. Biol. Chem.* **276,** 45470–45475.

Thakur, A. K., and Wetzel, R. (2002). Mutational analysis of the structural organization of polyglutamine aggregates. *Proc. Natl. Acad. Sci. USA* **99,** 17014–17019.

Trottier, Y., Lutz, Y., Stevanin, G., Imbert, G., Devys, D., Cancel, G., Saudou, F., Weber, C., David, L., Tora, L., *et al.* (1995). Polyglutamine expansion as a pathological epitope in Huntington's disease and four dominant cerebellar ataxias. *Nature* **378,** 403–406.

Van Stokkum, I. H., Spoelder, H. J., Bloemendal, M., van Grondelle, R., and Groen, F. C. (1990). Estimation of protein secondary structure and error analysis from circular dichroism spectra. *Anal. Biochem.* **191,** 110–118.

[2] Preparation of Amyloid β-Protein for Structural and Functional Studies

By DAVID B. TEPLOW

Abstract

Amyloid proteins cause a number of progressive, degenerative diseases. Among these is Alzheimer's disease (AD), the etiology of which is linked to the formation of neurotoxic assemblies by the amyloid β-protein (Aβ). The clinical importance of AD has stimulated intense interest in the mechanisms of Aβ folding and self-assembly. Studying these phenomena *in vitro* requires the preparation of Aβ peptide stocks that are well defined and display reproducible biophysical and biological behaviors. Unfortunately, the propensity of Aβ to self-assemble has made this goal difficult. I discuss here a biphasic strategy for preparing Aβ for structural and functional studies. The strategy involves sodium hydroxide pretreatment of synthetic Aβ, followed by size fractionation procedures. This approach produces Aβ solutions that have been used successfully in a variety of *in vitro* and *in vivo* experimental systems.

Background

Aβ comprises the fibrils found in the senile plaques that are pathognomonic for AD (Selkoe, 1991). Genetic, physiological, and biochemical data support the hypothesis that Aβ is a causative agent of AD (Selkoe, 2001). This may be a direct result of fibril neurotoxicity (Pike *et al.*, 1991, 1993). However, continuing structure-activity studies have revealed that fibril intermediates and many other types of Aβ assemblies also are neurotoxic (Hoshi *et al.*, 2003; Kirkitadze *et al.*, 2002; Klein *et al.*, 2004; Taylor *et al.*, 2003; Walsh and Selkoe, 2004; Walsh *et al.*, 1999), emphasizing the importance of a full elucidation of Aβ assembly. In particular, recognition of the clinical and biological importance of oligomeric assemblies (Klein *et al.*, 2004) has made determination of conformational and assembly states populated in the structural space between monomer and fibril especially significant. As one might predict (Murphy, 1949), it is within this structural space that the intrinsic propensity of Aβ to self-associate creates the largest experimental impediments. Aβ assembly is a complex process (Buxbaum, 2003; Teplow, 1998; Thirumalai *et al.*, 2003) that produces an array of metastable structures (for a recent review, see Lazo *et al.*, 2005). Metastability and polydispersity largely obviate the use of

METHODS IN ENZYMOLOGY, VOL. 413
0076-6879/06 $35.00
DOI: 10.1016/S0076-6879(06)13002-5

solution nuclear magnetic resonance (NMR) and X-ray crystallographic techniques that have been applied so effectively to determine the tertiary structure of homogeneous preparations of natively folded proteins. To overcome these problems, site-specific labeling techniques have been employed to minimize problems arising from population heterogeneity. These include site-directed spin-labeling coupled with electron paramagnetic resonance [EPR] (Torok *et al.*, 2002), hydrogen-deuterium exchange (Kheterpal *et al.*, 2000, 2003), intrinsic fluorescence (Maji *et al.*, 2005), and NMR on isotopically labeled samples (Antzutkin *et al.*, 2000; Balbach *et al.*, 2002; Benzinger *et al.*, 1998; Petkova *et al.*, 2002; Tycko, 2004). In each case, preparation of protein samples with maximal primary, secondary, tertiary, and quaternary structural homogeneity was critical. I discuss here approaches for producing such Aβ preparations.

Theoretical and Practical Aspects of Amyloid Protein Preparation

Pre Facto *Considerations for Studies of Aβ and Other Amyloid Proteins*

Aβ is produced by solid phase peptide synthesis (SPPS) or recombinant DNA techniques. Unfortunately, substantial compositional variation has been reported among Aβ preparations, resulting in experimental irreproducibility (Howlett *et al.*, 1995; Simmons *et al.*, 1994; Soto *et al.*, 1995). It is important that the experimentalist verifies that the peptide itself is chemically pure and that nonpeptide components are inert or absent. In practice, this is difficult. Most peptide lyophilizates do contain salts and other components. For example, fluoren-9-ylmethoxycarbonyl (FMOC)-mediated SPPS coupled with reverse phase high-performance liquid chromatography (HPLC) purification typically produces trifluoroacetic acid (TFA) salts of the resulting peptides. These salts, as well as chemical scavengers, often are present in lyophilized peptide preparations and can complicate the initial solvation and preparation of peptide stock solutions. In addition, these nonpeptide components can alter the biophysical and biological behavior of the peptide. Technical errors also can be made in calculation of nominal peptide concentration if the weights of nonpeptide components are not taken into account.

SPPS cannot produce an Aβ product that is 100% pure. Failure sequences—peptides missing one or more amino acids—are unavoidable, although with proper synthesis chemistry, their relative amounts can be minimized. Oxidation of Met35 to its sulfoxide form is a common side reaction that can occur during peptide workup and purification. Synthesis-related amino acid racemization and side reactions during peptide cleavage and deprotection also may be observed, but these generally occur infrequently. Most peptide suppliers perform quantitative amino acid analysis and mass spectrometry to characterize their products. However, because amino acid and simple mass analyses cannot determine primary structure, Edman or mass spectrometric sequence analysis can be used to prove formally that the peptide structure is correct.

For recombinantly–derived $A\beta$, primary structure changes are rare because of the high fidelity of the protein expression systems and the physiological conditions under which these systems operate. In systems in which $A\beta$ is produced as a fusion protein, posttranslational processing with specific endoproteases releases the free peptide. It is important to ensure that the $A\beta$ component of the fusion protein is not contaminated by uncleaved fusion protein, the enzyme itself, or peptide fragments produced through adventitious proteolysis. Because fusion protein cleavage is done under conditions necessary for efficient endoprotease action and these conditions are unlikely to be identical to those desired for subsequent experimentation, buffer exchange or removal may be necessary prior to peptide use.

Metastability and Polydispersity

Metastability and polydispersity are two important factors complicating the study of $A\beta$. The methods discussed here have primary application in the minimization of polydispersity. The reader should note, however, that metastability is an equally important issue. Metastability is an intrinsic property that exists in solution studies of $A\beta$ and other amyloid proteins done under quasiphysiological conditions (e.g., neutral pH, isotonic biologic buffers). As discussed above, $A\beta$ forms a variety of monomeric, oligomeric, and polymeric structures. The precursor-product relations and equilibria among these assemblies are not entirely understood and are areas of active investigation. Some controversy exists as to the oligomerization state of aggregate-free starting preparations of $A\beta$. Here, I refer to $A\beta$ solutions prepared at micromola peptide concentrations as low molecular weight (LMW) $A\beta$ (Walsh et al., 1999). I do so because techniques, including quasielastic light scattering (Walsh et al., 1997), chemical cross-linking (Bitan et al., 2001, 2003; LeVine, 1995, 2004), fluorescence resonance energy transfer (Garzon-Rodriguez et al., 1997), and ultracentrifugation (Huang et al., 2000; Seilheimer et al., 1997), have shown that $A\beta$ exists in these solutions as a mixture of monomers and low-order oligomers in quasiequilibrium rather than solely as monomers. The reader should note that for many experimental needs, this issue may be academic, because the primary concern may be the ability to prepare peptide stock solutions identical in their distribution of peptide assembly states, whatever that distribution may be. The importance of *pre facto* consideration of which assembly state(s) is being monitored cannot be overstated.

Irreproducibility

To conduct studies of amyloid protein assembly, one seeks ideally to produce a homogeneous protein solution within which protein folding and self-association can be initiated synchronously and then monitored. Unfortunately, this has not been possible in studies of Aβ. Instead, irreproducible behavior of the peptide commonly has been observed, especially among different laboratories. Causes of irreproducibility include the initial structure and aggregation state of the peptide, both in the solid state (Fezoui *et al.*, 2000) and immediately after solvation, the presence in the solvent of nidi for heterogeneous nucleation, and rapid (<1 s) milieu-dependent protein oligomerization (Bitan *et al.*, 2001, 2003; LeVine, 1995, 2004). The structure of an amyloid protein in the solid state is difficult to control because it depends upon the solution conditions prior to dehydration and the dehydration procedure itself. The issue becomes moot, however, if appropriate solvation procedures are used. These procedures should eliminate preexisting aggregates and create monomeric Aβ stock solutions from peptide lyophilizates.

Heterogeneous nucleation can be minimized by scrupulous preparation procedures that utilize high-purity water filtered to eliminate nanoparticulates, minimize the adherence of particulate matter to glassware and plasticware, and use ultrafiltration methods to eliminate particulates in buffers. These procedures have proven to be effective for biological assays and biophysical studies. It should be noted, however, that electrostatic charging of surfaces and the presence of ionic airborne particulates make complete elimination of particulates almost impossible in the absence of "clean room" procedures of the type used in the semiconductor industry. This fact is particularly relevant for studies using techniques, such as quasielastic light scattering, that have sensitivities directly proportional to analyte molecular weight (Lomakin *et al.*, 1999, 2005). Protein solutions of exceptional clarity (particulate-free) are required if such techniques are to be used successfully.

A "magic formula" for manipulating Aβ does not exist. Chaotropic agents (dimethylsulfoxide [DMSO]) (Lambert *et al.*, 2001; Stine *et al.*, 2003; Wang *et al.*, 2002), organic acids (TFA) (Zagorski *et al.*, 1999), organic cosolvents (trifluoroethanol [TFE] and hexafluoroisopropanol [HFIP]) (Zagorski and Barrow, 1992), and sodium hydroxide (NaOH) (Fezoui *et al.*, 2000) all have been used either singly or in combination

(Hou *et al.*, 2004) to solubilize and disaggregate Aβ lyophilizates. For example, examination of the secondary structure of Aβ dissolved in neat DMSO revealed no β-sheet (Shen and Murphy, 1995). Aβ treated in this way prior to initiation of fibril formation in biological buffers displayed the slowest polymerization rates relative to samples treated with lower concentrations of DMSO, 0.1% (v/v) TFA, or 0.1% TFA in acetonitrile (Shen and Murphy, 1995). Low (0.1%) TFA concentrations were not effective in disaggregating Aβ or preventing its self-association. In contrast, pretreatment of Aβ with concentrated TFA, followed by lyophilization, produced preparations that yielded solutions of protein monomers displaying "random coil" secondary structure following solubilization in biological buffers (Jao *et al.*, 1997). HFIP and TFE disrupt hydrophobic interactions in aggregated amyloid preparations and stabilize α-helical structure (Buck, 1998; Wood *et al.*, 1996; Zagorski and Barrow, 1992), leading to disruption of preexistent β-sheet structure. HFIP pretreatment of Aβ has been shown to yield peptide solutions of uniform globular morphology with predominantly α-helical and random coil secondary structure and less than 1% β-sheet (Stine *et al.*, 2003).

Preparing Aβ for Biophysical and Biological Study

In this section, I discuss the technique and benefits of pretreating synthetic Aβ with NaOH. I then present procedures using membrane filtration or gel permeation chromatography that produce LMW Aβ from the pretreated lyophilized peptide stocks.

Alkaline Pretreatment of Aβ

Introduction. The majority of biophysical work done on Aβ involves synthetic peptides. The chemical synthesis procedures involve cleavage and deprotection of peptide-resins with TFA or hydrogen fluoride (HF), depending on whether FMOC or *t-BOC* (tert-butyloxycarbonyl) chemistry, respectively, is employed. Peptide purification then is accomplished by reverse phase HPLC, typically using gradient elution with TFA in acetonitrile. The result of these procedures is a peptide lyophilizate containing residual, avidly bound trifluoroacetate or fluoride ions. The bound ions produce an acidic milieu following solubilization of the peptide in water. This low pH ($\approx$3–4) can facilitate peptide solubilization and has proven useful in spectroscopic studies of fibril nucleation and elongation (Lomakin *et al.*, 1996, 1999). However, solvation of Aβ lyophilizates in buffers of neutral pH produces a pH transition from acidic ($\approx$2) to neutral as residual TFA or HF is neutralized. During this transition, the solution pH passes through the isoelectric point of Aβ (5.5), at which Aβ aggregation propensity is maximal and solubility is minimal (Barrow *et al.*, 1992; Wood *et al.*, 1996). The result is conversion of Aβ monomers into a polydisperse population of interacting low-order oligomers and higher order polymers. This population has irreproducible assembly behavior characterized by significant pH-dependent morphologic and kinetic differences in fibril formation (Wood *et al.*, 1996). We reasoned that if initial solvation of Aβ could be done without causing pH transitions through the peptide pI, the problems described above could be mitigated. This was accomplished through "presolvation" of the Aβ lyophilizates with NaOH (pH $\approx$10.5–11) followed by relyophilization (Fezoui *et al.*, 2000). Solubilization of the pretreated peptide in neutral buffers then produced a pH shift from the alkaline regimen to neutrality, avoiding the Aβ pI. Comparative analyses of treated and untreated Aβ preparations from a variety of commercial and noncommerical sources showed that NaOH pretreatment produced peptide solutions with higher yields of LMW Aβ and with lower levels of preexistent aggregates (Fezoui *et al.*, 2000). The treated Aβ preparations reproducibly formed fibrils with conformational and tinctorial properties typical of amyloid fibrils, which were toxic to cultured neurons.

Historical precedent exists for the use of alkali in solubilizing Aβ. The effects of alkalis on amyloids were examined as early as 1898 (Krakow, 1898). In this work as well as in subsequent studies (Dubois *et al.*, 1999; Hass and Schulz, 1940; Perry *et al.*, 1981; Pras *et al.*, 1969; Shirahama and Cohen, 1967), alkalis, including barium hydroxide, calcium hydroxide, ammonium hydroxide, NaOH, and alkaline borate, and phosphate buffers were used to solubilize amyloids and amyloid proteins from *ex vivo* tissue samples. Ammonium hydroxide (Shirahama and Cohen, 1967) and NaOH (Pras *et al.*, 1969) were found to be particularly effective. Strong alkali treatment thus has proven useful for Aβ disaggregation and solubilization at both extremes of the polymerization state–monomer and fibril aggregate.

The Method. Dissolve lyophilized Aβ in 2 mM NaOH to produce a peptide concentration of <1 mg/ml. This may be accomplished conveniently in a 1.5-ml conical microcentrifuge tube. Ensure that the pH of the resulting solution is ≥10.5. It is very important to add the NaOH and gently agitate the tube only enough to wet the peptide lyophilizate entirely. Allow the solvation to proceed without additional agitation until a visually constant appearance is achieved ($\approx$1–3 min). The solution may be clear or remain somewhat turbid. Sonicate the peptide solution in a bath-type sonicator (e.g., a Branson model 1200-R; Branson Ultrasonics Corp., Danbury, CT) for 1 min. Lyophilize the solution and store the lyophilizate at $-20°$. This NaOH-treated peptide lyophilizate will be the starting material for all subsequent experimental uses.

Filtration and Size Exclusion Chromatography

Introduction. Alkali treatment of $A\beta$ prior to dissolution is effective in increasing peptide yields and decreasing the number and size of preexistent aggregates (Fezoui *et al.*, 2000). However, this approach alone does not yield peptide solutions amenable to detailed mechanistic study, because aggregates can form during peptide solvation. To produce fully disaggregated peptide solutions, a second preparation step is necessary. This step generally involves size fractionation, which eliminates aggregates, leaving a LMW $A\beta$ preparation. Filtration methods have been used to remove large and small aggregates from starting solutions of $A\beta$ (Fezoui *et al.*, 2000). Filtration of $A\beta$ through a 0.2-μm nylon microspin Whatman filter at 5000g for 10 min will remove fibrils, fibril aggregates, and other structures larger than 200 nm. However, for most experimental needs, this filtration is insufficient, because assemblies of 200-nm size are relatively large and generally are already fibrillar. Filtration through filters of 20-nm porosity (Anotop 10 Plus; Whatman Inc., Clifton, NJ) or with 10-kDa exclusion limits is a superior method. The latter procedure initially yields monomers and dimers. A second procedure for preparing aggregate-free $A\beta$ is size exclusion chromatography (SEC). An advantage of SEC relative to simple filtration is its fractionation capability, which allows collection of different populations of oligomers, including protofibrils (Walsh *et al.*, 1997) as well as relatively pure populations of monomer, dimers, and trimers (Bitan and Teplow, 2005).

It is important to reiterate that in peptide concentration regimens of micromola and higher, $A\beta$ monomers exist in rapid equilibrium with higher order oligomers and that this equilibrium is established within seconds. Most aggregate-free $A\beta$ solutions therefore comprise an oligodisperse population of assemblies. Nevertheless, because this population can be prepared reproducibly and is not polydisperse, consistent peptide assembly behavior can be observed.

Preparation of LMW Aβ by Filtration. Filtration is a simple, rapid method that requires few instrumental resources. Microcon YM-10 filters are washed with 200 μl of 10 mM phosphate buffer, pH 7.4, by adding the buffer to the filter and centrifuging at 16,000g for 20 min. The filtrate is discarded, and the washing is repeated once. The washed filter unit then is placed in a new 1.5 ml microcentrifuge tube. Lyophilized, NaOH-treated $A\beta$ is dissolved in water at a concentration of $\approx$4 mg/ml. An equal volume of 20 mM phosphate buffer, pH 7.4, is added to this solution, which then is sonicated for 1 min and transferred to the filter assembly. The filter unit is centrifuged at 16,000g for 30 min. The filtrate containing LMW $A\beta$ is collected and used immediately. For experiments involving extended incubation times, 0.01% (w/v) sodium azide is added to the buffers to prevent microbial growth.

Preparation of LMW Aβ by Size Exclusion Chromatography. Prepare 10 m*M* sodium phosphate, pH 7.4, using high-purity water. We have found that a Milli-Q Synthesis system (Millipore Corp., Billerica, MA) is an excellent source of water suitable for biophysical and biological studies. Filter the buffer solution through a 0.22 μm polyethersulfone (or equivalent) membrane to remove bacteria and any other large particulates. Use this buffer to wash and equilibrate a 10/30 Superdex 75 HR (Amersham Biosciences, Piscataway, NJ) column at a flow rate of 0.5 ml/min until a flat ultraviolet (UV) trace is observed. The chromatographic pumping system, *per se*, is not relevant to the procedure. It need only provide appropriate flow rates and include a detector capable of determining absorbance in the UV range (200–300 nm). Dissolve 350–400 μg of lyophilized Aβ in DMSO at a concentration of 2 mg/ml in a 1.5 ml

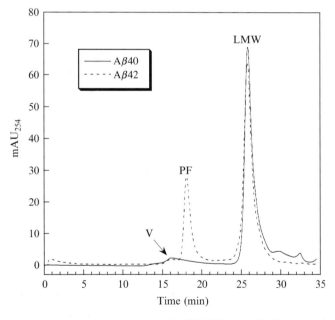

Fig. 1. Preparation of low molecular weight (LMW) amyloid β–protein (Aβ) by size exclusion chromatography (SEC). Aβ40 (solid line) and Aβ42 (dotted line) were fractionated by SEC using a Superdex 75 matrix and 10 m*M* sodium phosphate, pH 7.4, as the mobile phase. A small void volume peak (V) is observed in both samples, as is a major peak (LMW) in the included volume, which corresponds to LMW Aβ. A protofibril peak (PF) is shown in the Aβ42 sample. Adapted and reproduced with permission from Bitan *et al.* (2003).

microcentrifuge tube, and then sonicate the tube for 1 min in a bath sonicator. Centrifuge the peptide solution at 16,000g for 10 min to remove any large aggregates. Inject 160–180 μl of the supernate onto the equilibrated column and monitor the eluate using a UV detector. Protein peaks eluting from the column may be detected at a number of wavelengths, including 215 nm (peptide bonds), 254 nm (mercury line), or 280 nm (tyrosine absorbance). Column calibration with globular and polymeric standards will provide the most accurate indication of apparent molecular weight. Large aggregates elute first in a void volume peak that is followed by a peak comprising protofibrils (Fig. 1). Alkali-treated Aβ preparations should produce little or none of these components. LMW Aβ elutes later, with a retention time consistent with globular standards of molecular mass 5–15 kDa. Greatest homogeneity is obtained by collection only of the apex (the middle third, based on collection time) of the LMW peak. The LMW Aβ fraction should be used immediately after its isolation if structure activity correlations are to be made. Time delays and sample manipulation allow assembly of larger structures, which can complicate interpretation of the experimental data.

LMW Aβ40 produced by SEC is qualitatively similar to that produced by filtration (Fig. 2). The oligomerization state of Aβ42 differs within LMW fractions prepared by the two methods (Bitan *et al.*, 2003). In

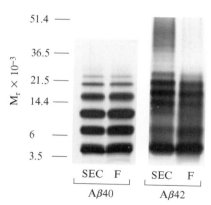

Fig. 2. Amyloid β–protein (Aβ) oligomer size distributions. Low molecular weight (LMW) Aβ40 and Aβ42 were isolated either by size exclusion chromatography (SEC) or by filtration through a 10,000 molecular weight cut-off [MWCO] filter. The peptides were photochemically cross-linked to produce a quantitative "snapshot" of the oligomer size distribution (Bitan *et al.*, 2001), and the products were analyzed by sodium dodecyl sulfate polyacrylamide gel electrophoresis. The mobilities of molecular mass markers are shown on the left. The oligomer size distributions obtained using the two methods are identical for Aβ40 but differ for Aβ42 (Bitan *et al.*, 2003). Adapted and reproduced with permission from Bitan *et al.* (2001) and Bitan *et al.* (2003).

addition to the relatively narrow (predominantly monomer through hepta-mer) distribution of oligomers observed with filtered preparations, SEC-isolated LMW Aβ42 produces higher order oligomers (Fig. 2).

Preparation of LMW Aβ by In Situ *Filtration.* The use of spectroscopic techniques, such as quasielastic light scattering, imposes stringent require-ments for sample "cleanliness." Successful measurement of the hydrody-namic radius of LMW molecules depends on the optical purity of the sample. Any high-molecular-weight particulate matter can prevent acqui-sition of useful spectra. In studies of Aβ, we have used the intrinsic filtering potential of chromatography column packing materials and a continuous flow procedure for washing the collection vessel (a cuvette) to produce optically pure samples (Fig. 3). Cuvettes first are prepared by heating the top 20 mm of standard 6 × 50 mm glass test tubes in the flame of a small Bunsen burner or torch. The tops of each tube then are pulled to form

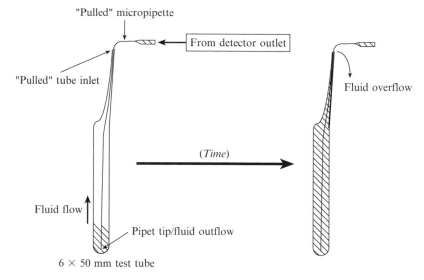

FIG. 3. Direct collection of low molecular weight (LMW) amyloid β–protein (Aβ) into cuvettes. One end of a 6 × 50 mm round-bottomed glass test tube is placed in a Bunsen burner, softened, and then pulled to create a narrow inlet. An identical procedure is performed on a glass micropipette. The "pulled" end of the micropipette then is inserted into the bottom of the test tube, and the nonpulled end is connected to the outlet of an ultraviolet detector flow cell. Size exclusion chromatography column eluate then washes the inside of the test tube from the bottom to the top, with the fluid and any dust particles continuously being washed over the lip of the test tube. After a peak of interest is collected, the capillary is removed from the end of the test tube, which is then heat-sealed to create a permanently "clean" environment.

narrow capillaries. A similar procedure is performed on disposable glass micropipettes to form a junction between the HPLC detector and the cuvette. In this case, the untreated end of the micropipette is inserted into the HPLC detector outflow line while the pulled tip of the micropipette is positioned inside the cuvette at its bottom. In this way, "filtered" buffer is constantly washing the interior of the cuvette from the bottom up through the narrow capillary top. To prepare the $A\beta$ sample, alkali-pretreated peptide is fractionated by SEC. When the peak of interest flows through the detector and fills the cuvette, the micropipette is removed from the cuvette and the end of the cuvette is fire-sealed immediately. This procedure, although somewhat cumbersome, provides excellent dust-free samples. The peak volumes of different $A\beta$ fractions are large relative to the volume of the cuvette; therefore, the cuvette behaves analogously to the detector flow cell in that cross-contamination among different fractions is not problematic.

Acknowledgments

This work was supported by grants NS38328, NS44147, AG18921, and AG027818 from the National Institutes of Health and by the Foundation for Neurologic Diseases. The author thanks Drs. Gal Bitan and Samir Maji for constructive criticism and comments on the manuscript. He also acknowledges the past and continuing efforts of members of the Teplow laboratory, who have developed and continue to refine $A\beta$ preparation techniques. Special recognition is accorded to Dr. Youcef Fezoui, who developed the NaOH procedure, and to Drs. Aleksey Lomakin and Dominic Walsh, who developed the *in situ* filtration technique.

References

Antzutkin, O. N., Balbach, J. J., Leapman, R. D., Rizzo, N. W., Reed, J., and Tycko, R. (2000). Multiple quantum solid-state NMR indicates a parallel, not anti parallel, organization of β-sheets in Alzheimer's β-amyloid fibrils. *Proc. Natl. Acad. Sci. USA* **97,** 13045–13050.

Balbach, J. J., Petkova, A. T., Oyler, N. A., Antzutkin, O. N., Gordon, D. J., Meredith, S. C., and Tycko, R. (2002). Supramolecular structure in full-length Alzheimer's β-amyloid fibrils: Evidence for a parallel β-sheet organization from solid-state nuclear magnetic resonance. *Biophys. J.* **83,** 1205–1216.

Barrow, C. J., Yasuda, A., Kenny, P. T. M., and Zagorski, M. (1992). Solution conformations and aggregational properties of synthetic amyloid β peptides of Alzheimer's disease. *J. Mol. Biol.* **225,** 1075–1093.

Benzinger, T. L. S., Gregory, D. M., Burkoth, T. S., Millerauer, H., Lynn, D. G., Botto, R. E., and Meredith, S. C. (1998). Propagating structure of Alzheimer's β-amyloid(10–35) is parallel β-sheet with residues in exact register. *Proc. Natl. Acad. Sci. USA* **95,** 13407–13412.

Bitan, G., Kirkitadze, M. D., Lomakin, A., Vollers, S. S., Benedek, G. B., and Teplow, D. B. (2003). Amyloid β-protein (Aβ) assembly: Aβ40 and Aβ42 oligomerize through distinct pathways. *Proc. Natl. Acad. Sci. USA* **100**, 330–335.

Bitan, G., Lomakin, A., and Teplow, D. B. (2001). Amyloid β-protein oligomerization— Prenucleation interactions revealed by photo-induced crosslinking of unmodified proteins. *J. Biol. Chem.* **276**, 35176–35184.

Bitan, G., and Teplow, D. (2005). Preparation of aggregate-free, low molecular weight amyloid-β for assembly and toxicity assays. *In* "Amyloid Proteins—Methods and Protocols" (D. E. M. Sigurdsson, ed.), pp. 3–9. Humana Press, Totowa, NJ.

Buck, M. (1998). Trifluoroethanol and colleagues: Cosolvents come of age. Recent studies with peptides and proteins. *Q. Rev. Biophys.* **31**, 297–355.

Buxbaum, J. N. (2003). Diseases of protein conformation: What do *in vitro* experiments tell us about *in vivo* diseases? *Trends Biochem. Sci.* **28**, 585–592.

Dubois, J., Ismail, A. A., Chan, S. L., and Ali-Khan, Z. (1999). Fourier transform infrared spectroscopic investigation of temperature- and pressure-induced disaggregation of amyloid A. *Scand. J. Immunol.* **49**, 376–380.

Fezoui, Y., Hartley, D. M., Harper, J. D., Khurana, R., Walsh, D. M., Condron, M. M., Selkoe, P. T., Lansbury, P. T., Fink, A. L., and Teplow, D. B. (2000). An improved method of preparing the amyloid β-protein for fibrillogenesis and neurotoxicity experiments. *Amyloid* **7**, 166–178.

Garzon-Rodriguez, W., Sepulveda-Becerra, M., Milton, S., and Glabe, C. G. (1997). Soluble amyloid Aβ-(1-40) exists as a stable dimer at low concentrations. *J. Biol. Chem.* **272**, 21037–21044.

Hass, G., and Schulz, R. Z. (1940). Amyloid. 1. Methods of isolating amyloid from other tissue elements. *Arch. Pathol.* **30**, 240–259.

Hoshi, M., Sato, M., Matsumoto, S., Noguchi, A., Yasutake, K., Yoshida, N., and Sato, K. (2003). Spherical aggregates of β-amyloid (amylospheroid) show high neurotoxicity and activate tau protein kinase I/glycogen synthase kinase-3β. *Proc. Natl. Acad. Sci. USA* **100**, 6370–6375.

Hou, L. M., Shao, H. Y., Zhang, Y. B., Li, H., Menon, N. K., Neuhaus, E. B., Brewer, J. M., Byeon, D. G., Ray, D. G., Vitek, M. P., Iwashita, T., Makula, R. A., Przybyla, A. B., and Zagorski, M. G. (2004). Solution NMR studies of the Aβ(1-40) and Aβ(1-42) peptides establish that the Met35 oxidation state affects the mechanism of amyloid formation. *J. Am. Chem. Soc.* **126**, 1992–2005.

Howlett, D. R., Jennings, K. H., Lee, D. C., Clark, M. S., Brown, F., Wetzel, R., Wood, S. J., Camilleri, P., and Roberts, G. W. (1995). Aggregation state and neurotoxic properties of Alzheimer β-amyloid peptide. *Neurodegeneration* **4**, 23–32.

Huang, T. H., Yang, D. S., Plaskos, N. P., Go, S., Yip, C. M., Fraser, P. E., and Chakrabartty, A. (2000). Structural studies of soluble oligomers of the Alzheimer β-amyloid peptide. *J. Mol. Biol.* **297**, 73–87.

Jao, S. C., Ma, K., Talafous, J., Orlando, R., and Zagorski, M. G. (1997). Trifluoroacetic acid pretreatment reproducibly disaggregates the amyloid β peptide. *Amyloid* **4**, 240–252.

Kheterpal, I., Lashuel, H. A., Hartley, D. M., Walz, T., Lansbury, P. T., Jr., and Wetzel, R. (2003). Aβ protofibrils possess a stable core structure resistant to hydrogen exchange. *Biochemistry* **42**, 14092–14098.

Kheterpal, I., Zhou, S., Cook, K. D., and Wetzel, R. (2000). Aβ amyloid fibrils possess a core structure highly resistant to hydrogen exchange. *Proc. Natl. Acad. Sci. USA* **97**, 13597–13601.

Kirkitadze, M. D., Bitan, G., and Teplow, D. B. (2002). Paradigm shifts in Alzheimer's disease and other neurodegenerative disorders: The emerging role of oligomeric assemblies. *J. Neurosci. Res.* **69,** 567–577.

Klein, W. L., Stine, W. B., Jr., and Teplow, D. B. (2004). Small assemblies of unmodified amyloid β-protein are the proximate neurotoxin in Alzheimer's disease. *Neurobiol. Aging* **25,** 569–580.

Krakow, N. P. (1898). Beitdige zur chemie der amyloidenartung. *Arch. Exp. Pathol. Pharmakol.* **40,** 195–220.

Lambert, M. P., Viola, K. L., Chromy, B. A., Chang, L., Morgan, T. E., Yu, J. X., Venton, G. A., Krafft, G. A., Finch, C. E., and Klein, W. L. (2001). Vaccination with soluble Aβ oligomers generates toxicity-neutralizing antibodies. *J. Neurochem.* **79,** 595–605.

Lazo, N. D., Maji, S. K., Fradinger, E. A., Bitan, G., and Teplow, D. B. (2005). The amyloid β-protein. *In* "Amyloid Proteins—The Beta Sheet Conformation and Disease" (J. C. Sipe, ed.), pp. 385–492. Wiley-VCH, Weinheim.

LeVine, H., III. (1995). Soluble multimeric Alzheimer β(1-40) pre-amyloid complexes in dilute solution. *Neurobiol. Aging* **16,** 755–764.

LeVine, H., III. (2004). Alzheimer's β-peptide oligomer formation at physiologic concentrations. *Anal. Biochem.* **335,** 81–90.

Lomakin, A., Benedek, G. B., and Teplow, D. B. (1999). Monitoring protein assembly using quasielastic light scattering spectroscopy. *Methods Enzymol.* **309,** 429–459.

Lomakin, A., Chung, D. S., Benedek, G. B., Kirschner, D. A., and Teplow, D. B. (1996). On the nucleation and growth of amyloid β-protein fibrils: Detection of nuclei and quantitation of rate constants. *Proc. Natl. Acad. Sci. USA* **93,** 1125–1129.

Lomakin, A., Teplow, D. B., and Benedek, G. B. (2005). Quasielastic light scattering for protein assembly studies. *In* "Amyloid Proteins: Methods and Protocols" (E. M. Sigurdsson, ed.), pp. 153–174. Humana Press, Totowa, NJ.

Maji, S. K., Amsden, J. J., Rothschild, K. J., Condron, M. M., and Teplow, D. B. (2005). Conformational dynamics of amyloid β-protein assembly probed using intrinsic fluorescence. *Biochemistry* **44,** 13365–13376.

Murphy, E. A. (1949). Murphy's law. Available at: http://www.edwards.af.mil/history/docs_html/tidbits/murphy's_law.html.

Perry, E. K., Oakley, A. E., Candy, J. M., and Perry, R. H. (1981). Properties and possible significance of substance P and insulin fibrils. *Neurosci. Lett.* **25,** 321–325.

Petkova, A. T., Ishii, Y., Balbach, J. J., Antzutkin, O. N., Leapman, R. D., Delaglio, F., and Tycko, R. (2002). A structural model for Alzheimer's β-amyloid fibrils based on experimental constraints from solid state NMR. *Proc. Natl. Acad. Sci. USA* **99,** 16742–16747.

Pike, C. J., Burdick, D., Walencewicz, A. J., Glabe, C. G., and Cotman, C. (1993). Neurodegeneration induced by β-amyloid peptides *in vitro*: The role of peptide assembly state. *J. Neurosci.* **13,** 1676–1687.

Pike, C. J., Walencewicz, A. J., Glabe, C. G., and Cotman, C. W. (1991). *In vitro* aging of β-amyloid protein causes peptide aggregation and neurotoxicity. *Brain Res.* **563,** 311–314.

Pras, M., Zucker-Franklin, D., Rimon, A., and Franklin, E. C. (1969). Physical, chemical, and ultrastructural studies of water-soluble human amyloid fibrils. *J. Exp. Med.* **130,** 777–796.

Seilheimer, B., Bohrmann, B., Bondolfi, L., Muller, F., Stuber, D., and Dobeli, H. (1997). The toxicity of the Alzheimer's β-amyloid peptide correlates with a distinct fiber morphology. *J. Struct. Biol.* **119,** 59–71.

Selkoe, D. J. (1991). The molecular pathology of Alzheimer's disease. *Neuron* **6,** 487–498.

Selkoe, D. J. (2001). Alzheimer's disease: Genes, proteins, and therapy. *Physiol. Rev.* **81,** 741–766.

Shen, C. L., and Murphy, R. M. (1995). Solvent effects on self-assembly of β-amyloid peptide. *Biophys. J.* **69,** 640–651.

Shirahama, T., and Cohen, A. S. (1967). Reconstitution of amyloid fibrils from alkaline extracts. *J. Cell Biol.* **35,** 459–464.

Simmons, L. K., May, P. C., Tomaselli, K. J., Rydel, R. E., Fuson, K. S., Brigham, E. F., Wright, I., Lieberburg, I., Becker, G. W., Brems, D. N., and Li, W. Y. (1994). Secondary structure of amyloid β peptide correlates with neurotoxic activity *in vitro. Mol. Pharmacol.* **45,** 373–379.

Soto, C., Castano, E. M., Kumar, R. A., Beavis, R. C., and Frangione, B. (1995). Fibrillogenesis of synthetic amyloid-β peptides is dependent on their initial secondary structure. *Neurosci. Lett.* **200,** 105–108.

Stine, W. B., Dahlgren, K. N., Krafft, G. A., and LaDu, M. J. (2003). *In vitro* characterization of conditions for amyloid-β peptide oligomerization and fibrillogenesis. *J. Biol. Chern.* **278,** 11612–11622.

Taylor, B. M., Sarver, R. W., Fici, G., Poorman, R. A., Lutzke, B. S., Molinari, A., Kawabe, T., Kappenman, A. E., Buhl, A. E., and Epps, D. E. (2003). Spontaneous aggregation and cytotoxicity of the β-amyloid Aβ1–40: A kinetic model. *J. Prot. Chem.* **22,** 31–40.

Teplow, D. B. (1998). Structural and kinetic features of amyloid β-protein fibrillogenesis. *Amyloid* **5,** 121–142.

Thirumalai, D., Klimov, D. K., and Dima, R. I. (2003). Emerging ideas on the molecular basis of protein and peptide aggregation. *Curr. Opin. Struct. Biol.* **13,** 146–159.

Torok, M., Milton, S., Kayed, R., Wu, P., McIntire, T., Glabe, C., and Langen, R. (2002). Structural and dynamic features of Alzheimer's Aβ peptide in amyloid fibrils studied by site-directed spin labeling. *J. Biol. Chem.* **277,** 40810–40815.

Tycko, R. (2004). Progress towards a molecular-level structural understanding of amyloid fibrils. *Curr. Opin. Struct. Biol.* **14,** 96–103.

Walsh, D. M., Hartley, D. M., Kusumoto, Y, Fezoui, Y, Condron, M. M., Lomakin, A., Benedek, D. J., Selkoe, D. J., and Teplow, D. B. (1999). Amyloid β-protein fibrillogenesis—Structure and biological activity of protofibrillar intermediates. *J. Biol. Chem.* **274,** 25945–25952.

Walsh, D. M., Lomakin, A., Benedek, G. B., Condron, M. M., and Teplow, D. B. (1997). Amyloid β-protein fibrillogenesis—Detection of a protofibrillar intermediate. *J. Biol. Chem.* **272,** 22364–22372.

Walsh, D. M., and Selkoe, D. J. (2004). Deciphering the molecular basis of memory failure in Alzheimer's disease. *Neuron* **44,** 181–193.

Wang, H. W., Pasternak, J. F., Kuo, H., Ristic, H., Lambert, M. P., Chromy, B., Viola, K. L., Klein, W. B., Stine, W. B., Krafft, G. A., and Trommer, B. L. (2002). Soluble oligomers of β amyloid (1-42) inhibit long-term potentiation but not long-term depression in rat dentate gyrus. *Brain Res.* **924,** 133–140.

Wood, S. J., Maleeff, B., Hart, T., and Wetzel, R. (1996). Physical, morphological and functional differences between pH 5.8 and 7.4 aggregates of the Alzheimer's amyloid peptide Aβ. *J. Mol. Biol.* **256,** 870–877.

Zagorski, M. G., and Barrow, C. J. (1992). NMR studies of amyloid β-peptides: Proton assignments, secondary structure, and mechanism of an α-helix→β-sheet conversion for a homologous, 28-residue, N-terminal fragment. *Biochemistry* **31,** 5621–5631.

Zagorski, M. G., Yang, J., Shao, H., Ma, K., Zeng, H., and Hong, A. (1999). Methodological and chemical factors affecting amyloid-β peptide amyloidogenicity. *Methods Enzymol.* **309,** 189–204.

[3]   Kinetics and Thermodynamics of Amyloid Assembly Using a High-Performance Liquid Chromatography–Based Sedimentation Assay

By Brian O'Nuallain, Ashwani K. Thakur, Angela D. Williams, Anusri M. Bhattacharyya, Songming Chen, Geetha Thiagarajan, and Ronald Wetzel

Abstract

Nonnative protein aggregation has been classically treated as an amorphous process occurring by colloidal coagulation kinetics and proceeding to an essentially irreversible endpoint often ascribed to a chaotic tangle of unfolded chains. However, some nonnative aggregates, particularly amyloid fibrils, exhibit ordered structures that appear to assemble according to ordered mechanisms. Some of these fibrils, as illustrated here with the Alzheimer's plaque peptide amyloid β, assemble to an endpoint that is a dynamic equilibrium between monomers and fibrils exhibiting a characteristic equilibrium constant with an associated free energy of formation. Some fibrils, as illustrated here with the polyglutamine repeat sequences associated with Huntington's disease, assemble via highly regular mechanisms exhibiting nucleated growth polymerization kinetics. Here, we describe a series of linked methods for quantitative analysis of such aggregation kinetics and thermodynamics, focusing on a robust high-performance liquid chromatography (HPLC)–based sedimentation assay. An integrated group of protocols is provided for peptide disaggregation, setting up the HPLC sedimentation assay, the preparation of fibril seed stocks and determination of the average functional molecular weight of the fibrils, elongation and nucleation kinetics analysis, and the determination of the critical concentration describing the thermodynamic endpoint of fibril elongation.

Introduction

Some of the first analytical techniques applied to the understanding of purified individual proteins involved monitoring their aggregation in response to solvent changes, heat, or other conditions of stress (Neurath *et al.*, 1944). Despite this history, our understanding of the mechanisms of formation, structures, and structural dynamics of protein aggregates has lagged well behind our appreciation of the behavior of globular proteins in

METHODS IN ENZYMOLOGY, VOL. 413
Copyright 2006, Elsevier Inc. All rights reserved.

solution. To some extent, this is attributable to a historical lack of interest in protein aggregation (Neurath *et al.*, 1944) that only began to change as the importance of the process to basic cellular biology (Kopito, 2000; Meijer and Codogno, 2004), protein folding (Stefani and Dobson, 2003), biotechnology (Mitraki and King, 1989; Wetzel, 1994), and human disease (Falk *et al.*, 1997; Martin, 1999; Merlini and Bellotti, 2003) became apparent. An additional constraint, however, has been the lack of development of analytical technologies of sufficient precision to allow extraction of interpretable mechanistic information. As interest in protein aggregation has increased over the past 30 years, newer methods have been developed. This is especially true over the past decade, a time of rapid growth in the number of laboratories working on protein aggregation in general and on amyloid in particular.

To approach a mechanistic understanding of any physicochemical transformation, it is important to monitor molecular conversions quantitatively—to know how many molecular different species exist along the reaction profile, to be able to assign numbers of molecules to each of these molecular bins, and to know how these change with reaction time. This has been a problem in protein aggregation studies. Turbidity assays have been used effectively to develop some basic understanding of aggregation kinetics (Andreu and Timasheff, 1986; Mulkerrin and Wetzel, 1989) but suffer from the tendency of the largest aggregates to dominate the signal. More sophisticated static and dynamic light-scattering methods (Lomakin *et al.*, 1999) offer significant improvements over turbidity assays but also suffer from a dominant contribution to the signal from larger aggregates, which can make it difficult to get reliable data on small components, particularly the monomer pool. Dye-binding assays, such as with Congo red (Klunk *et al.*, 1999) and thioflavin T (ThT) (LeVine, 1999) offer some advantages (e.g., discrimination between fibrils and protofibrils) but also some limitations (e.g., widely varying fluorescence yields per weight of fibril). In favorable cases, circular dichroism (CD) performs so well that it can distinguish different conformations of amyloid (Yamaguchi *et al.*, 2005), but CD also can suffer from interference from light scattering by large aggregates (Colon, 1999) and is less useful if the native state is rich in β-sheet. Fourier transform infrared (FTIR) spectroscopy has also been used to follow amyloid formation (Nilsson, 2004) but is not convenient for monitoring reaction time points at the relatively low concentrations normally used for fibril formation reactions. These and other methods are capable of qualitatively monitoring aggregate formation but are less useful at rigorously quantifying both aggregates and unreacted monomer. This limits their utility in serving detailed kinetics analysis (Chen *et al.*, 2002b; O'Nuallain *et al.*, 2005).

This chapter is devoted to studies focusing on the amount of material remaining in solution as low-molecular-weight species during the aggregation reaction. The major advantage of this approach is that it is relatively straightforward to quantify the soluble material with great accuracy and reproducibility. A disadvantage is that by only focusing on the soluble, nonsedimentable fraction, one risks oversimplifying the aggregation process. This is an important concern, because it is now clear there are multiple aggregated states, including multiple amyloid states in some cases (Petkova *et al.*, 2005; Tanaka *et al.*, 2004), which are available to many proteins and peptides and can coexist at intermediate times during amyloid formation. It is therefore important to conduct preliminary studies of any new aggregation system to convince oneself that one understands the nature of the reactants, intermediates, and products using methods, such as hydrogen–deuterium exchange protection (see Chapter 8 by Kheterpal and colleagues in this volume) that allow accurate discrimination and quantitation of various molecular states. Such studies should establish the feasibility of studying aggregation reaction flux using a monomer-centric method, such as the sedimentation assay described here. Where it is not possible to quantify intermediates, the ability to conduct detailed kinetics analysis is diminished and one is constrained to focus on the reaction endpoint (see section on thermodynamics of amyloid fibril elongation).

The examples used in this chapter to illustrate the use of high-performance liquid chromatography (HPLC) sedimentation assays are relatively simple peptides from the point of view of their exhibiting no detectible stable structure in native buffer. This can be a tremendous advantage, because it removes the elements of precursor unfolding or misfolding, which are otherwise required initiating steps for the aggregation of many globular proteins (Hurle *et al.*, 1994; McCutchen *et al.*, 1993). One approach to studying aggregation of globular proteins is to study the aggregation reaction under conditions where native globular structure has been destabilized (Ignatova and Gierasch, 2005) (see Chapter 4 by Chiti and colleagues in this volume).

This chapter describes our laboratory's techniques built around the use of the HPLC sedimentation assay to study amyloid formation by peptides. These methods were developed for studies on the Alzheimer's plaque peptide amyloid β (Aβ) (Cannon *et al.*, 2004; O'Nuallain *et al.*, 2004, 2005; Shivaprasad and Wetzel, 2004, 2006; Williams *et al.*, 2004, 2005, 2006; Wood *et al.*, 1996) and the polyglutamine (polyGln) sequence important in Huntington's disease and other expanded CAG repeat diseases (Bhattacharyya *et al.*, 2005, 2006; Chen *et al.*, 2001, 2002b; Sharma *et al.*, 2005; Thakur and Wetzel, 2002; Yang *et al.*, 2002). The various elements include: (1) disaggregation of peptides prior to initiation of aggregation,

(2) preparation of seeds for use in elongation studies, (3) the basic HPLC sedimentation assay, (4) determination of the average functional molecular weight of an aggregate population, (5) elongation kinetics, (6) nucleation kinetics, and (7) thermodynamics of amyloid fibril elongation. The chapter is necessarily lengthy, and it might be argued that it could have been divided into multiple chapters. However, because the methods are all interlinked, the most efficient way of describing them all is to describe them all together.

Disaggregation of Peptides

The realization that many peptide preparations can be contaminated with small but functionally important levels of preexisting aggregates capable of acting as cytotoxic agents (Howlett et al., 1995), as seeds to accelerate the aggregation of the monomeric peptides (Evans et al., 1995), or as inhibitors to poison spontaneous aggregation (Wood et al., 1996) and the description of methods for the effective removal of these aggregates were important advances allowing the field to achieve some degree of control and reproducibility of in vitro studies. A number of disaggregation methods have been described. These include filtration of dimethyl sulfoxide (DMSO) solutions (Evans et al., 1995), sequential, transient treatment with trifluoroacetic acid (TFA) and hexafluoroisopropanol (HFIP) (Zagorski et al., 1999), transient treatment with HFIP (Wood et al., 1996), ultracentrifugation of aqueous solutions (Zhang et al., 2000), dissolution in basic aqueous solution (see Chapter 2 by Teplow in this volume), and transient treatment in a 1:1 mixture of TFA and HFIP followed by ultracentrifugation in aqueous solution (Chen and Wetzel, 2001; Wetzel, 2005). Each has potential advantages and disadvantages. Some workers successfully use material directly from the lyophilized powder, without even a centrifugation step to remove any aggregates. For some applications, disaggregation may not be necessary, for example, if one is making aggregates for a subsequent use and is less interested in details of assembly kinetics and mechanism. However, even in this case, the decision to skip the disaggregation step may carry with it unappreciated consequences. For example, the structure of any adventitious aggregate seed present in the peptide solution will self-propagate and therefore determine the structure of the aggregated product; in some cases, this might differ from the structure of an aggregate formed by spontaneous aggregation from fully disaggregated monomer under the same conditions. Until recently, this concern was only hypothetical, but it has become real with the description of different conformers of amyloid fibrils being formed from the same peptide sequence (Petkova et al., 2005; Tanaka et al., 2004).

Experimental Results

The basic method described here involves two steps: (1) the dissolution and breakdown of aggregated structures through treatment with TFA or TFA-HFIP mixtures, followed by removal of these volatile solvents, and (2) aqueous dissolution of the resulting disaggregated peptide film, followed by high-speed centrifugation to remove trace aggregates. For peptide samples that are obtained in highly aggregated form, both steps are important. For peptide samples that appear to dissolve readily in water or phosphate-buffered saline (PBS), it is possible that the centrifugation step might be sufficient. We routinely carry out both steps, however. Although this method in outline is general, if this approach is applied to other amyloidogenic peptides, it must be kept in mind that modifications, for example, based on awareness of peptide's pI, may be required.

The sometimes subtle importance of aggregates in apparently soluble peptides is illustrated by work with a short polyGln peptide $D_2Q_{15}K_2$ (Chen and Wetzel, 2001). When this peptide is dissolved from the lyophilized powder in pH 3 water, it appears to dissolve completely, giving a clear solution with very low Rayleigh light scattering. Adjusting this "solution" to PBS conditions, however, and incubating at 37° lead to rapid aggregation that is over after 5–10 min (Fig. 1). After pretreatment of another sample of lyophilized peptide with a TFA/HFIP/centrifugation protocol,

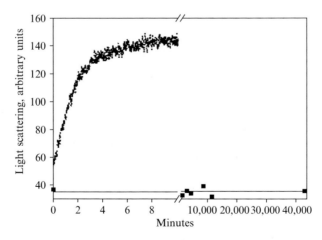

Fig. 1. Time course of aggregation of a $7\text{-}\mu M$ solution of D2Q15K2 at pH 7.5 and 37° when synthetic peptide was dissolved directly in pH 3 water before adjusting to pH 7.5 (top curve, rapid aggregation) or pretreated with the trifluoroacetic acid-hexafluoroisopropanol (TFA-HFIP) protocol before dissolving in pH 3 water and then adjusting to pH 7.5 (bottom curve, ■; no aggregation). Reproduced with permission from Chen and Wetzel (2001).

however, the same concentration of the peptide in PBS and 37° does not aggregate appreciably after hours or even days of incubation, only reaching an approximate equilibrium position after about 6 months. Thus, even when peptide solutions appear to be soluble, they may harbor levels of seed aggregates that can dramatically change the course of aggregation.

As peptides become less soluble, disaggregation conditions must be made more stringent. The Q_{15} peptide discussed above can be disaggregated by sequential treatments of TFA and HFIP. The TFA has the combined ability of a strong organic acid to dissolve H-bonded protein aggregates, on the one hand, although being relatively volatile and therefore easy to remove on the other hand. The primary function of the HFIP is normally to facilitate the removal of TFA (Zagorski et al., 1999), although it has some ability itself to break down aggregates (Wood et al., 1996). A 1:1 mixture of TFA and HFIP was found, surprisingly, to have superior disaggregation properties to TFA alone, such that long polyGln peptides that are not disaggregated by TFA alone can be handled by the 1:1 mixture (Chen and Wetzel, 2001). Incubation time is also important. Optimal incubation time can be determined empirically by occasionally removing aliquots for a microscale workup, followed by a light-scattering test for aggregate formation (Chen and Wetzel, 2001; Wetzel, 2005). The ultimate measure of success of a disaggregation protocol is the presence of a lag phase of reproducible and maximal length when the treated peptide is incubated under appropriate conditions. Thus, application of a modification of the 1:1 TFA/HFIP protocol to the problematic islet amyloid polypeptide (IAPP) led to spontaneous aggregation reactions featuring a measurable lag time (O'Nuallain et al., 2004), which is generally not observed when IAPP is disaggregated by other protocols.

It should be pointed out that although nucleation is a stochastic process at the molecular level, in a solution containing many molecules and exhibiting multiple independent nucleation events (i.e., multiple fibrils), one expects lag times to be reproducible. Variation in observed lag times in multiple ensemble experiments is not evidence for a nucleated growth mechanism but rather an indication of lack of experimental control over the system.

Whatever the details of the TFA/HFIP treatment, the manner by which these volatile solvents are removed can be extremely important. It is convenient to remove the bulk of the volatile solvent under a stream of gas in a chemical fume hood. For peptides containing easily oxidizable amino acids like methionine or cysteine, it is probably important to use an oxygen-free gas. Oxygen-free nitrogen is available, or argon can be used. Once the peptide has been disaggregated and the bulk of the organic solvent has been removed so that only a peptide film remains, it is important to remove traces of HFIP under high vacuum. This is because buffers containing low percentages of HFIP can radically alter the aggregation pathway of amyloidogenic peptides (Nichols *et al.*, 2005). Perhaps in part because of the importance of removing HFIP, when scaling the TFA/HFIP step for larger amounts of peptide, it is important to retain approximately equivalent surface-to-volume ratios. One simple way to scale up a disaggregation protocol reliably is to set up parallel, multiple small reactions rather than place all the material in a single large vessel.

It may be possible to achieve satisfactory results using a rotary evaporator or speed-vac to remove the bulk of the TFA and HFIP; we have not attempted to do this. Some cautions are in order for those attempting one of these adaptations. Using a speed-vac in place of gentle volatilization under a stream of nitrogen or argon may lead to all of the protein being concentrated in a small volume of gum at the base of the flask rather than being spread over a large area. This may affect the efficiency of dissolution in the aqueous solvent and lead to insoluble material. A rotary evaporator may produce the desired thin film of peptide residue, but it is not advisable to use a water aspirator to provide the vacuum. Exposure of dry films of an amyloidogenic peptide to water vapor can lead to rapid conversion to a β-sheet-rich structure (Heller *et al.*, 1996), most likely indicative of aggregation.

In addition to TFA treatment conditions, other parameters that affect peptide solubilities include the pH of the aqueous solvent used to dissolve the peptide after removal of TFA/HFIP, the aqueous solvent used for the centrifugation step, and the details of the centrifugation conditions. The pH is important, because undue exposure of an amyloidogenic peptide to a pH near its pI can lead to rapid isoelectric precipitation to an alternatively structured aggregate that may even poison amyloid growth (Wood *et al.*, 1996). For example, weak acid is used for dissolving polyGln peptides containing flanking Lys residues, but this is contraindicated for the peptide $A\beta$ (pI $\approx$6), for which the preferred aqueous solvent is dilute aqueous sodium hydroxide quickly followed by addition of $2\times$ PBS. Centrifugation conditions are important, because some highly aggregation-prone peptides, such as $A\beta(1\text{-}42)$, are completely lost to solution on prolonged high-speed centrifugation. The centrifugation buffer can also be adjusted to minimize the possibility of aggregation during centrifugation; for example, if aggregation is greatly favored by salt, the centrifugation can be conducted in low-salt conditions and the buffer adjusted to the desired salt concentration after the supernatant has been harvested.

Using this protocol, we obtain reproducible spontaneous aggregation kinetics for various $A\beta(1\text{–}40)$ molecules and for polyGln peptides of various repeat lengths. Difficult peptides, such as $A\beta(1\text{–}42)$, IAPP, and polyGln peptides of repeat lengths in the range of 50, have also been success-fully treated. Whether organic solvent dissolution can be an effective protocol for amyloidogenic globular proteins is not clear, because it requires the ability of the protein to refold efficiently when resuspended in aqueous buffer. With appropriate concern for the variables discussed above, however, the method should be applicable to many amyloidogenic peptides.

Because biological activity normally requires solubility, the ability of a bioactive peptide to aggregate, facilitated by small amounts of seeds, can compromise any experiment exploring the activity or properties of pep-tides. These methods may thus prove useful for working with synthetic peptides in all sorts in applications other than aggregation studies (Chen and Wetzel, 2001).

Protocols

Disaggregation of Polyglutamine Peptides

1. Suspend approximately 2 mg dry peptide in 4 ml of a 1:1 mixture (v/v) of TFA (Pierce) and HFIP (Acros) in a sealed 20-ml glass Erlenmeyer flask and incubate at room temperature (RT) for 5 h or until a test for residual aggregates is negative.

2. In a chemical fume hood, dry the peptide solution under a stream of argon gas, and then immediately place it under vacuum to remove any residual volatile solvents. In humid climates, it may be advisable to vent the vacuum with dry air or gas.

3. Dissolve the residue in pH 3 aqueous TFA to a concentration in the range of 100 μM and centrifuge at 386,000g for 3 h (or overnight if necessary) at 4° in a table-top ultracentrifuge.

4. Carefully remove the top two thirds of the solution, place on ice, and determine the polyGln concentration using the HPLC assay (see section on HPLC sedimentation assay for amyloid transformations). Add an appropriate volume of a buffer concentrate, such as 10× PBS, to adjust the peptide to buffered reaction conditions, and dilute as necessary with buffer to adjust to the desired concentration.

5. Disaggregated peptide should normally be prepared fresh, because aggregates build up even when the peptide is stored at −80°. If storage is attempted, avoid exposing the polyGln to frozen conditions above the eutectic points of the solutes (typically above −30°), because this will likely lead to aggregation attributable to freeze concentration (Chen *et al.*, 2002a) (see section on preparation of stocks of aggregate seeds). Thus, snap-freeze in dry ice and ethanol or liquid nitrogen, and store at −80°. It may be possible to remove trace aggregates from stored samples by

repeating the centrifugation step after thawing. If this is done, it may be advisable to redetermine the peptide concentration in the supernatant.

6. Aggregate test. To test TFA/HFIP-incubated peptide for degree of disaggregation, remove a small aliquot, dry under argon followed by brief exposure to vacuum, dissolve in pH 3 water, and add a 1:9 volume of 10× PBS (or buffer concentrate later to be used in the reaction). This solution should be analyzed in a fluorometer in which emission and excitation wavelengths are set at 450 nm. The Rayleigh light-scattering signal obtained should be comparable to scattering from buffer alone and significantly lower than dilutions of an equivalent weight concentration of completely aggregated peptide in the same buffer. Analysis of aliquots should be continued at longer incubation times until the signal is that of buffer alone or, failing that, has reached a minimum not far from buffer alone.

Disaggregation of Aβ(1–40) Peptides

1. Suspend 1 mg lyophilized peptide powder in 1 ml TFA in a small glass vial. Cap the vial, and sonicate the sample in a water bath sonicator for 10 min.

2. In a chemical fume hood, dry the peptide solution under a gentle stream of argon gas that distributes material to the flask walls as it reduces volume. When done properly, you should see a thin film on the walls of the vial.

3. Dissolve the peptide in 1 ml HFIP. Swirl carefully to make sure all the monomer has gone into solution and is not clinging to the sides of the vial. Incubate the closed solution of peptide at 37° for 1 h.

4. Transfer a small, measured aliquot of the HFIP solution into 0.05% TFA in water and chromatograph on HPLC to determine the amount of peptide in the HFIP solution (see section on HPLC sedimentation assay for amyloid transformations). This value will be approximate, because the low surface tension of HFIP makes accurate pipetting difficult (transfer rapidly). However, because lyophilized peptide contains significant and variable amounts of water, it will be more accurate than the powder weight of the peptide from step 1.

5. In a chemical fume hood, dry the peptide solution under a stream of argon.

6. For a second time, dissolve the residue in 2 ml HFIP. This step is done to ensure that all of the TFA is removed from the sample.

7. Based on the HPLC determination of the amount of peptide in the sample, aliquot sample into glass tubes at 0.25 mg per tube. Do not put more peptide in any one tube so as to avoid precipitation of Aβ during step 10.

8. Evaporate the HFIP in tubes under a stream of argon in a chemical fume hood.

9. Immediately further dry the peptide under vacuum (a standard lyophilizer is fine) for 30–60 min to ensure that all of the TFA and HFIP has been removed from the peptide. It is critical to move without delay from step 8 to step 9 and from step 9 to step 10. Delays can lead to aggregate formation upon addition of the 2 mM NaOH. This is probably attributable to the adsorption of water from the atmosphere by the dry peptide films, leading to essentially very concentrated aqueous solutions of peptide that are prone to aggregate. In humid environments, vent the vacuum with dry air or dry gas.

10. Slowly add 0.5 ml fresh 2 mM NaOH per tube (do not maintain a stock solution of 2 mM NaOH because it will be neutralized over time from carbonic acid formed from atmospheric CO_2; this will lead to incomplete neutralization of TFA and isoelectric precipitation of Aβ). Let it stand 5 min undisturbed, and then slowly add 0.5 ml 2× PBS with 0.1% sodium azide per tube. When adding each of these solvents, do not agitate. If needed, roll the liquid in the tube gently around the sides. Any aggressive agitation of the solution (e.g., vortexing, shaking vigorously) will induce precipitation. Centrifuge the peptide at 386,000g overnight at 4° in a table-top ultracentrifuge.

11. Carefully remove the peptide solution supernatant. If desired, keep it at 4° while determining the Aβ(1–40) concentration using the HPLC assay.

12. Disaggregated monomer solutions can be preserved by first snap-freezing in dry ice and ethanol or liquid nitrogen and then storing at −80°.

Preparation of Stocks of Aggregate Seeds

Defined preparations of amyloid fibrils and other aggregates have a number of uses. For example, a seed stock is useful to ensure uniformity when amyloid preparations are made for other studies via seeded elongation. In addition, preformed aggregates serve as the basis of some elongation assays for compound library screening (Williams et al., 2005) and otherwise can be used to initiate solution phase elongation assays (Bhattacharyya et al., 2005). Aggregates can also be administered to cells (Bucciantini et al., 2002; Howlett et al., 1995; Tanaka et al., 2004; Yang et al., 2002) for determination of toxicity or other biological activities. Growth, stabilization, and characterization of the specific elongation activity of an aggregate seed preparation are critical to the calculation of the nucleation equilibrium constant for a nucleated growth polymerization mechanism (Bhattacharyya et al., 2005). Well-characterized aggregates are also important in cross-seeding experiments that can inform about structure–function aspects of amyloid fibril elongation, and hence amyloid structure (Jones and Surewicz, 2005; O'Nuallain *et al.*, 2004).

Experimental Results

The particular conformation of an amyloid fibril appears to "breed true" when fibrils are used as seeds for stimulating growth of additional aggregates by elongation (Petkova *et al.*, 2005), and this appears to be the basis of strain effects and species barriers in prion diseases (Tanaka *et al.*, 2004). Seeded fibril elongation thus provides a way of maintaining individual "lines" of fibrils distinguished by their conformations and unique properties. At the same time, this places a premium on conformational purity in the seed stock, because, in principle, the conformational purity of the aggregated product is unlikely to be any better than that of the seeds used to produce it. In some cases, the ratio of amyloid conformations within an aggregate population can shift during sequential rounds of elongation (Yamaguchi *et al.*, 2005). Under special circumstances, a particular amyloid can be prepared in a conformationally pure state from aggregates prepared *in vitro* by using microbial cloning methods (Tanaka *et al.*, 2004). Except for such methods, however, the field is currently limited in its ability to create absolutely pure conformational states. New separation methods (see Chapter by Silveira and colleagues and Chapter by Bagriantsev and co-workers, in this volume) or a cloning strategy (see Chapter by Tanaka and Weissman in this volume) may ultimately facilitate preparation of pure or enriched states of conformational variants.

The specific conformational features of aggregates formed by spontaneous growth from pure monomeric (disaggregated) populations can vary significantly with growth conditions. Traditionally, amyloid fibrils of the $A\beta(1-40)$ peptide have been grown either with shaking (Esler *et al.*, 1997; Evans *et al.*, 1995) or under quiescent conditions (Hilbich *et al.*, 1992; Wood *et al.*, 1996). The Tycko group showed that the amyloid fibrils formed by these two approaches differ both in macroscopic structure and in the details of how $A\beta$ is folded within the fibril (Petkova *et al.*, 2005). Although fibrils formed under a particular set of conditions may be dominated by certain conformational types, they may still harbor a variety of minor states as well, which presumably also breed true and will continue to contaminate the elongated aggregation products. It may eventually be possible to estimate the conformational purity of a preparation with the aid of convenient physical measurements.

The polyGln peptides also exhibit a wide range of amyloid-like aggregate morphologies, many of which are capable of seeding aggregate growth by monomer addition. Product structure ranges from typical amyloid fibrils, to ribbons and bundles of long rigid protofilaments, to smaller protofibril-like structures (Chen *et al.*, 2002a). Aggregate structure is controlled both by polyGln repeat length and growth conditions (Chen *et al.*, 2002a). Longer polyGln sequences favor amyloid fibril formation, as do the low-salt growth conditions that tend to slow down aggregate assembly. Formation of protofibril-like structures requires the unusual growth condition of freezing a PBS solution in the $-20°$ to $-5°$ range. Aggregates form efficiently in the space of 24 h because of the process of freeze concentration (Franks, 1993), which forces solutes (both the polyGln and the salts) into small fissures within the imperfect ice lattice and thereby creates very high concentrations that drive aggregate formation ($A\beta$ solutions appear not to be susceptible to this effect). Freezing monomer at $-70°$, below the eutectic points of the solutes, preserves the monomer state.

Seeding efficiency, which is related to the average number of growth sites per weight of aggregate, can be increased by sonication (Jarrett and Lansbury, 1992). Effective shearing of fibril products normally requires a probe (as opposed to a bath) sonicator. The ability of sonication to magnify the seeding potential of a preparation is the basis of a method for increasing the sensitivity of detection methods for mammalian prions (see Chapter by Soto and colleagues in this volume). Sonication apparently shears fibrils, altering their conformational integrity, although more work is required to confirm this. Sonication is normally used to enhance seeding potential, but it also can be used to increase the uniformity of a fibril preparation (O'Nuallain *et al.*, 2004) and enhance the ability of cells to take up aggregates (Yang *et al.*, 2002). Although electron microscopy (EM) analysis confirms that isolated fibrils have a shorter average length after sonication, it appears that sonication does not necessarily significantly reduce the average particle size of fibril particles in suspension phase, presumably because of fibril clustering (see Chapter 11 by Mok and Howlett in this volume).

For many applications, it is useful to be able to characterize a fibril preparation according to weight concentration and the number of growth sites per unit weight. Weight concentration can be indirectly determined by measuring the amount of monomer left unreacted at the end of a fibril formation reaction. The assumption that all peptide not accounted for in the monomer pool must be in the aggregate pool is fine, so long as the aggregate has not been subjected to any transfers or concentration steps in the interim. If that is the case, however, it is better to assess the aggregate weight concentration of the final aggregate stock suspension directly, as described below. Methods for titrating fibril growth sites are described in the section on titration of fibril growing ends.

Stock aggregate suspensions are best stored by snap-freezing in liquid nitrogen or dry ice and ethanol, followed by storage at $-80°$. Even under these conditions, their properties, such as seeding potential, may degrade somewhat over a period of months. There is not enough experience to formulate general rules; aggregate stability under storage should be confirmed for each system and aggregate type.

Protocols

Growth and Storage of Aβ(1–40) Aggregates Under Conditions of Quiescent Growth

1. For spontaneous growth of $A\beta(1$–40) fibrils, incubate 20–60 μM freshly disaggregated $A\beta(1$–40) in PBS with 0.05% sodium azide in an Eppendorf or Falcon tube at 37°. The samples are kept undisturbed for the duration of the growth. This procedure also works for many mutants of $A\beta$ (1–40), although some require higher starting concentrations of peptide because of their high critical concentration (C_r) values. Other buffers and temperatures can also be used but may lead to alternate fibril conformations (Petkova *et al.*, 2005).

2. Assay time points should be taken for the HPLC sedimentation assay and/or ThT assay by gently inverting the reaction several times to mix before removing aliquots to be tested.

3. When the aggregation is complete as determined by the ThT and/or HPLC sedimentation assay, collect the $A\beta(1$-40) aggregates by centrifugation for 30 min at 20,800g (14,000 rpm in an Eppendorf 5417R centrifuge). Resuspend in buffer with gentle swirling or vortexing. This step removes the azide and buffer components as well as any unpolymerized monomer and allows for generation of a concentrated aggregate suspension, if desired. Determine the weight concentration of aggregates, if desired. Preserve the stock suspensions by snap-freezing in dry ice and ethanol or liquid nitrogen and then placing the stocks at $-80°$ for long-term storage.

4. If desired, sonicate the aggregate stock suspension using a probe sonicator. We typically sonicate on ice five times using a cycle of a 30-s pulse followed a by 1-min delay. We normally include 1 mM DTT as a precaution against air oxidation during the sonication, but, of course, this is not advisable, for example, for mutated $A\beta$ fibrils containing disulfide bonds. We have no data indicating that the DTT is required, but Met 35 of $A\beta$ is the most chemically labile amino acid residue in wild-type $A\beta$ under most conditions.

5. It should be noted that removal of unpolymerized monomer in step 3 may create a situation where aggregate is prone to dissociate partially if allowed to stand for extended periods under dissociation conditions. This is more of a concern with aggregates that exhibit significant C_r values and

dissociation kinetics. In some cases, where the projected use of the aggregate is not compromised by small amounts of azide or monomer, it may be better simply to store the reaction mixture without taking steps to remove these soluble molecules.

Growth and Storage of polyGln Aggregates (37°)

1. Dilute the disaggregated peptide stock in a Falcon tube to a 10-μM final concentration in PBS containing 0.05% sodium azide. Cap and seal the Falcon tube with parafilm, and incubate at 37°.

2. Monitor by HPLC sedimentation assay and/or ThT assay until the signal plateaus. This may take several to many days, depending on repeat length (Chen *et al.*, 2002a) and features of the sequence.

3. Collect aggregates by centrifugation, wash, and concentrate in fresh buffer, if desired, as described previously. Determine the weight concentration of aggregates if desired, snap-freeze, and store at −80°.

4. Different growth conditions at 37° give different morphologies (Chen *et al.*, 2002a). The above protocol leads to ribbon-like parallel assemblies of protofilament-like structures with intermediate repeat length polyGln peptides in PBS. Amyloid-like twisted ropes can be prepared by growing the same peptides in Tris HCl, pH 7, without added salts, but this requires significantly longer incubation times. Longer polyGln peptides (45 or more Gln) grow into fibrils even in PBS.

Growth and Storage of polyGln Aggregates (−20°)

1. Another polymorphism of polyGln aggregates with many interesting properties and uses is the short, curvilinear filaments with amyloid-like seeding activity and properties formed in frozen PBS solutions (Chen *et al.*, 2002a).

2. Dilute the peptide from a disaggregated stock solution to a 10-μM final concentration in PBS with 0.05% sodium azide in an Eppendorf or Falcon tube. Incubate the aggregates for 24 h at 37°, and then snap-freeze in liquid nitrogen and store at −20° for 24 h. It may be necessary to eliminate the 37° preincubation step for peptides that begin to aggregate over this 24-h period. The role of this preincubation step is not clear, but we have observed somewhat different properties of aggregated product if it is eliminated.

3. Thaw on the bench-top and check for aggregation by ThT, light scattering, or HPLC sedimentation assay. Collect aggregates at 20,800g, and resuspend the pellet in PBS to prepare a suspension of the desired approximate weight concentration. Determine the exact weight concentration by the protocol below.

4. Aliquot if desired, snap-freeze, and store at −80°.

Determination of Aggregate Weight Concentration

1. Remove an aliquot of the aggregate stock suspension sufficient to deliver at least 10 μg (or, alternately, the smallest volume of suspension that can be accurately transferred), and add 100% formic acid to give 70% formic acid. Vortex vigorously, and incubate at room temperature for 10 min.

2. Add water to give a final volume of 120 μl and vortex.

3. Inject 100 μl onto the HPLC sedimentation assay using a reverse-phase column and gradient sufficient for resolution of the monomeric form of the aggregate component (see section on HPLC sedimentation assay for amyloid transformations).

4. Integrate the peak of the correct retention time, and obtain the micrograms of peptide from the standard curve (see section on HPLC sedimentation assay for amyloid transformations). Based on the dilution scheme, calculate the weight concentration of aggregates in the stock suspension.

An HPLC Sedimentation Assay for Amyloid Transformations

The dilemma in trying to rigorously track amyloid transformations is that there is no reproducible, convenient method that allows quantitation of all the reaction species, ranging from monomers to various types of aggregates. The most discriminating assays for amyloid, for example, are EM and atomic force microscopy (AFM), but they are not quantitative for aggregates and give no concentration information at all on small monomeric peptides. ThT gives a relatively strong response to amyloid, but it can vary over at least a 20-fold range depending on the fibril (Shivaprasad and Wetzel, 2006) (Fig. 2) and does not appear to give a significant response to some Aβ protofibrils (Williams et al., 2005). Sedimentation assays can give information on concentrations of unreacted monomer but can provide little information on the nature of the aggregated, sedimented fraction. If the aggregates exhibit radically different sedimentation coefficients, some gross features of the aggregate product profile can be assessed by varying centrifugation conditions (see Chapter 11 by Mok and Howlett in this volume). Other techniques must be used to confirm the nature of the aggregates along the reaction coordinate.

There are a number of ways of measuring protein in the supernatant after sedimentation of the aggregates. The advantage of HPLC is that it allows one to monitor multiple monomeric species independently, and thus, for example, observe chemical degradation during the aggregation process, or to monitor the parallel aggregation of multiple peptides in the same reaction. Disadvantages include a modest sensitivity, a relatively lengthy

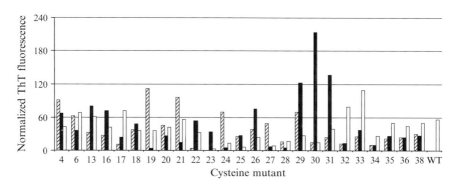

Fig. 2. Comparisons of weight-normalized thioflavin T (ThT) fluorescence signals for amyloid fibrils of amyloid β (Aβ)(1-40) point mutants containing single Cys point mutations in various chemical states. Free Cys, cross-hatched bars; carboxymethyl-Cys, filled bars; methyl-Cys, open bars. Reprinted with permission from Shivaprasad and Wetzel (2006).

analysis time, and the need to ensure that supernatants in the HPLC analysis queue do not degrade (e.g., continue to aggregate) before injection.

Experimental Results

The sedimentation assay requires a standard curve relating the integrated area of a peptide HPLC peak to the mass of the peptide. The assay requires reproducible recovery of injected peptide throughout the dynamic range of the assay and, preferably, nearly quantitative recovery. Wavelengths in the range of 215 to 220, where the peptide bond (and other amides) strongly absorb, are normally used for quantitation. Figure 3 shows the excellent linearity of response for Aβ(1-40) and for the peptide $K_2Q_{25}K_2$ using peptide stock solutions that were independently calibrated for concentration using amino acid composition analysis. Figure 3 also shows that because of somewhat different extinction coefficients of different amino acids, different peptides can exhibit different standard curves. In a practical compromise, we typically use the same repeat length polyGln peptide standard for all polyGln peptides regardless of repeat length and the existence of occasional mutations. Similarly, we use a wild-type Aβ (1–40) stock solution for point mutants of Aβ.

It is recommended that one establish the appropriateness of the HPLC sedimentation assay as a faithful reporter of a reaction by carrying out a preliminary validation in which the aggregation reaction is monitored by multiple measures. Figure 4A shows the time course of a spontaneous aggregation reaction of a polyGln peptide as monitored by the sedimentation assay, the ThT assay, a Rayleigh light-scattering assay, and CD monitoring of

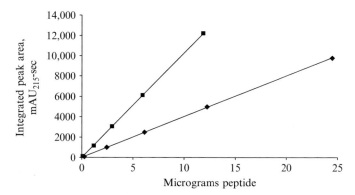

FIG. 3. Standard curves for high-performance liquid chromatography (HPLC) determination of peptide mass as described in section on the HPLC sedimentation assay for amyloid transformations (amyloid β (Aβ)(1-40), ■; K2Q25K2, ◆).

the random coil to β-sheet transition (Chen *et al.*, 2002b). Figure 4B shows a contrasting case, in which the HPLC sedimentation assay of Aβ aggregation picks up a significant loss of monomer from solution during the ThT lag phase, a discontinuity explained by the early generation of ThT-unresponsive oligomers and/or protofibrils (Williams *et al.*, 2005).

Protocols

Construction of the Standard Curve

1. Disaggregate a sample of purified peptide, and dissolve the dried film in a volume of 1% TFA in water sufficient to produce a solution of approximately 100 μg/ml. Aliquot this solution into volumes of about 500 μl, snap-freeze, and store at −80°. This is the standard solution to be used for routine construction of the standard curve. For the polyGln standard, we use the peptide $K_2Q_{25}K_2$ so that there are no major concerns about the peptide aggregating in the autosampler.

2. To determine the peptide concentration in the standard solution accurately, dilute with the highest possible accuracy (by serial dilutions and/or by determining aliquot volumes gravimetrically) into 1% aqueous TFA to a concentration of approximately 10 μg/ml. Seal in an ampoule, and submit (including duplicate or triplicate independent dilutions from the same stock solution) for amino acid composition analysis (e.g., Commonwealth Biotechnologies, Inc., 601 Biotech Drive, Richmond, VA 23235). Use the returned mean value to calculate the exact concentration of the standard solution.

3. In separate runs, inject (preferably with escalating peptide amounts) aliquots of the standard solution on an analytical HPLC instrument with analytical-quality gradient reproducibility and ultraviolet (UV) detector sensitivity (we use the Agilent 1100 binary gradient system with either a diode array or a variable wavelength detector), using an analytical reverse-phase column (C3 works well for $A\beta$(1-40) and for many polyGln

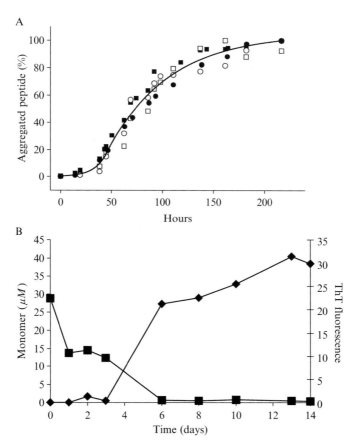

FIG. 4. Testing for aggregation reaction complexity with parallel assays. (A) Aggregation progress of 0.39 mg/ml K2Q42K2 in 10 mM Tris-trifluoroacetic acid (TFA), pH 7.0, with monitoring by circular dichroism (CD; $[\Theta]_{200}$) ($\blacksquare$), high-performance liquid chromatography (HPLC) ($\bullet$), light scattering ($\square$), and thioflavin T (ThT) ($\bigcirc$) methods, consistent with one class of aggregated product. Reprinted with permission from Chen *et al.* (2002b). (B) Solution phase spontaneous aggregation of disaggregated amyloid β ($A\beta$)(1-40) in phosphate-buffered saline (PBS) at 37°, monitored by ThT ($\blacklozenge$) and HPLC of high-speed centrifugation supernatants ($\blacksquare$), suggests the existence of a ThT-insensitive aggregate populated before fibril formation ensues on days 4–6. Modified and reprinted with permission from Williams *et al.* (2005).

peptides) and a gradient of acetonitrile in 0.05% TFA. Monitor at a wavelength of 215 nm, and obtain the integral of the peak corresponding to the peptide. To judge peptide recovery and ensure against inaccuracies caused by sample carryover, blank runs should be interspersed with runs on injected standard, at least for injections of higher masses of peptide. After the standard curve runs are complete, discard the remaining standard solution (do not refreeze).

4. Other peptide bond wavelengths can be used for detection, but it is obviously critical to always use the same wavelength for unknowns and for the standard curve. Other reverse-phase columns may be used. Chain lengths greater than C3 may be required for good behavior of short polyGln peptides but may give loss of peptide and/or broadening for $A\beta$ peptides. Better resolution may be obtained for $A\beta(1–42)$ by using elevated column temperatures.

5. Prepare a standard curve by plotting peak areas with respect to the injected mass of peptide (Fig. 3). By using peptide weights rather than moles, one can use one polyGln standard for determining concentrations of other polyGln peptides of other repeat lengths without incurring any significant errors. Alternatively, standard curves can be generated by the above protocol for each peptide to be analyzed.

Determination of Autosampler Storage Conditions

Obtaining accurate kinetics data may require analysis of many time points, and perhaps replicate samples, leading to a requirement for an autosampler. Care must be taken to ensure that the samples do not degrade (particularly by continuing to aggregate) on the autosampler tray, however. It is strongly advised to conduct preliminary studies to determine sample stability as follows. Stability against aggregation in the autosampler tube should be probed not with simple dilution of a monomer solution but with an example of the type of sedimentation supernatant that will eventually be the subject of an aggregation study. This is because of the possibility that a small amount of residual microaggregates will contaminate the sample in the vial, with the potential to give a much more aggressive aggregation in the autosampler queue via seeded elongation.

1. Prepare an example of a test aggregation kinetics reaction, and allow it to reach 25–50% complete aggregation.
2. Remove an aliquot of the suspension, and centrifuge under conditions required for doing the kinetics analysis.
3. Carefully decant an aliquot, and transfer it into the candidate stabilization/assay buffer. For $A\beta(1–40)$ samples, we have found that 0.5% TFA prevents aggregation in the vial indefinitely. For polyGln

samples, a final concentration of 20% formic acid was required to prevent aggregation in the vial for up to 20 h at RT.

4. Analyze injections from this test vial at periodic intervals using a standard HPLC gradient and appropriate standard curve to establish whether the test solvent is adequate to preserve sample concentration over time.

Sedimentation Assay Protocol

1. Take a 100-μl aliquot of the sample, and centrifuge to remove aggregates. If it can be established that amyloid-scale aggregates are the only aggregated product, 30 min at 20,400g is sufficient. If microaggregates, such as protofibrils, are a possibility, higher speeds, such as 386,000g for 1 h at 4° in a table-top ultracentrifuge, may be required. Smaller aggregates may remain behind in the supernatant if shorter times and slower speeds are used (Fig. 4B) (Chen *et al.*, 2002b; Williams *et al.*, 2005; see Chapter 11 by Mok and Howlett in this volume). At the same time, in some buffers, protofibrils appear to clump to give superaggregates that pellet well at 20,400g.

2. Remove 75 μl of supernatant carefully from the centrifuge tube, and place it in an Eppendorf tube along with any reagents found necessary to stabilize the supernatant against aggregation in the vial.

3. Transfer the major portion of the sample onto a reverse-phase HPLC column (directly or by queuing in an autosampler), and develop with a linear gradient of acetonitrile in aqueous 0.05% TFA.

4. Integrate the peak of the correct retention time, and obtain the micrograms of peptide from the standard curve. If automated software integration is used, it is best to confirm that it is giving appropriate numbers, especially as the signal-to-noise ratio decreases toward the end of the aggregation reaction, by manual integration.

5. Be vigilant for the time-dependent appearance of new peaks in the HPLC assay that may indicate oxidation or hydrolysis of side chains in the peptide. Significant formation of chemical degradation products severely complicates kinetic and thermodynamic interpretations of the data.

Titration of Fibril Growing Ends

Analysis of many aspects of fibril growth reactions requires knowledge of the average number of growth/elongation sites per unit weight of aggregate. Although fibril elongation reactions can be easily fit to pseudo-first-order kinetics, the rate constant obtained is only valid for the particular preparation of fibrils used as seeds, because the reaction is of the first order in both monomer and aggregate (O'Nuallain *et al.*, 2004) and aggregates can differ considerably in their average molecular weights within and between preparations. A more universal value, the second-order elongation rate constant, which takes into account the concentration of fibril growing ends, is characteristic of the fibril type and the monomer used in the reaction as well as other reaction conditions. However, it can only be calculated with knowledge of the number of growing ends per weight of the fibrils.

Such second-order rate constants are very useful in characterizing amyloid systems. For example, they can be used to compare the cross-seeding efficiencies of an aggregate in the elongation of a variety of monomers (O'Nuallain et al., 2004) or the efficiencies of a number of aggregate seeds for the same monomer. Second-order rate constants are also required for deconvolution of the complex kinetic parameter obtained in the analysis of nucleation kinetics. In principle, one might obtain estimates of the average molecular weight of a fibril preparation using EM analysis, although this requires a number of assumptions, including the viability of all observed fibril ends. Molecular rate constants can also be obtained by microscopic monitoring of the growth of individual fibrils (DePace and Weissman, 2002; see Chapter 5 by Ban and Goto in this volume); extending such values to a fibril population, however, still would require knowledge of average fibril molecular weight. We describe here a solution phase-binding assay using tagged monomer to determine the number of productive elongation sites in a fibril preparation (Bhattacharyya et al., 2005).

Experimental Results

In principle, a variety of tags can be used. Radioactive tags are the most straightforward to assay but require extra precautions and can have limited lifetimes as reagents as well as leading to free radical damage to the peptide on storage. We used a nonradioactive strategy involving biotin-tagged peptide and a highly sensitive time-resolved fluorescence (DELFIA) approach to developing the signal. Biotin-tagged polyGln peptides with a time-resolved fluorescence workup have been used to follow aggregate elongation in a microplate assay suitable for inhibitor screening (see Chapter 16 by Berthelier and colleagues in this volume; Berthelier et al., 2001) and have also been used to stain elongation-competent sites in Huntington's disease tissue (see Chapter by Osmand and co-workers in this volume). The structure of the tagged peptide used in experiments described here is shown in Fig. 5A.

The amyloid fibril elongation cycle comprises several steps, including an initial "docking" step, followed by one (Esler et al., 2000) or more (Cannon et al., 2004) rearrangement ("locking") steps that consolidate amyloid structure and create a new binding site for addition of the next monomer. If the locking steps are relatively slow compared with the docking step, it is theoretically possible to count the number of growth sites using a tagged version of the monomer. Appropriately timed addition of unlabeled monomer during the workup can stimulate additional cycles of elongation and thus prevent the loss of label (Fig. 5B). Alternatively, it may be possible to identify a temperature and binding conditions at which the docking step is favored, whereas the subsequent steps are not. We found that conducting the binding assay on polyGln aggregates at 25° allowed us to observe saturable binding without the need for a "cold chase" with unlabeled polyGln (Bhattacharyya et al., 2005). It cannot be assumed that this will be true for other aggregate systems, however. Using polyGln aggregates prepared as described above, we conducted a preliminary time course of binding at 37° (Fig. 5C). The observed binding is lost if unlabeled monomer is left out of the workup buffers (Fig. 5C), consistent with the unstable nature of peptide docked onto the growing end of the fibril. Consistent with the molecular specificity of amyloid elongation (O'Nuallain et al., 2004), significant binding does not occur using tagged $A\beta(1–40)$ (Fig. 5C). Significant binding also does not occur using a tagged aggregation-incompetent proline mutant of polyGln (Fig. 5C). Binding is saturable (Fig. 5D) and amounts to only a fraction of a percent of the total tagged monomer in solution (Bhattacharyya et al., 2005).

With the determination of the molar concentration of aggregate elongation sites, it is possible to convert the pseudo-first-order elongation rate constant obtained in reactions seeded with these aggregates (see section on elongation kinetics) into a second-order rate constant. For example, the rate constant for elongation (see below) of a Q_{47} peptide by a Q_{47} aggregate at 37° in PBS is in the range of $10^4 \ M^{-1}s^{-1}$ (Bhattacharyya et al., 2005).

Two phase kinetics associated with the dock and lock mechanism are observed in microtiter plate elongation assays (Berthelier et al., 2001; Esler et al., 1997); because of the relative convenience of these assays, it is tempting to attempt to determine growth site concentrations of aggregates by extrapolation of the kinetic plot (Bhattacharyya et al., 2005). However, although relatively good agreement with suspension phase-binding assays can be obtained in some cases, more generally, the values obtained in the microplate assay can be significantly lower than in the suspension phase assay (A. K. Thakur, unpublished data), possibly because some sites are masked in the binding to plastic. Our recommendation is to use the suspension phase assay described below rather than a microtiter plate–based method.

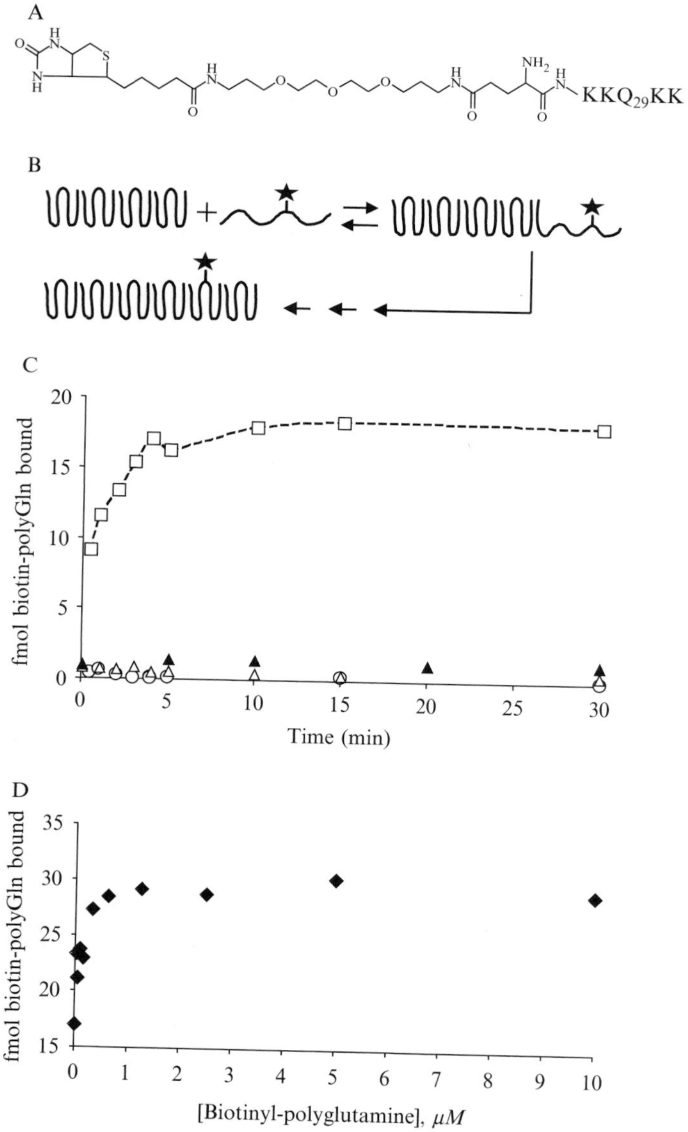

FIG. 5. Titration of growth sites on amyloid fibrils. (A) Structure of the biotin-tagged polyGln molecule. (B) Schematic of a multistep dock-and-lock fibril elongation reaction shows that a tagged monomer freshly added to the growing end of the aggregate will tend to dissociate unless stabilized by subsequent rounds of elongation. Modified and reprinted with permission from Bhattacharyya et al. (2005). (C) Time dependence of addition of biotinylated-Q29 to sonicated Q47 aggregates at 25° using a workup procedure, including subsequent incubation with excess unlabeled Q30 (□); control in which unlabeled Q30 was

Protocol

Determination of the Growth Site Concentration of an Aggregate Suspension

1. Following purification, disaggregation, and ultracentrifugation steps, place 500 μl of different concentrations (0.05–10 μM) of biotin-PEG-Q_{29} (Fig. 5A) (obtained by custom synthesis from the Keck Center at Yale University) in PBS in 1.5-ml Eppendorf tubes.

2. Prepare an appropriate volume of a stock of a known concentration of aggregates (e.g., 1.9 mg/ml).

3. From this suspension, deliver 200-ng aggregates into each Eppendorf tube containing 500 μl biotin-PEG-Q_{29} from step 1. Vortex tubes gently at RT.

4. As a control, process 200-ng aggregates in 500 μl PBS with no biotinyl peptide. Data obtained from the control provide the background.

5. Immediately transfer all the tubes to a water bath set at 25°, and incubate for 30 min.

6. Immediately transfer all the tubes to a precooled 4° Eppendorf centrifuge, and centrifuge at 14,000 rpm (20,400g) for 1 h. These conditions should be followed exactly in order to produce a tight pellet.

7. Gently remove the supernatant with the help of gel loading tips, and put 100 μl Eu-streptavidin (use 1:1000 dilution from the solution provided by the manufacturer, PerkinElmer) to each pellet, vortex gently to resuspend, and incubate at RT in the dark for 1 h. Because the pellet is not visible, it is best to leave around 5% supernatant in each wash cycle to avoid disturbing the pellet. Signal can be lost if the pellet is disturbed.

8. Wash the pellet three times by centrifugation for 1 h at 4°, followed by careful decantation (leaving 5% of the supernatant each time) and resuspension in PBS.

9. Resuspend the pellet in 100 μl of enhancement solution (Perkin-Elmer). Keep the suspension for 10 min at RT in the dark.

10. Transfer the 100-μl suspension from each tube to enzyme-linked immunosorbent assay (ELISA) plate wells, and determine the europium counts by DELFIA/time-resolved fluorescence (Diamandis, 1988) in a PerkinElmer/EG&G Wallac Victor^2 microtiter plate reader.

left out (O); experiment adding biotin-linked amyloid β (Aβ)(1–40) to the Q47 aggregate, followed by incubation with unlabeled Aβ (▲); and experiment adding biotin-linked, elongation incompetent proline mutant of polyGln, followed by incubation with unlabeled version of the same peptide. Modified and reprinted with permission from Bhattacharyya *et al.* (2005). (D) Concentration dependence of binding of biotinyl-Q30 to Q47 aggregates in 30 min at 25°. Reprinted with permission from Bhattacharyya *et al.* (2005).

11. Convert counts to femtomoles of europium by using a standard curve established with a calibrated europium solution (PerkinElmer). Convert femtomoles of europium to femtomoles of bound biotin-peptide using the manufacturer's determination of europium atoms per streptavidin molecule and an assumption of 1 M biotin bound per 1 M streptavidin tetramer.

12. Subtract the background signal obtained in the control experiment (step 4).

13. Convert femtomoles of biotin-peptide bound per 200-ng aggregate into femtomoles bound per microgram of aggregate.

14. This number can be used to calculate the molar concentration of growth sites for a given aggregate preparation used in an elongation reaction.

15. Note that this protocol is for polyGln aggregates, for which it has been validated that there is only one addition of biotinyl-polyGln to each growth site when incubation is at 25°. For other aggregate systems, until it is established to be unnecessary, it is strongly recommended that unlabeled peptide chaser be added during the workup (Fig. 5B,C). This promotes elongation with unlabeled peptide and thus stabilizes the embedded biotinylated peptides.

Elongation Kinetics

The reversible addition of a monomer to a fibril to extend the fibril by one unit is shown in Fig. 6A. Assuming that there is no multiplication of fibril growing ends over the course of the reaction attributable to secondary nucleation pathways (Ferrone, 1999), the forward elongation kinetics for an amyloid fibril in a seeded reaction are then described by the simple second-order rate expression shown in Eq. (1), in which $^2k_+$ is the second-order rate constant. Again, assuming no change in fibril growth site concentration over the course of the reaction, this expression reduces to one describing a pseudo-first-order reaction, as shown in Eq. (2), where $^{\psi 1}k_+$ is the pseudo-first-order rate constant. This rate constant is only valid for the reaction mixture in which it was determined, however, because the concentration of growing ends normally is not known. If this concentration can be determined (see section on titration of fibril growing ends), the more robust second-order rate constant can be calculated according to Eq. (3).

Some amyloid fibrils are sufficiently unstable that their dissociation can be easily observed. The dissociation rate is of the first order in fibril growth sites and is described by Eq. (4), where $^2k_-$ is the first-order dissociation rate constant. This expression should be valid for early stages of fibril dissociation but only to the point where fibrils begin to dissolve completely and disappear; at that point, the concentration of growth sites begins to decline. Assuming an unchanging concentration of fibrils, Eq. (4) reduces to the pseudo-zero-order expression shown in Eq. (5), in which the pseudo-zero-order rate constant $^{\psi 0}k_-$ is equivalent to the observed dissociation rate and, again, is only valid for the fibril dissociation reaction studied.

The existence of a significant fibril dissociation rate has two consequences. First, the observed fibril elongation rate will be the sum of the forward and reverse rates. Second, fibril formation will arrive at a point of dynamic equilibrium described by an equilibrium constant that is the ratio of forward and reverse rate constants, as shown in Eq. (6).

$$\text{Elongation rate} = {}^{2}k_+[\text{Fibril}][\text{Monomer}] \tag{1}$$

$$\text{Elongation rate} = {}^{\psi 1}k_+[\text{Monomer}] \tag{2}$$

$${}^{2}k_+ = {}^{\psi 1}k_+/[\text{Fibril}] \tag{3}$$

$$\text{Dissociation rate} = {}^{1}k_-[\text{Fibril}] \tag{4}$$

$$\text{Dissociation rate} = {}^{\psi 0}k_- \tag{5}$$

$$k_a = {}^{2}k_+/{}^{1}k_- = {}^{\psi 1}k_+/{}^{\psi 0}k_- \tag{6}$$

In spite of the many technical difficulties and caveats, it is possible to obtain good approximations of these rate constants for robust fibril elongation and dissociation reactions. Experimental determination of such rate constants is important for a number of reasons. As will be seen in the section on nuclear kinetics analysis, knowledge of the second-order elongation rate constant allows a fuller understanding of nucleation kinetics and a determination of the equilibrium constant controlling nucleation. In addition, the second-order elongation rate constants describing homologous and heterologous seeded elongation reactions are the key parameters for making comparisons in cross-seeding experiments, which, for example, can provide information on fibril structure (O'Nuallain et al., 2004).

Elongation kinetics can be monitored by a number of methods. ThT generally works well, especially for fibril formation reactions that go essentially to completion. Because ThT fluorescence varies with instrumentation conditions as well as fibril structure (Shivaprasad and Wetzel, 2006), the absolute mass of fibrils associated with a particular ThT signal has to be independently determined, for example, by also doing a sedimentation assay. For such reactions, the pseudo-first-order rate constant determined

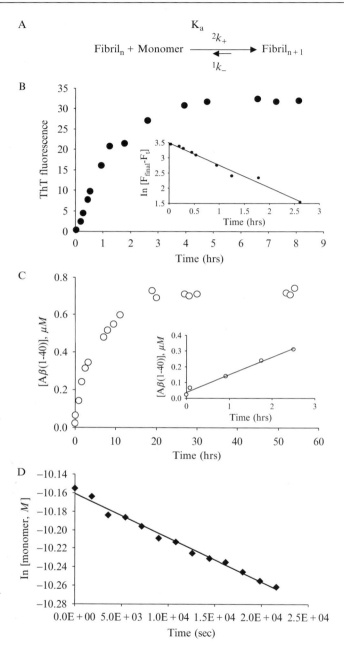

FIG. 6. Amyloid elongation reactions. (A) Model for fibril growth by monomer addition; (B) 25 μM amyloid β (Aβ)(1–40) seeded with 8.4% by weight of Aβ(1–40) fibrils and

for fibril elongation using ThT fluorescence is a composite of the elongation and dissociation rate constants (O'Nuallain et al., 2005). An HPLC sedimentation assay focusing on the amount of monomer left unreacted at different times allows determination of kinetics in more complex reactions, such as when one is monitoring the simultaneous aggregation of several molecules in a mixture of molecules (Bhattacharyya et al., 2006).

Experimental Results

Freshly disaggregated Aβ(1–40) was seeded with previously prepared fibrils and incubated in PBS at 37°. Aliquots were removed and subjected to ThT analysis, and the results were plotted by first-order kinetics treatment to determine $^{\psi 1}k_+$ (Fig. 6B). This could equally well be monitored and analyzed using the HPLC sedimentation assay to give the same rate constant (O'Nuallain et al., 2005). After the reaction reached completion (a position of dynamic equilibrium), it was gently mixed and an aliquot was diluted by a known amount into reaction buffer, being careful to keep the total Aβ concentration above the previously determined C_r. The rate of dissociation, equivalent to the apparent $^{\psi 0}k_+$, was determined (Fig. 6C) using the HPLC sedimentation assay (see section on HPLC sedimentation assay for amyloid transformations) and then multiplied by the dilution factor used to prepare this reaction in order to calculate the apparent dissociation rate constant for a fibril concentration equivalent to that in the forward elongation reaction. An estimate of the elongation equilibrium constant of 3.8 μM was obtained from the $^{\psi 1}k_+$ and $^{\psi 0}k_-$ values substituted into Eq. (6) (O'Nuallain et al., 2005).

In another example of the use of elongation kinetics analysis, to facilitate the analysis of polyGln nucleation kinetics of a Q$_{47}$ peptide, a Q$_{47}$ aggregate elongation reaction was carried out using an aliquot of a stock suspension of Q$_{47}$ aggregate previously grown in PBS at 37° and titrated for growth site concentration as described in the section on titration of fibril growing ends. Aggregates were diluted into a solution of freshly disaggregation Q$_{47}$ monomers in PBS, and the reaction was incubated at 37°.

monitored by thioflavin T (ThT) fluorescence (inset: pseudo-first-order plot of initial rate). Reprinted with permission from O'Nuallain et al. (2005). (C) Equilibrium reaction mixture from the above reaction was diluted 25-fold and incubated at 37°, and the Aβ(1–40) released was monitored by high-performance liquid chromatography (HPLC) sedimentation assay (inset: initial rate). Reprinted with permission from O'Nuallain et al. (2005). (D) Pseudo-first-order elongation of Q47 aggregates by monomeric Q47 in phosphate-buffered saline (PBS) at 37°, which was monitored by the HPLC sedimentation assay. Reprinted with permission from Bhattacharyya et al. (2005).

Aliquots were removed and analyzed by centrifugation, followed by HPLC analysis of the supernatant, yielding the aggregation kinetics profile shown in Fig. 6D. The pseudo-first-order rate constant for elongation, $^{\psi 1}k_+$, was determined to be 4.7×10^{-6} s^{-1}. Dividing this value by the concentration of fibril growth sites available in the seeded reaction yielded a second-order elongation rate constant, $^{2}k_+$, equal to 11,400 M^{-1}s^{-1} (Bhattacharyya et al., 2005).

Protocol

Determination of the Pseudo-First-Order Rate Constant $^{\psi 1}k_+$

1. Prepare aggregates of Q_{47} at 37° in PBS, and determine the weight concentration as described above.

2. If desired, sonicate the aggregate stock suspension using a probe sonicator five times with 30-s pulses, with each pulse followed by a 1-min delay, on ice. If required, preserve some of the sonicated stock for independent determination of growth site concentration as described above. Because the number of growth sites in an aggregate, including sonicated aggregates, can decay on storage, it is best to conduct all measurements on a particular aggregate in the same time frame or to store the aggregates at −80° after snap-freezing.

3. Prepare disaggregated monomers of Q_{47} peptide as described in the section on disaggregation of peptides. Prepare a volume (chosen to provide the desired number of data points) of a 20-μM solution of disaggregated peptide in PBS in an Eppendorf or Falcon tube, and preincubate at 37°. Add 5% weight (fibril seeds to monomer) of sonicated aggregates, and incubate without agitation at 37° to initiate elongation.

4. At different time intervals, remove a 100-μl aliquot from the gently mixed reaction and centrifuge at 14,000 rpm for 30 min at 4° in a 5417R Eppendorf centrifuge. Transfer 70 μl of the supernatant to an autosampler vial, and add formic acid to give 20% formic acid. Inject 50 μl of supernatant from each processed time point. If injection is immediate (no autosampler), the formic acid can be eliminated.

5. Determine the concentration of Q_{47} monomer at each time point from the HPLC trace by reading the polyGln mass from the standard curve (see section on HPLC sedimentation assay for amyloid transformations), converting to moles using the molecular weight of the Q_{47} peptide, correcting for volume manipulations during assay, and calculating the molar concentration.

6. Plot the data according to a first-order kinetic model, and determine $^{\psi 1}k_+$.

Nucleation Kinetics Analysis

The assumed high degree of structural order of amyloid fibrils, plus the observation of a lag phase in the spontaneous growth of fibrils from monomeric proteins, led to speculation that fibril growth might be mediated by a process of nucleation-dependent polymerization (Jarrett and Lansbury, 1992) similar to how protein crystallization is thought to be initiated. As pointed out by Ferrone (Ferrone, 1999), however, the presence of a lag phase, even one that can be eliminated by providing exogenous seeds, is not sufficient to prove nucleated growth by a monomer addition mechanism. In fact, for many amyloids, the situation must be more complex, because significant amounts of nonfibrillar aggregates are observed to form relatively rapidly, often coinciding with the observed lag phase in the development of amyloid-dependent ThT fluorescence. These have been implicated in alternative mechanisms to explain amyloid nucleation and growth (Harper et al., 1997; Serio et al., 2000).

The nature of the nucleation event in such mechanisms is not clear. According to one formulation, nucleation involves the association of several protofibrils to form a nascent fibril with a much enhanced ability, compared with the isolated protofibrils, to grow (Harper et al., 1997). In contrast, others have obtained data for some amyloid fibril formation reactions more consistent with models, in which the nonamyloid aggregates formed early in spontaneous amyloid formation reactions are off-pathway assemblies (Collins et al., 2004; Goldsbury et al., 2005; Gosal et al., 2005), such that their role in amyloid formation is to deplete reversibly the environment of monomers. The question of whether oligomers are on-pathway or off-pathway would appear to be experimentally approachable, albeit challenging. For the purpose of this chapter, the discussion of the analysis of nucleation kinetics will be limited to the formation of amyloid-like (Chen et al., 2002a) aggregates from simple polyGln peptides, which is an apparently relatively rare case of amyloid formation that does not involve the early formation of nonamyloid aggregates.

Determining this mechanism involves the collection of aggregation data representing the earliest portions of the spontaneous aggregation reaction. A variety of methods can be used, each with advantages and disadvantages. In this section, we show the application of data obtained using the HPLC-linked sedimentation assay discussed above. The data are then analyzed by equations describing nucleated growth by monomer addition. A number of approaches can be taken to model such data (Ferrone, 1999). Here, we use an analytic approach based on a "thermodynamic" model for nucleation, in which the nucleus is considered to be less stable than the monomer ground state and, in fact, the least stable species on the aggregation

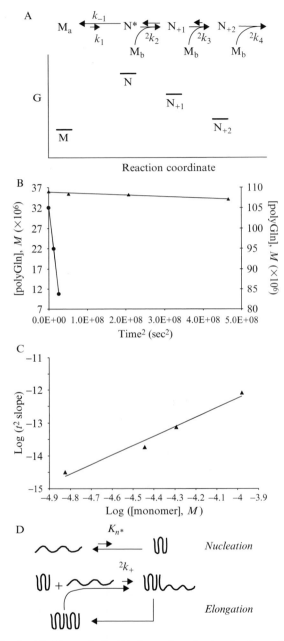

FIG. 7. Nucleation kinetics analysis. (A) General mechanism for the nucleation phase of a nucleated growth polymerization reaction, based on a thermodynamic, pre-equilibrium model for nucleation (M = monomer; N* = aggregation nucleus; N$_{+1}$ and N$_{+2}$, elongated nuclei/

reaction profile (Ferrone, 1999). This is illustrated in the reaction scheme shown in Fig. 7A. The kinetics expression for nucleation kinetics that comes out of this scheme is shown in Eq. (7), where Δ is the molar concentration of monomers that has been converted to aggregates at time t, $^2k_+$ is the second-order elongation rate constant for nucleus elongation and aggregate elongation (which are assumed to be identical), c is the molar concentration of monomers at the start of the reaction, n^* is the critical nucleus (the number of monomers that associate together to form the nucleus), and K_{n^*} is the nucleation equilibrium constant. Ideally, a full description of the nucleation kinetics will include determination of the nucleation equilibrium constant, the critical nucleus, and the elongation second-order rate constant; together, these control the nucleation and overall aggregation reaction.

$$\Delta = \{1/2\}(^2k_+{}^2)(K_{n^*})\, c^{(n^*+2)}t^2 \tag{7}$$

$$\log(\text{slope A}) = \log[\{1/2\}(^2k_+{}^2)(K_{n^*})] + (n^* + 2)\log(c) \tag{8}$$

$$\text{x-intercept} = \log[\{1/2\}(^2k_+{}^2)(K_{n^*})] \tag{9}$$

Experimental Results

The polyGln aggregation passes the first two tests of a nucleated growth polymerization reaction: (1) it exhibits a lag phase that (2) is shortened or abrogated when a small amount of previously formed aggregate is included in the reaction mixture (Chen et al., 2002a). Further tests are required, however, before the mechanism is convincingly established. These are described here. Equation (7) predicts that in a nucleated growth polymerization reaction, the increase in the amount of monomer converted into aggregate should give a linear plot with respect to time^2 with a slope (slope A) equal to $\{1/2\}(^2k_+{}^2)(K_{n^*})\, c^{(n^*+2)}$ (Chen et al., 2002b); Fig. 7B (Bhattacharyya et al., 2005) shows that for a Q_{47} peptide, this is indeed the case. Equation (8), derived from Eq. (7), further predicts that a log-log plot of the slopes of these time^2 plots with respect to the starting monomer concentrations should give a straight line of positive slope (i.e., an increase

---

nascent aggregates). (B) Representative time^2 plots of the early portion of Q47 aggregation in phosphate-buffered saline (PBS), 37°, at two different concentrations (▲ , 36 μM, left y-axis; ●, 105 μM, right y-axis). (C) Log-log plot of the slopes of the time^2 plots *versus* Q47 concentration. (D) Mechanism of nucleated growth polymerization of amyloid-like polyGln aggregates, featuring a monomeric nucleus. Figures B–D reprinted with permission from Bhattacharyya et al. (2005).

in aggregation with increasing concentration) (Chen et al., 2002b); Fig. 7C (Bhattacharyya et al., 2005) shows that this is also the case.

These additional tests not only confirm the mechanism as being nucleated growth with monomer addition but provide experimental parameters that provide key details of the nucleation mechanism. Thus, the slope of the log-log plot (Fig. 7C) is equal to $n^* + 2$. Classically, n^* has been visualized as an integer greater than 1. In contrast, n^* for polyGln aggregation, as determined from plots like that shown in Fig. 7C, consistently (Bhattacharyya et al., 2005, 2006; Chen et al., 2002b; Thakur and Wetzel, 2002) gives values close to 1. That is, nucleus formation in polyGln aggregation involves the formation from bulk phase monomer of a highly unfavorable folded form of the monomer required for aggregation initiation (Fig. 7D) (Chen et al., 2002b). The x-intercept of the log-log plot is an equally important parameter containing the second-order rate constants for elongation of the nucleus and mature aggregate, along with the nucleation equilibrium constant K_{n^*}. Making the assumption that the two elongation rate constants are identical, the expression reduces to that shown in Eq. (9). If one can independently obtain an experimental value for the second-order elongation rate constant $^2k_+$ for the type of aggregate formed early in the Q_{47} aggregation process, K_{n^*} can then be calculated from $^2k_+$ and the x-intercept using Eq. (9). Our ability to determine $^2k_+$ by determining both $^{\psi 1}k_+$ (see section on HPLC sedimentation assay for amyloid transformations) and the concentration of aggregation growth sites in the aggregate (see section on titration of fibril growing ends) allows us to solve Eq. (9) for K_{n^*}. For Q_{47} aggregation in PBS at 37°, the calculated $K_{n^*} = 2.6 \times 10^{-9}$ (Bhattacharyya et al., 2005).

Protocol

Nucleation Kinetics Analysis

1. Make at least five different concentrations (e.g., 40, 20, 10, 5, and 2.5 μM) of freshly disaggregated polyGln in PBS as described in earlier protocols, and set up aggregation reactions at 37° in Falcon tubes.

2. For each time point, remove a 100-μl aliquot and centrifuge at 14,000 rpm (20,400g) for 30 min at 4°. Carefully remove 70 μl of the supernatant, and determine the concentration of residual monomer using the HPLC assay (see section on HPLC sedimentation assay for amyloid transformations). Strive to obtain as much data as possible for the first 20% of the aggregation reaction final amplitude. It may be helpful to use the ThT assay to follow the reaction time course semiquantitatively in real time so as to judge better the number and frequency of time points required for the HPLC assay.

3. For the data analysis, tabulate all the data for each reaction in terms of monomer concentration (moles) and reaction time (seconds).

4. Using only the portion of the data covering the initial 20% (or less) of the reaction amplitude, and construct plots of remaining monomer concentration (moles) *versus* time^2 (s^2) for each reaction.

5. Calculate slopes of the t^2 plots for all concentrations.

6. Plot the log of the starting concentrations (moles) of each reaction in the series *versus* the log of the t^2 plot slopes, and fit to a straight line. From the slope of this line, calculate n* using the expression "slope = n* + 2."

7. From the x-intercept of the log-log plot, calculate the nucleation equilibrium constant (K_n*) using Eq. (9) and the second-order elongation rate constant value ($^2k_+$) obtained as described in the section on elongation kinetics.

8. It should be noted that although we defined Δ in Eq. (7) as the molar concentration of monomers converted into aggregates at time t, the data analysis shown here consistently plots the molar concentration of unaggregated monomer. This is legitimate only because [soluble monomer] = [soluble monomer]$_{t=0}$ − Δ, such that the absolute value of the slope is the same for either treatment of the data.

9. The centrifugation conditions recommended in this protocol are valid for reactions in which there is essentially no development of small oligomeric aggregates in the early stages of the aggregation reaction. This hypothesis should be rigorously confirmed to validate the use of these equations. For example, one can compare the amount of monomer remaining in the supernatant of low-speed (e.g., 20,400g for 30 min) and high-speed (e.g., 350,000g for 3 h) centrifugation runs at several time points in the aggregation kinetics; mature aggregates as formed in even the early phases of polyGln aggregation will pellet at 20,400g (Chen *et al.*, 2002b), whereas some Aβ(1–40) early aggregates require 350,000g (Williams *et al.*, 2005). An additional test is whether the earliest formed aggregates are good seeds for elongation of monomer solutions. If they are not, the equations used here, which were specifically derived for ensemble data for nucleated growth with monomer addition, are not valid.

Thermodynamics of Amyloid Fibril Elongation

As discussed in the section on elongation kinetics, at least some amyloid fibril formation reactions arrive at an endpoint at which there remains a measurable amount of monomer remaining in solution. Although this is consistent with an endpoint represented by a dynamic equilibrium between monomer additions and dissociations, steps must be taken to confirm that the residual monomer does not result from an artifact. If further tests confirm the equilibrium, the K_a describing it can be converted to a standard free energy describing the fibril elongation reaction. This value corresponds to the difference in free energy between fibril plus monomer, before the addition step, and fibril elongated by one additional monomer, after addition. K_a is simply the reciprocal of C_r, the concentration of monomer measured when fibril formation has reached equilibrium. Figure 6A shows a single, reversible step in the fibril elongation reaction. Because, near equilibrium, fibril molar concentration does not change as the fibril elongates by one monomer (and because the fibrils differing in length by one monomer differ negligibly in activity coefficients), the fibril concentration terms in the expression for K_a cancel, leaving only the concentration of monomer, which, at equilibrium, is equal to C_r:

$$K_a = [\text{Fibril}_{n+1}]/[\text{Fibril}_n][\text{Monomer}] = 1/[\text{Monomer}] = 1/C_r \qquad (10)$$

One important use of the derived ΔG values for amyloid fibril elongation is the derivation of $\Delta \Delta G$ values reflecting the effect of point mutations on fibril elongation energetics. As in the parallel derivation of $\Delta \Delta G_{\text{folding}}$ values associated with mutational changes in folding stability of globular proteins, there are a number of caveats associated with this interpretation (Williams et al., 2006). For example, formally, the $\Delta \Delta G$ is a measure of the difference in net free energy change between reactants and products with and without mutation, such that the value will reflect mutational effects not only on fibril stability but on monomer stability. In addition, although it is desirable to interpret the $\Delta \Delta G$ value in terms of structural effects local to the site of mutation, it is possible for the mutational effect to be propagated distally from the site of mutation, allowing for compensation and underestimation of the local destabilizing effect of the mutation in some cases. Much more work will be required to probe the utility of these fibril elongation ΔG and $\Delta \Delta G$ values, but preliminary results suggest that these values, if interpreted conservatively, can provide significant insights into amyloid structure and the energetics of amyloid formation (Williams et al., 2006).

Experimental Results

Whether fibril formation by $A\beta(1-40)$ takes place spontaneously from disaggregated monomer or via seeded elongation, the reaction in PBS at $37°$ arrives at a plateau value in the range of $0.6–1.0 \ \mu M$ monomer. The residual monomer is chemically unchanged and, in fact, is capable of further fibril growth if concentrated and seeded with $A\beta$ fibrils (O'Nuallain et al., 2005). The fibrils at plateau also are active seeds for further $A\beta(1-40)$

elongation. In addition, fibrils diluted into PBS dissociate to an equilibrium position with the same C_r (O'Nuallain et al., 2005). Both the K_a estimated from the ratio of elongation and dissociation rate constants and the K_a propagated from microscopic forward and reverse rate constants determined by surface plasmon resonance (Cannon et al., 2004) are in good agreement with the K_a from the C_r value (O'Nuallain et al., 2005).

$A\beta(1–40)$ fibril formation reactions allowed to incubate for extended periods (longer than 1 week) appear to undergo a further assembly reaction, perhaps involving fibril superassociation, that leads to a substantial further decrease in the measured monomer concentration (O'Nuallain et al., 2005). This further assembly is fragile, however, because the C_r for dissociation of these aged fibrils is identical ($\approx 0.8~\mu M$) to that of "fresh" fibrils if the aged fibrils are vortexed to initiate the dissociation reaction (O'Nuallain et al., 2005). We assume that the C_r value at the initial plateau represents the stability of individual fibrils or small clusters of fibrils and is therefore a better indication of the energetics of structure formation within the individual fibril. It is thus important to be able to distinguish this initial plateau and not to be confused by the second phase of further diminution of monomer concentration. To make observation of this initial plateau more straightforward, we grow fibrils using a relatively high level of seeds, which eliminates the lag time and promotes a rapid approach to the equilibrium position (O'Nuallain et al., 2005). When analyzing the stability of fibrils from mutant peptides, seeding with wild-type fibrils may also help to bias fibril structure toward the wild-type conformer, allowing for cleaner interpretation of the derived $\Delta\Delta G$ (Williams et al., 2006).

C_r-derived ΔG and $\Delta\Delta G$ values have been used to quantify the effects on the $A\beta(1–40)$ fibril stability of proline (Williams et al., 2004) and alanine (Williams et al., 2006) replacements. In addition, the effect of side chain charge and hydrophobicity has been explored by determining the C_r values for fibril stability of a series of Cys point mutants alkylated with either iodoacetic acid or methyl iodide (Shivaprasad and Wetzel, 2006). The compatibility of disulfide bonds engineered into the fibril has also been assessed by Gs derived from C_r determination (Shivaprasad and Wetzel, 2004) (see Chapter 10 by Shivaprasad and Wetzel in this volume). C_r-derived ΔG and $\Delta\Delta G$ values have also been used to assess the effect of a C-terminal oligoproline sequence on fibril stability, an effect that appears to be attributable to an effect on the free energy of the monomer pool rather than on the free energy of the fibril (Bhattacharyya et al., 2006).

Support for the significance of such $\Delta\Delta G$ values comes from a comparison of the energetic consequences to fibril stability and globular protein stability by identical Ala replacements. The similarity of effects of mutations within the parallel β-sheets of an amyloid fibril (Williams et al., 2006) and the small, globular protein Gβ1 (Merkel *et al.*, 1999) suggest that the fibril $\Delta\Delta$G values obtained as described here are valid. At the same time, proline replacements within the Aβ-sheet (Williams *et al.*, 2004) are somewhat less devastating than expected based on Gβ1 studies, perhaps because of the formation of additional stabilizing backbone H-bonds within the fibril in response to proline mutations (Williams *et al.*, 2004).

Protocol

Seeded and Unseeded Quiescent Growths of Aβ(1–40) to Calculate the Critical Concentration

1. For both seeded and unseeded growth of Aβ(1–40) fibrils, 20–60 μM freshly disaggregated Aβ(1–40) is prepared in 1× PBS containing 0.05% sodium azide (see section on disaggregation of peptides).

2. Confirm the concentration of the PBS solution of Aβ(1–40) obtained after disaggregation, by diluting 20 μl Aβ(1–40) sample with 100 μl 1% TFA aqueous solution and analyzing 100 μl by the HPLC assay to determine the micrograms of Aβ peptide (see section on HPLC sedimentation assay for amyloid transformations). The starting concentration is adjusted as desired using PBS.

3. Appropriate volumes of the Aβ(1–40) are incubated undisturbed at 37° with and without the addition of 0.1% by weight of wild-type Aβ (1–40)-sonicated seed fibrils.

4. Time points are taken periodically for the HPLC sedimentation assay and ThT assay during the course of growth. Depending on the Aβ mutant, the seeded reaction should proceed quickly, with little or no lag time. Once the reaction appears to be complete by a lack of further increases in the ThT signal, aliquots are analyzed daily by the HPLC sedimentation assay to confirm that the remaining level of monomer is no longer dropping. This plateau value in the residual monomer concentration is the C_r.

Concluding Remarks

Many technically sophisticated methods have been applied to the study of protein aggregation reactions and have provided valuable information. Because many of these methods are more suited for analysis of aggregates than quantitation of the small molecular-weight pool of starting material, they are of limited use in putting together quantitative descriptions of some aspects of aggregation reactions, such as the early phases of aggregation and equilibrium positions at the reaction endpoint, where monomer concentrations are low. Quantitation by HPLC is reasonably sensitive and has the added major advantage of being able to quantify multiple chemical species independently in a mixture of peptides, whether this mixture is arranged by design or is the result of chemical decay of the starting material. In the context of other methods done in parallel, sedimentation results can also provide information on the existence of multiple aggregated species on (or off) the reaction coordinate. The power of this simple set of methods is illustrated in the examples described here, which include the determination of reliable macroscopic rate constants, including true second-order rate constants for elongation, nucleation kinetics analysis, and the determination of equilibrium constants and derived free energy changes associated with fibril nucleation and elongation reactions. These tools should allow the determination of reaction parameters for other amyloid systems in addition to $A\beta(1-40)$ and polyGln, which will help to improve our understanding of the general features of amyloid structure and assembly.

Acknowledgment

This work was supported by grants R01 AG18416 and R01 AG19322 from the National Institutes of Health.

References

Andreu, J. M., and Timasheff, S. N. (1986). The measurement of cooperative protein self-assembly by turbidity and other techniques. *Methods Enzymol.* **130,** 47–59.

Berthelier, V., Hamilton, J. B., Chen, S., and Wetzel, R. (2001). A microtiter plate assay for polyglutamine aggregate extension. *Anal. Biochem.* **295,** 227–236.

Bhattacharyya, A. M., Thakur, A., and Wetzel, R. (2005). Polyglutamine aggregation nucleation: Thermodynamics of a highly unfavorable protein folding reaction. *Proc. Natl. Acad. Sci. USA* **102,** 15400–15405.

Bhattacharyya, A. M., Thakur, A. K., Hermann, V. M., Thiagarajan, G., Williams, A. D., Chellgren, T. P., Creamer, T. P., and Wetzel, R. (2006). Oligoproline effects on polyglutamine conformation and aggregation. *J. Mol. Biol.* **355,** 524–535.

Bucciantini, M., Giannoni, E., Chiti, F., Baroni, F., Formigli, L., Zurdo, J., Taddei, N., Ramponi, C. M., Dobson, C. M., and Stefani, M. (2002). Inherent toxicity of aggregates implies a common mechanism for protein misfolding diseases. *Nature* **416,** 507–511.

Cannon, M. J., Williams, A., Wetzel, R., and Myszka, D. G. (2004). Kinetic analysis of $A\beta$ fibril elongation. *Anal. Biochem.* **328,** 67–75.

Chen, S., Berthelier, V., Hamilton, J. B., O'Nuallain, B., and Wetzel, R. (2002a). Amyloid-like features of polyglutamine aggregates and their assembly kinetics. *Biochemistry* **41,** 7391–7399.

Chen, S., Berthelier, V., Yang, W., and Wetzel, R. (2001). Polyglutamine aggregation behavior *in vitro* supports a recruitment mechanism of cytotoxicity. *J. Mol. Biol.* **311,** 173–182.

Chen, S., Ferrone, F., and Wetzel, R. (2002b). Huntington's disease age-of-onset linked to polyglutamine aggregation nucleation. *Proc. Natl. Acad. Sci. USA* **99,** 11884–11889.

Chen, S., and Wetzel, R. (2001). Solubilization and disaggregation of polyglutamine peptides. *Protein Sci.* **10,** 887–891.

Collins, S. R., Douglass, A., Vale, R. D., and Weissman, J. S. (2004). Mechanism of prion propagation: Efficient amyloid growth in the absence of oligomeric intermediates. *PLoS* **2,** 1582–1590.

Colon, W. (1999). Analysis of protein structure by solution optical spectroscopy. *Methods Enzymol.* **309,** 605–632.

DePace, A. H., and Weissman, J. S. (2002). Origins and kinetic consequences of diversity in Sup35 yeast prion fibers. *Nat. Struct. Biol.* **9,** 389–396.

Diamandis, E. P. (1988). Immunoassays with time-resolved fluorescence spectroscopy: Principles and applications. *Clin. Biochem.* **21,** 139–150.

Esler, W. P., Stimson, E. R., Ghilardi, J. R., Felix, A. M., Lu, Y. A., Vinters, H. V., Mantyh, P. W., and Maggio, J. E. (1997). A beta deposition inhibitor screen using synthetic amyloid. *Nat. Biotechnol.* **15,** 258–263.

Esler, W. P., Stimson, E. R., Jennings, J. M., Vinters, H. V., Ghilardi, J. R., Lee, J. P., Mantyh, P. W., and Maggio, J. E. (2000). Alzheimer's disease amyloid propagation by a template-dependent dock-lock mechanism. *Biochemistry* **39,** 6288–6295.

Evans, K. C., Berger, E. P., Cho, C.-G., Weisgraber, K. H., and Lansbury, P. T., Jr. (1995). Apolipoprotein E is a kinetic but not a thermodynamic inhibitor of amyloid formation: Implications for the pathogenesis and treatment of Alzheimer disease. *Proc. Natl. Acad. Sci. USA* **92,** 763–767.

Falk, R. H., Comenzo, R. L., and Skinner, M. (1997). The systemic amyloidoses [see comments]. *N. Engl. J. Med.* **337,** 898–909.

Ferrone, F. (1999). Analysis of protein aggregation kinetics. *Methods Enzymol.* **309,** 256–274.

Franks, F. (1993). Storage stabilization of proteins. *In* "Protein Biotechnology: Isolation, Characterization and Stabilization" (F. Franks, ed.), pp. 489–531. Humana Press, New York.

Goldsbury, C., Frey, P., Olivieri, V., Aebi, U., and Muller, S. A. (2005). Multiple assembly pathways underlie amyloid-beta fibril polymorphisms. *J. Mol. Biol.* **352,** 282–298.

Gosal, W. S., Morten, I. J., Hewitt, E. W., Smith, D. A., Thomson, N. H., and Radford, S. E. (2005). Competing pathways determine fibril morphology in the self-assembly of beta2-microglobulin into amyloid. *J. Mol. Biol.* **351,** 850–864.

Harper, J. D., Lieber, C. M., and Lansbury, P. T., Jr. (1997). Atomic force microscopic imaging of seeded fibril formation and fibril branching by the Alzheimer's disease amyloid-beta protein. *Chem. Biol.* **4,** 951–959.

Heller, J., Kolbert, A. C., Larsen, R., Ernst, M., Bekker, T., Baldwin, M., Prusiner, S. B., Pines, A., and Wemmer, D. E. (1996). Solid-state NMR studies of the prion protein H1 fragment. *Protein Sci.* **5,** 1655–1661.

Hilbich, C., Kisters-Woike, B., Reed, J., Masters, C. L., and Beyreuther, K. (1992). Substitutions of hydrophobic amino acids reduce the amyloidogenicity of Alzheimer's disease bA4 peptides. *J. Mol. Biol.* **228,** 460–473.

Howlett, D. R., Jennings, K. H., Lee, D. C., Clark, M. S., Brown, F., Wetzel, R., Wood, S. J., Camilleri, P., and Roberts, G. W. (1995). Aggregation state and neurotoxic properties of Alzheimer beta-amyloid peptide. *Neurodegeneration* **4,** 23–32.

Hurle, M. R., Helms, L. R., Li, L., Chan, W., and Wetzel, R. (1994). A role for destabilizing amino acid replacements in light chain amyloidosis. *Proc. Natl. Acad. Sci. USA* **91,** 5446–5450.

Ignatova, Z., and Gierasch, L. M. (2005). Aggregation of a slow-folding mutant of a beta-clam protein proceeds through a monomeric nucleus. *Biochemistry* **44,** 7266–7274.

Jarrett, J. T., and Lansbury, P. T., Jr. (1992). Amyloid fibril formation requires a chemically discriminating nucleation event: Studies of an amyloidogenic sequence from the bacterial protein OsmB. *Biochemistry* **31,** 12345–12352.

Jones, E. M., and Surewicz, W. K. (2005). Fibril conformation as the basis of species- and strain-dependent seeding specificity of Mammalian prion amyloids. *Cell* **121,** 63–72.

Klunk, W. E., Jacob, R. F., and Mason, R. P. (1999). Quantifying amyloid by Congo red spectral shift assay. *Methods Enzymol.* **309,** 285–305.

Kopito, R. R. (2000). Aggresomes, inclusion bodies and protein aggregation. *Trends Cell. Biol.* **10,** 524–530.

LeVine, H. (1999). Quantification of β-sheet amyloid fibril structures with thioflavin T. *Methods Enzymol.* **309,** 274–284.

Lomakin, A., Benedek, G. B., and Teplow, D. B. (1999). Monitoring protein assembly using quasielastic light scattering spectroscopy. *Methods Enzymol.* **309,** 429–459.

Martin, J. B. (1999). Molecular basis of the neurodegenerative disorders [published erratum appears in *N. Engl. J. Med.* [1999] **341,** 1407]. *N. Engl. J. Med.* **340,** 1970–1980.

McCutchen, S. L., Colon, W., and Kelly, J. W. (1993). Transthyretin mutation Leu-55-Pro significantly alters tetramer stability and increases amyloidogenicity. *Biochemistry* **32,** 12119–12127.

Meijer, A. J., and Codogno, P. (2004). Regulation and role of autophagy in mammalian cells. *Int. J. Biochem. Cell. Biol.* **36,** 2445–2462.

Merkel, J. S., Sturtevant, J. M., and Regan, L. (1999). Sidechain interactions in parallel beta sheets: The energetics of cross-strand pairings. *Structure Fold Des.* **7,** 1333–1343.

Merlini, G., and Bellotti, V. (2003). Molecular mechanisms of amyloidosis. *N. Engl. J. Med.* **349,** 583–596.

Mitraki, A., and King, J. (1989). Protein folding intermediates and inclusion body formation. *Biotech.* **7,** 690–697.

Mulkerrin, M. G., and Wetzel, R. (1989). pH dependence of the reversible and irreversible thermal denaturation of γ interferons. *Biochem.* **28,** 6556–6561.

Neurath, H., Greenstein, J. P., Putnam, F. W., and Erickson, J. O. (1944). The chemistry of protein denaturation. *Chem. Rev.* **34,** 157–265.

Nichols, M. R., Moss, M. A., Reed, D. K., Cratic-McDaniel, S., Hoh, J. H., and Rosenberry, T. L. (2005). Amyloid-beta protofibrils differ from amyloid-beta aggregates induced in dilute hexafluoroisopropanol in stability and morphology. *J. Biol. Chem.* **280,** 2471–2480.

Nilsson, M. R. (2004). Techniques to study amyloid fibril formation *in vitro. Methods* **34,** 151–160.

O'Nuallain, B., Shivaprasad, S., Kheterpal, I., and Wetzel, R. (2005). Thermodynamics of abeta(1-40) amyloid fibril elongation. *Biochemistry* **44,** 12709–12718.

O'Nuallain, B., Williams, A. D., Westermark, P., and Wetzel, R. (2004). Seeding specificity in amyloid growth induced by heterologous fibrils. *J. Biol. Chem.* **279,** 17490–17499.

Petkova, A. T., Leapman, R. D., Guo, Z., Yau, W. M., Mattson, M. P., and Tycko, R. (2005). Self-propagating, molecular-level polymorphism in Alzheimer's β-amyloid fibrils. *Science* **307,** 262–265.

Serio, T. R., Cashikar, A. G., Kowal, A. S., Sawicki, G. J., Moslehi, J. J., Serpell, L., Arnsdorf, M. F., and Lindquist, S. L. (2000). Nucleated conformational conversion and the replication of conformational information by a prion determinant. *Science* **289,** 1317–1321.

Sharma, D., Shinchuk, L. M., Inouye, H., Wetzel, R., and Kirschner, D. A. (2005). Polyglutamine homopolymers having 8–45 residues form slablike beta-crystallite assemblies. *Proteins* **61,** 398–411.

Shivaprasad, S., and Wetzel, R. (2004). An intersheet packing interaction in Aβ fibrils mapped by disulfide crosslinking. *Biochemistry* **43,** 15310–15317.

Shivaprasad, S., and Wetzel, R. (2006). Scanning cysteine mutagenesis analysis of Aβ(1-40) amyloid fibrils. *J. Biol. Chem.* **281,** 993–1000.

Stefani, M., and Dobson, C. M. (2003). Protein aggregation and aggregate toxicity: New insights into protein folding, misfolding diseases and biological evolution. *J. Mol. Med.* **81,** 678–699.

Tanaka, M., Chien, P., Naber, N., Cooke, R., and Weissman, J. S. (2004). Conformational variations in an infectious protein determine prion strain differences. *Nature* **428,** 323–328.

Thakur, A., and Wetzel, R. (2002). Mutational analysis of the structural organization of polyglutamine aggregates. *Proc. Natl. Acad. Sci. USA* **99,** 17014–17019.

Wetzel, R. (1994). Mutations and off-pathway aggregation. *Trends Biotechnol.* **12,** 193–198.

Wetzel, R. (2005). Protein folding and aggregation in the expanded polyglutamine repeat diseases. *In* "The Protein Folding Handbook. Part II." (J. Buchner and T. Kiefhaber, eds.), pp. 1170–1214. Wiley-VCH, Weinheim.

Williams, A. D., Portelius, E., Kheterpal, I., Guo, J. T., Cook, K. D., Xu, Y., and Wetzel, R. (2004). Mapping abeta amyloid fibril secondary structure using scanning proline mutagenesis. *J. Mol. Biol.* **335,** 833–842.

Williams, A. D., Sega, M., Chen, M., Kheterpal, I., Geva, M., Berthelier, V., Kaleta, D. T., Cook, K. D., and Wetzel, R. (2005). Structural properties of Aβ protofibrils stabilized by a small molecule. *Proc. Natl. Acad. Sci. USA* **102,** 7115–7120.

Williams, A. D., Shivaprasad, S., and Wetzel, R. (2006). Alanine scanning mutagenesis of Aβ (1-40) amyloid fibril stability. *J. Mol. Biol.* in press.

Wood, S. J., Maleeff, B., Hart, T., and Wetzel, R. (1996). Physical, morphological and functional differences between pH 5.8 and 7.4 aggregates of the Alzheimer's peptide Ab. *J. Mol. Biol.* **256,** 870–877.

Yamaguchi, K., Takahashi, S., Kawai, T., Naiki, H., and Goto, Y. (2005). Seeding-dependent propagation and maturation of amyloid fibril conformation. *J. Mol. Biol.* **352,** 952–960.

Yang, W., Dunlap, J. R., Andrews, R. B., and Wetzel, R. (2002). Aggregated polyglutamine peptides delivered to nuclei are toxic to mammalian cells. *Hum. Mol. Genet.* **11,** 2905–2917.

Zagorski, M. G., Yang, J., Shao, H., Ma, K., Zeng, H., and Hong, A. (1999). Methodological and chemical factors affecting amyloid-β amyloidogenicity. *Methods Enzymol.* **309,** 189–204.

Zhang, S., Iwata, K., Lachenmann, M. J., Peng, J. W., Li, S., Stimson, E. R., Lu, Y., Felix, J. E., Maggio, J. E., and Lee, J. P. (2000). The Alzheimer's peptide a beta adopts a collapsed coil structure in water. *J. Struct. Biol.* **130,** 130–141.

[4]   Protein Aggregation Starting From The Native Globular State[1]

By Giordana Marcon, Georgia Plakoutsi, and Fabrizio Chiti

Abstract

Amyloid formation by globular proteins that normally adopt a compact folded structure is generally induced *in vitro* under harsh conditions involving low pH, high temperature, high pressure, or in the presence of organic solvents. Under these conditions, folded proteins are generally unfolded, at least partially. The approach described here shows a rationale and two detailed examples as to how the mechanism of aggregation of a globular protein can be probed under conditions in which it is initially in its folded conformation, and hence relevant to a physiological environment.

Introduction

Many of the peptides and proteins that have been recognized to deposit into ordered protein aggregates under pathological conditions have a disordered structure in their soluble state under physiological conditions. Examples include the amyloid β (Aβ) peptide, α-synuclein, and amylin (Uversky and Fink, 2004). By contrast, other disease-related proteins have a well-defined globular structure in their soluble state, such as lysozyme, β2-microglobulin, cystatin c, and superoxide dismutase (Stefani and Dobson, 2003; Uversky and Fink, 2004). In addition, it is increasingly evident that many of the natural proteins that normally adopt a folded structure can form, under appropriate conditions, amyloid-like fibrils morphologically and structurally similar to those associated with diseases (Stefani and Dobson, 2003; Uversky and Fink, 2004). These may even be cytotoxic in the form of precursor prefibrillar aggregates (Bucciantini *et al.*, 2002). To understand the pathogenesis of protein deposition diseases associated with globular proteins and the potential conversion of folded proteins into toxic oligomers that the cell needs to combat actively, we need to clarify the mechanism by which a normally folded protein aggregates *in vivo*.

The hypothesis that a folded protein needs to unfold, at least partially, to form a conformational state that is susceptible to aggregate is generally accepted (Kelly, 1998). The finding that many natural mutations associated

---

[1] G. Marcon and G. Plakoutsi contributed equally to this work.

METHODS IN ENZYMOLOGY, VOL. 413
0076-6879/06 $35.00
DOI: 10.1016/S0076-6879(06)13004-9

with inherited forms of amyloid diseases cause the native state to be destabilized provides support to this hypothesis. However, the generalization of this notion has been challenged by a number of observations. Some proteins, such as Ure2p and lithostatin, are found to aggregate under nondestabilizing physiological conditions and form fibrils in which the individual protein molecules retain a native-like structure (Baxa *et al.*, 2003; Bousset *et al.*, 2002; Laurine *et al.*, 2003). Human insulin forms at least precursor aggregates with a native-like α-helical content (Bouchard *et al.*, 2000). In the S6 protein from *Thermus thermophilus*, the propensity to aggregate does not correlate with the destabilization of the native state by mutations (Pedersen *et al.*, 2004). These and other observations suggest that although some proteins do unfold to generate aggregates, at least partially and transiently, others can assemble to form at least precursor aggregates in the absence of or before undergoing a major conformational change. The complexity and multiplicity of pathways are shown in Fig. 1. In this work, we will outline a methodology designed to determine the aggregation mechanism of a folded protein into early oligomeric species. We will describe the application of this procedure to two sample proteins that aggregate with different mechanisms.

Basic Features of Our Approach

Identification of Conditions That Promote Aggregation

The approach for investigating the mechanism of protein aggregation under conditions in which a folded protein is initially in its native conformation consists of three steps. First, we need to identify solution conditions that promote the conversion of the protein of interest into amyloid-like fibrils or protofibrils. Second, we select those conditions in which the protein is initially native prior to aggregation. Finally, we determine the rate constants for basic processes (e.g., folding, unfolding, aggregation, and disaggregation) and proceed to a kinetic inspection to elucidate the underlying aggregation mechanism.

Folded proteins are generally very stable in conditions close to physiological. One common strategy is therefore to choose solution conditions that destabilize the native state (i.e., that facilitate its flexibility, favoring the population of quasi-native states, or reduce the free energy of unfolding). Examples are solutions with small amounts of organic solvents, a mild acidic or basic pH, a high temperature, or a high pressure. Combinations of two or more of these destabilizing agents can also be useful (e.g., small decreases of pH concomitantly with small increases of temperature). If small solvent perturbations are not sufficient to induce aggregation, the

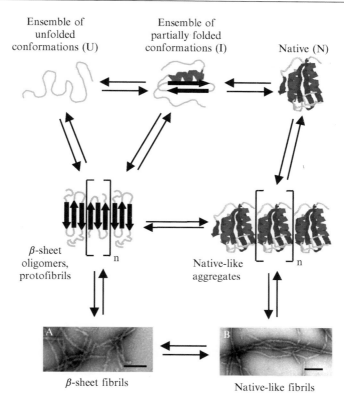

FIG. 1. Possible pathways of aggregation for a globular protein. The negative stained electron micrographs show heat-treated amyloid-like fibrils (A) and native-like fibrils (B) from Ure2p (bar = 0.2 μm). The electron micrographs were reproduced with permission from Bousset *et al.* (2003).

tests can be repeated at higher protein concentrations, because aggregation events are multimolecular reactions and are generally facilitated by increases in protein concentration. If the protein of interest has a high net charge, either positive or negative, high salt concentrations can shield the electrostatic repulsion between the protein molecules and increase the probability to observe aggregation. The same result can be achieved by bringing the pH of the solution close to the isoionic point of the protein.

Selection of Conditions in Which the Protein Is Initially in Its Native Conformation

Once the appropriate solvent conditions have been identified, one needs to select those in which the protein is initially in its folded state before aggregating. The protein can be destabilized, but it must not be denatured. In thermodynamic terms, the free energy of unfolding into a totally or partially unfolded state can be decreased but has to remain substantially higher than zero. A free energy of unfolding higher than 10 kJ mol^{-1} ensures that the protein is over 98% folded at room temperature. It is recommendable, however, that the solvent perturbation is as small as possible so that the stability of the native state is only marginally affected. The ideal condition is the smallest perturbation that still promotes aggregation.

Several techniques can, in principle, be utilized to evaluate the conformational structure of a protein under conditions that favor aggregation when the aggregation process has not yet started significantly. Intrinsic fluorescence and near-ultraviolet (UV) circular dichroism (CD) spectroscopy allow the chemical environment around the tryptophan side chains and other aromatic residues to be evaluated. Far-UV CD and Fourier-transform infrared (FT-IR) spectroscopies can be used to probe the percentages of secondary structure types. Fluorescence of 1-anilino-8-naphthalenesulfonic acid (ANS) defines the solvent accessibility of hydrophobic clusters.

The procedure consists of determining the spectra of the protein sample immediately after incubation under aggregating conditions and filtration with 0.02-μm pore-sized filters. The aim is obviously to make sure that they are similar to those obtained for a protein sample under conditions in which the protein does not aggregate and is stable in its native conformation. Nuclear magnetic resonance (NMR) spectroscopy is the best technique to define the overall structure in solution and to detect even subtle changes following the transfer of the protein from native to aggregating conditions. If the protein behaves as an enzyme or binds to specific ligands or target proteins, enzymatic activity or binding assays can also be utilized to confirm the correct folding of the protein of interest.

Although this analysis is apparently simple (it is based on the comparison of spectroscopic or biochemical features of the same protein under two different conditions), some care is necessary. Our first recommendation is to utilize more than one probe to detect possible conformational changes. In fact, many of the techniques mentioned above can only monitor specific levels of structure when they are not used in conjunction with others. For example, the observation that a protein has apparently identical far-UV CD spectra in different conditions does not by itself rule out the occurrence of a conformational change: the protein can just relax the packing of residues in the hydrophobic core while maintaining the overall topology.

Another complicating factor is that spectroscopic or enzymatic activity changes are not necessarily caused by conformational modifications of the protein. These can result from baseline effects associated with changes of solvent conditions rather than arising from real conformational changes. This is particularly evident when using intrinsic tryptophan fluorescence as a probe, because different solution conditions can have different fluorescence quenching properties. This problem can be sorted out by making measurements using stepwise changes of conditions. For example, if the native and aggregating conditions that one needs to compare are solutions containing 0 and 2 M urea, the spectra can be acquired within this range using 0.2 M intervals. If the resulting plot of fluorescence emission *versus* urea concentration is linear, the change is likely to arise from solvent effects; if a nonlinear transition is observed, the observed spectroscopic change is more likely caused by a conformational change.

The ongoing aggregation process under the selected conditions can be viewed as a further complexity in the analysis. The presence of newly formed aggregates during the measurements would obviously interfere with the biophysical and biochemical analysis of the initial state. This problem can be particularly serious, because oligomers and protofibrils can form rapidly in some cases and escape experimental detection with probes specific for amyloid fibrils, such as thioflavin T (ThT) or Congo red (CR) binding. To obviate this problem, rapid techniques that require small protein concentrations are preferentially selected. Enzymatic activity assays are very useful, because the enzymes are generally highly specific and efficient and the assays require extremely low protein concentrations. A stopped- or continuous-flow analysis to measure the rates of folding and unfolding is also preferred in some cases. These reactions may occur on the millisecond or submillisecond time scale; if folding is largely more rapid than unfolding, the protein is folded under the selected conditions. It is advisable, however, to check that protein aggregates do not form significantly during the time required for the conformational analysis of the initial state. This can be achieved using, for example, dynamic light scattering (DLS), ultracentrifugation, size-exclusion chromatography, or photo-induced cross-linking of unmodified proteins (PICUP).

Elucidating the Mechanism of Aggregation

After identifying the conditions that promote formation of amyloid-like fibrils and in which the protein is initially in its folded structure, we are ready to investigate the aggregation pathway. The first parameter that needs to be determined is the apparent rate constant for aggregation (k_{app}) under the selected conditions. The probe to determine the kinetic profile of aggregation can be selected on the grounds of the specific aggregation phase in which one is interested (e.g., oligomer formation, protofibril elongation, fibril elongation) and on the size of the forming aggregates. The time-dependent build-up of the aggregates can be followed with FT-IR spectroscopy, CD spectroscopy, ThT fluorescence, size-exclusion chromatography, light scattering, and other techniques. The plot of the spectroscopic signal versus time can be analyzed with a procedure of best-fit using an exponential function (this is reported below as Eq. 1). The procedure of best-fit yields the value of k_{app}. If a lag phase is present, it is advisable to acquire the kinetic plot in the presence of preformed fibrils to act as seeds. In the presence of a sufficiently high concentration of seeds, the lag phase is no longer present and the fitting procedure yields the rate constant of the exponential phase, which we again call k_{app}. For an ideal analysis of the data in the presence of a lag phase, it will be necessary to develop equations that allow fitting of both the lag and growth phases.

The folding and unfolding rate constants (k_{IN} and k_{NI}) also need to be determined under the selected conditions. These measurements generally need a stopped-flow apparatus that allows fluorescence or CD changes to be detected starting from a few milliseconds after placing the protein under folding or unfolding conditions. If folding and/or unfolding occurs on the submillisecond time scale, other instruments, such as a continuous-flow apparatus of the type described by Roder and co-workers, are required (Welker et al., 2004). Finally, conventional fluorometers or CD instruments are required if these processes occur slowly (on the time scale of a few seconds or minutes). The k_{NI} cannot be determined directly, because the protein is stable in its native state under the conditions selected for aggregation. Nevertheless, the k_{NI} can be determined under the selected conditions but in the presence of various concentrations of a denaturant, for example, urea or guanidine. The plot of the logarithm of the k_{NI} versus the denaturant concentration can be fitted to a straight line or to a second-order polynomial function to determine the k_{NI} in the absence of denaturant. This is exactly the k_{NI} value (i.e., the unfolding rate constant under the conditions selected for aggregation).

The k_{IN} can be measured directly under the selected conditions or by a similar extrapolation procedure. If more than one phase is observed during folding, a preliminary study is needed so that all the observed phases can be assigned to specific events during folding. Kinetic inspection of the problem will clarify if only the major phase for folding or all of them have to be considered in the analysis. In most cases, the relevant phase is the major folding phase, during which most of the protein molecules attain the native state, either from a partially folded state accumulating during the process or from the fully unfolded state (see the aggregation mechanism of the N-terminal domain of the *Escherichia coli* HypF [HypF-N] described below as an example). For simplicity, we will call this rate constant k_{IN}

regardless of whether the folding process starts from a partially or fully unfolded state.

In some cases, it is necessary to determine the rate of aggregation when the initial state is not the folded state but a partially or fully unfolded state (i.e., the same conformational state that represents the starting point of the investigated folding reaction). This aggregation rate constant, which we call k_{agg}, is not meant to be a microscopic rate constant but the observed rate constant for a complex process that probably leads to heterogeneity in aggregated species with a number of steps involving, for example, nucleation and elongation phases. This process can approximate to single exponential kinetics but understandably results from the combination of a number of microscopic rate constants. The k_{agg} can be technically difficult to measure, mainly because, under the selected conditions, the protein populates predominantly the folded state. Similar to k_{NI}, k_{agg} can be measured in conditions in which the protein is denatured and extrapolated back to the chosen conditions. The correctness of a linear extrapolation needs, however, to be assessed. In the absence of an accurate k_{agg} value determined experimentally, one can utilize a recently published and well-tested algorithm that allows the k_{agg} value of an unstructured polypeptide chain to be predicted with reasonable confidence (Dubay et al., 2004).

The rate constants of all these processes represent valuable pieces of information to determine the aggregation pathway followed by the folded protein. In some cases, it may be necessary to estimate additional rate constants depending on the complexity of the aggregation mechanism. A procedure of universal applicability as to how these parameters can be converted into a mechanistic description of the aggregation process cannot be presented here, because the utilization of the parameters depends on the numerical values of the rate constants themselves. In the following two paragraphs, we will provide two examples as to how the methodology described here and the resulting rate constants were used to gain information on the aggregation pathway of two proteins, HypF-N and the acylphosphatase from *Sulfolobus solfataricus* (Sso AcP).

Aggregation Mechanism of HypF-N

Selection of the Aggregating Conditions and Assessment of an Initial Native-Like Structure

The conditions selected to promote aggregation of HypF-N are 6–12% (v/v) 2,2,2-trifluoroethanol (TFE) in 5 or 50 mM acetate buffer, 2 mM dithiothreitol (DTT), pH 5.5, 25° (Marcon et al., 2005). In these media, HypF-N aggregates, at a concentration of 0.4 mg ml^{-1}, into spherical oligomers that subsequently convert into amyloid-like fibrils. Both the oligomers and fibrils bind and increase the fluorescence of ThT. By contrast, TFE concentrations lower than 6% (v/v) do not induce aggregation.

The far-UV CD spectra acquired for filtered samples (0.02 μm) in the presence of 6–12% TFE are very similar to the spectrum obtained for the native protein in its absence. This rules out any detectable change in secondary structure following the addition of small quantities of TFE and before aggregation has proceeded to any measurable degree (Marcon *et al.*, 2005). The intrinsic fluorescence spectra acquired in 0–10% TFE were also indistinguishable from each other. Moreover, the intrinsic fluorescence of HypF-N under native and aggregating conditions is quenched to a similar extent by acrylamide, indicating that the chemical environment around the two tryptophan side chains of the protein (Trp27 and Trp81) is not changed. The ANS fluorescence emissions in the presence of 6% TFE and in its absence are also highly superimposable, ruling out the possibility that hydrophobic clusters become superficial following the addition of small amounts of TFE.

Determination of the k_{app}

The ThT assay was chosen to analyze the time-dependent assembly of aggregates of HypF-N under the selected conditions. A total of 0.4 mg ml^{-1} HypF-N was incubated at 25° in 50 mM acetate buffer, 2 mM DTT, and 10% (v/v) TFE. At regular time intervals, 60-μl aliquots were added to 440 μl of solutions containing 25 μM ThT, 25 mM phosphate buffer, pH 6.0. The steady-state ThT fluorescence intensity, F(t), of the resulting samples, measured using a Perkin Elmer LS 55 spectrofluorometer (Wellesley, MA), was plotted versus time (Fig. 2). The plot was fitted to a single exponential function of the form:

$$F(t) = F_{eq} + A \exp(-k_{app}t) \tag{1}$$

where F_{eq} is the maximum ThT fluorescence measured at the end of the aggregation exponential phase, A is the amplitude of the fluorescence change, and k_{app} is the apparent rate constant for aggregation. The resulting k_{app} value is $1.17 \pm (0.40) \times 10^{-3}$ s^{-1} ($\ln k_{app} = -6.75 \pm 0.20$).

Determination of the Disaggregation Rate Constant (k_{dis})

The HypF-N aggregates, preformed in 10% (v/v) TFE, were isolated by centrifugation and resuspended in 20 mM Hepes buffer, 137 mM NaCl, pH 7.4, 25° in the absence of TFE. Disaggregation was followed using the ThT assay (Marcon *et al.*, 2005). The decrease of fluorescence intensity (Fig. 2, inset) occurred with an apparent disaggregation rate constant (k_{dis})

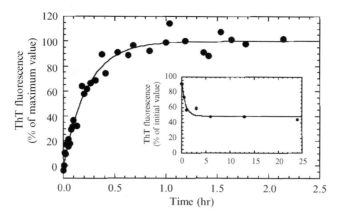

FIG. 2. Time course of HypF-N aggregation measured by thioflavin T (ThT) fluorescence. Conditions were 0.4 mg ml^{-1} N-terminal domain of the *Escherichia coli* HypF (HypF-N) in 10% (v/v) 2,2,2-trifluoroethanol (TFE), 50 mM acetate buffer, 2 mM dithiothreitol (DTT), pH 5.5, 25°. The solid line through the data represents the best fit to a single exponential function (Eq. 1). The fitting procedure yielded the apparent rate constant for aggregation (k_{app}). The inset shows the time-course of disaggregation of HypF-N in the absence of TFE after preincubating the protein for 1 h in 10% (v/v) TFE in the buffer described above. The equation of best fit provided the disaggregation rate constant (k_{dis}). Reproduced with permission from Marcon *et al.* (2005).

of 3.1×10^{-4} s^{-1} (ln$k_{dis} = -8.07 \pm 0.30$). From the observation of the inset in Fig. 2, it can be deduced that the aggregation process of HypF-N in 10% (v/v) TFE is partially reversible, because 50% of the aggregates are disrupted after removing the aggregating agent.

Determination of the k_{agg} from the Partially Folded State

The rate constant of aggregation under conditions in which only the partially folded state is present (k_{agg}) was also determined. A total of 0.4 mg ml^{-1} HypF-N was incubated at 25° in 50 mM acetate buffer, 2 mM DTT, 25% (v/v) TFE. TFE concentrations of 20–25% (v/v) ensure that the protein is denatured and that the native state is no longer present at the beginning of the aggregation process. A value of 1.0 ± 0.5 s^{-1} was obtained for k_{agg} (Calamai *et al.*, 2003). Nevertheless, we were interested in the k_{agg} value in 10% (v/v) TFE. We could not utilize the k_{agg} value obtained in 20–25% (v/v) TFE because of a dependence of k_{agg} on TFE concentration (Fezoui and Teplow, 2002). Results reported in literature for the unstructured Aβ peptide assess that k_{agg} value in 10% (v/v) TFE is approximately sevenfold lower than in 25% (v/v) TFE (Fezoui and

Teplow, 2002). We made the assumption that HypF-N behaves in an approximately analogous manner and attributed a value of 0.14 s^{-1} (ln $k_{agg} = -1.95 \pm 0.60$) to the k_{agg} parameter in 10% (v/v) TFE. Although the dependence of k_{agg} on TFE concentration may differ for different systems, this will not affect our conclusions, because alternative models are based on parameters that differ by orders of magnitude.

Determination of the k_{IN} and k_{NI}

Folding and unfolding traces of HypF-N were recorded with a Bio-logic SFM-3 stopped-flow device coupled to a fluorescence detection system (Bio-logic, Claix, France). All the kinetic traces were acquired in 50 mM acetate, 2 mM DTT, 10% TFE, pH 5.5, 25° at final protein concentrations of 0.015–0.04 mg ml^{-1} and in the presence of various urea concentrations. The unfolding and refolding traces were fitted to single and double exponential functions, respectively, of the form:

$$y(t) = q + \sum_{i=1}^{n} A_i \exp(-k_i t) \qquad (2)$$

where $y(t)$ is the fluorescence at time t; A_i and k_i are the amplitude and rate constant of the ith phase, respectively; q is the fluorescence value at equilibrium; and n is the number of observed phases. The second slow phase observed during folding of HypF-N has a small amplitude and arises from *cis-trans* proline isomerism (Marcon et al., 2005). Kinetic inspection of the system shows that the resulting model of the aggregation mechanism of HypF-N does not depend on whether or not such a slow phase is considered in the analysis (Marcon et al., 2005). For simplicity, we will therefore neglect this minor phase. The natural logarithm of the rate constant for the first major phase of folding or for unfolding (k_{IN} or k_{NI}, respectively) was plotted against urea concentration (Fig. 3), and the resulting plot was fitted to a polynomial function of the form:

$$\ln k = \ln(k^{H_2O}) + a[\text{urea}] + b[\text{urea}]^2 \qquad (3)$$

where k^{H_2O} is the rate constant in the absence of urea and a and b are the parameters related to the dependence of ln$k$ on urea concentration (Fig. 3). The values of the two rate constants k_{IN} and k_{NI}, extrapolated from the reported plots in the absence of urea, were determined to be 570 and 4.7 s^{-1} (ln$k_{IN} = 6.35 \pm 0.35$, ln$k_{NI} = 1.55 \pm 0.42$). Because folding and unfolding of HypF-N are both very fast with respect to aggregation, the latter process does not prevent the acquisition of satisfactory kinetic traces

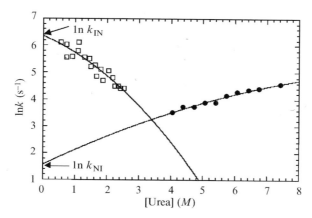

FIG. 3. Rate constants for folding and unfolding of the N-terminal domain of the *Escherichia coli* HypF (HypF-N). The unfolding (filled circles) and folding of the partially folded state into the fully native state (open squares) were measured in 10% 2,2,2,-trifluoroethanol (TFE), 50 mM acetate, 2 mM dithiothreitol (DTT), pH 5.5, 25°. The solid lines represent the best fits to Eq. (3). The folding and unfolding rate constants extrapolated in the absence of urea and used in the kinetic analysis are indicated with labels (k_{IN} and k_{NI}). Reproduced with permission from Marcon *et al.* (2005). k_{IN}, folding rate constant; k_{NI}, unfolding rate constant.

or the subsequent determination of the rate constants. The finding that $k_{IN} \gg k_{NI}$ is a further proof that HypF-N is predominantly in its native state under these conditions.

Utilization of the Relevant Rate Constants to Gain Information on the Mechanism of HypF-N Aggregation

The experimentally determined rate constants can be used to gain information on the pathway followed by native HypF-N to aggregate into ThT-binding oligomers among the various possibilities described in Fig. 1. We initially evaluated the pathway described in Fig. 4 (Scheme 1). Our aim was to ascertain whether the proposed model was consistent with the experimental data, thus assessing that the hypothesized model was correct. In particular, we were interested in proving the similarity between k_{app} measured in 10% (v/v) TFE and the k_{app} value derived from the model.

All the rate constants reported in Fig. 4 (Scheme 1) refer to the conditions selected for studying HypF-N aggregation. In this model, formation of aggregates [Agg] is assumed to occur from the partially unfolded state rather than from the native state. In the system considered here, I and N equilibrate very rapidly, with a k_{app} of $k_{NI} + k_{IN} = 575$ s^{-1}. This process

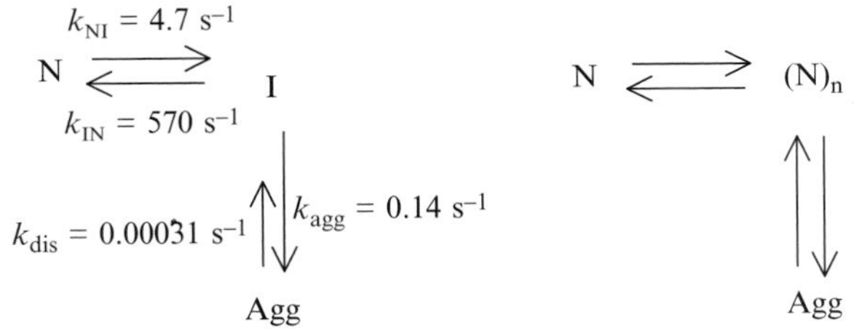

$$k_{NI} = 4.7 \text{ s}^{-1}$$

$$N \underset{k_{IN} = 570 \text{ s}^{-1}}{\overset{}{\rightleftarrows}} I \qquad\qquad N \rightleftarrows (N)_n$$

$$k_{dis} = 0.00031 \text{ s}^{-1} \Big\Updownarrow k_{agg} = 0.14 \text{ s}^{-1} \qquad\qquad \Updownarrow$$

Agg   Agg

Scheme 1 (HypF-N)   Scheme 2 (Sso AcP)

FIG. 4. Schematic representations of the aggregation mechanisms for HypF-N (Scheme 1) and *Sulfolobus solfataricus* acylphosphatase (Sso AcP) (Scheme 2). HypF-N aggregates via a partially unfolded species that is scarcely populated but in equilibrium with the fully native state. By contrast, Sso AcP forms native-like assemblies that consequently transform into ordered aggregates, without a preexisting unfolding phase. k_{app}, aggregation rate constant; k_{IN}, folding rate constant; k_{NI}, unfolding rate constant; k_{dis}, disaggregation rate constant.

occurs well before any significant degree of aggregation has taken place, because k_{agg} is over three orders of magnitude slower than the first relaxation rate. This simplifies the analysis considerably. In addition, disaggregation is much slower than aggregation ($k_{dis} \ll k_{agg}$); the latter can therefore be assumed to be an irreversible process. Taking advantage of these simplifications, the differential equations describing the change of concentration of the various species with time can be solved using matrix algebra to obtain the time courses of [N], [I], and [Agg]. Differential equations can be solved as described (Gutfreund, 1995). Alternatively, to solve systems of ordinary differential equations, a program like Berkeley Madonna™, developed by R. Macey and G. Oster, can be used (a free fully functional version is available on the Web at http://www.berkeleymadonna. com). The calculated time course of [Agg] is as follows:

$$[\text{Agg}](t) = [N]_0 - [N]_0 \exp[-k_{agg}k_{NI}/(k_{NI} + k_{IN})t] \qquad (4)$$

where [Agg](t) is the concentration of protein monomers in the aggregated form at time t and $[N]_0$ is the initial concentration of the protein monomers in the native form. According to Eq. (4), k_{app} corresponds to k_{agg} multiplied by the fraction of the intermediate species present in solution [$k_{app} = k_{agg} k_{NI}/(k_{NI} + k_{IN})$]. From the values of k_{agg}, k_{NI}, and k_{IN} in 10% TFE, we obtain $k_{app} = 0.00117 \text{ s}^{-1}$ (ln $k_{app} = -6.75 \pm 0.40$). This is effectively consistent with the experimental k_{app} value of 0.00117 s^{-1} (ln $k_{app} = -6.75 \pm 0.20$). Thus, the model depicted in Fig. 4 (Scheme 1) adequately describes the mechanism of aggregation of HypF-N under conditions in which the protein is initially predominantly in its native state. The small fraction of partially unfolded molecules (I) in rapid equilibrium with the fully folded state (N), with the former being approximately 1% of the total population, appears to drive the aggregation process of HypF-N. The reader is referred to Marcon *et al.* (2005) for more details on the biophysical investigation and kinetic analysis that have been described in this paragraph.

Aggregation Mechanism of Sso AcP

Appropriate Conditions Under Which the Aggregation Process Is Promoted

Sso AcP is a very stable protein with a free energy of unfolding of 48 ± 4 kJ mol^{-1} at pH 5.5, 37° (Bemporad *et al.*, 2004). It therefore represents an ideal model system for our studies, because this protein remains folded in its native structure under the relatively harsh conditions that generally promote amyloid formation. Sso AcP forms ordered aggregates when incubated for 30 min at concentrations of 0.1–1.0 mg ml^{-1} in 15–25% (v/v) TFE, 50 mM acetate, pH 5.5, 25° (Plakoutsi *et al.*, 2004; Plakoutsi *et al.*, 2005). These aggregates appear to have the morphology of globules or short and thin fibrils with approximate diameters of 3–5 nm (Plakoutsi *et al.*, 2004). They bind ThT, CR, and ANS and appear to possess an extensive β-sheet structure, as indicated by far-UV CD and FT-IR spectroscopies. Such properties are reminiscent of amyloid protofibrils.

Assessment of the Native-Like Structure of the Initial Conformation

Under the conditions found to promote protofibril formation, Sso AcP aggregates very rapidly, with the first oligomers forming with a kinetic phase that is complete within a few seconds. A detailed spectroscopic investigation of the initial monomeric state has therefore been very difficult to achieve. This technical difficulty has been circumvented using enzymatic activity measurements and a stopped-flow analysis (Plakoutsi *et al.*, 2004).

In a first experimental set, Sso AcP was preincubated at a concentration of 0.03 mg ml^{-1} for 30 min in 50 mM acetate, pH 5.5, 25° in different TFE concentrations. In each case, 50 μl of the protein samples was added to 950 μl of solutions containing 2.5 mM benzoylphosphate, 50 mM acetate, pH 5.5, and TFE concentrations identical to those present in the initial preincubation mixtures. The final protein concentration was 0.0015 mg ml^{-1}. At TFE concentrations higher than 12% (v/v), the activity was found to diminish abruptly relative to lower TFE concentrations. When the samples were not preincubated, in a second set of experiments, the activity decreased only at TFE concentrations higher than 26% (v/v), showing that when the aggregation process is at an early stage or not even started, native-like Sso AcP is still present in 15–25% (v/v) TFE. In a third set of experiments, the protein was fully unfolded by a 30-min preincubation in 6 M guanidine hydrochloride at 25°. As described above, 50-μl aliquots were added to 950-μl solutions containing various concentrations of TFE. Removal of the denaturant led to an immediate recovery of the enzymatic activity for TFE concentrations up to 26% (v/v), reaching levels similar to those of the second set of experiments.

These results imply that in the first set of experiments, the protein had enough time to aggregate and that before aggregating, the protein existed in a native-like and enzymatically active conformation that was in equilibrium with a scarcely populated denatured state. The measured activity of the protein in 15–20% TFE was lower than that measured in the absence of TFE. However, the dependence of activity on TFE concentration was found to decrease linearly and probably reflects the change of catalytic efficiency as a result of the solvent changes. This view is supported further by the observation that a mutant of Sso AcP that does not aggregate in 15–25% (v/v) TFE and remains stable in its native conformation has a dependence of enzymatic activity on TFE concentration similar to that observed for the wild-type protein (data not published).

For a further analysis of the protein's conformational state, the k_{IN} and k_{NI} of Sso AcP were measured under different TFE concentrations using a stopped-flow device, as described above for the HypF-N protein (Plakoutsi et al., 2004). For the unfolding experiments, 1 volume of 0.4 mg ml^{-1} Sso AcP in 50 mM acetate buffer, pH 5.5, was mixed with 19 volumes of solutions with TFE concentrations ranging from 45–60% (v/v). For the folding experiments, 1 volume of 0.4 mg ml^{-1} Sso AcP denatured in 6 M guanidine hydrochloride was mixed with 19 volumes of solutions with TFE concentrations ranging from 0 to 18% (v/v). The k_{IN} and k_{NI} were plotted versus TFE concentration to extrapolate the k_{IN} and k_{NI} values in 15–25% (v/v) TFE (Fig. 5). As shown clearly in Fig. 5, folding is much faster than unfolding under these conditions, indicating that the native state is thermodynamically more stable than the partially unfolded state (Plakoutsi et al., 2004).

Utilization of the Relevant Rate Constants to Gain Information on the Mechanism of Sso AcP Aggregation

The k_{app} was measured at various TFE concentrations using the ThT assay as described above for the HypF-N model. The aggregation rate is faster than unfolding under conditions in which protofibril formation occurs (Fig. 5). Therefore, the aggregation process does not involve a transition of native Sso AcP into a partially unfolded state but occurs

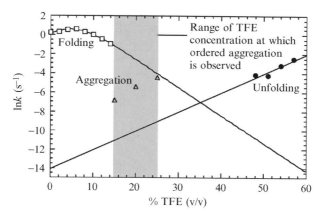

FIG. 5. Rate constants for folding (empty squares) and unfolding (filled circles) of *Sulfolobus solfataricus* acylphosphatase (Sso AcP). Aggregation rate constant values are also reported (empty triangles). The grey area represents the range of 2,2,2,-trifluoroethanol (TFE) concentration in which oligomers containing β-sheet structure and binding thioflavin T (ThT), Congo red (CR), and 1-anilo-8-naphthalenesulfonic acid (ANS) are formed. All measurements were performed in 50 m*M* acetate, pH 5.5, 25°. Reproduced with permission from Plakoutsi *et al.* (2004).

directly from a native-like conformation. On the other hand, the protofibrils forming at the end of the monitored process have a secondary structure content and ANS binding properties very different from those of the native state (Plakoutsi *et al.*, 2004). The proposed model for the Sso AcP aggregation mechanism consists in the assembly of the Sso AcP molecules in their folded state before they transform into the protofibril aggregates (Fig. 4). For more details, the reader is referred to Plakoutsi *et al.* (2004) and Plakoutsi *et al.* (2005).

Conclusions

    In summary, we have described a methodology to investigate the pathway followed by a folded protein to convert into those oligomeric species that are cytotoxic and potentially precursors to amyloid fibrils. The application of this procedure to the two sample proteins described here (HypF-N and Sso AcP) has allowed two different aggregation mechanisms to be outlined for these two cases. Of course, the procedure described here cannot describe all phases of the aggregation process; for instance, it cannot determine the thermodynamics and kinetics of nucleation and elongation phases. Nor can this method contribute to the acquisition of structural information on the oligomers or transiently formed species that precede their formation (e.g., the partially folded species for HypF-N or the native-like oligomer for Sso AcP) or clarify which residues or regions of the sequence can promote the process of amyloid formation. Nevertheless, it has the advantage of elucidating, with relatively simple biophysical and kinetic analyses, whether aggregation occurs via or before a major conformational change. It can show whether a partially folded conformation in a monomeric state or, instead, an assembly of molecules in a native-like conformation can be a key intermediate of the overall process. Once the key intermediate species have been identified, a more detailed investigation can be pursued to isolate these species and determine their structures and properties.

References

Baxa, U., Taylor, K. L., Wall, J. S., Simon, M. N., Cheng, N., Wickner, R. B., and Steven, A. C. (2003). Architecture of Ure2p prion filaments: The N-terminal domains form a central core fiber. *J. Biol. Chem.* **278**, 43717–43727.

Bemporad, F., Capanni, C., Calamai, M., Tutino, M. L., Stefani, M., and Chiti, F. (2004). Studying of the folding process of the acylphosphatase from *Sulfolobus solfataricus*. A comparative analysis with other proteins from the same superfamily. *Biochemistry* **43**, 9116–9126.

Bouchard, M., Zurdo, J., Nettleton, E. J., Dobson, C. M., and Robinson, C. V. (2000). Formation of insulin amyloid fibrils followed by FTIR simultaneously with CD and electron microscopy. *Protein Sci.* **9**, 960–967.

Bousset, L., Briki, F., Doucet, J., and Melki, R. (2003). The native-like conformation of Ure2p in fibrils assembled under physiologically relevant conditions switches to an amyloid-like conformation upon heat-treatment of the fibrils. *J. Struct. Biol.* **141**, 132–142.

Bousset, L., Thomson, N. H., Radford, S. E., and Melki, R. (2002). The yeast prion Ure2p retains its native alpha-helical conformation upon assembly into protein fibrils *in vitro*. *EMBO J.* **17**, 2903–2911.

Bucciantini, M., Giannoni, E., Chiti, F., Baroni, F., Formigli, L., Zurdo, J., Taddei, N., Ramponi, G., Dobson, C. M., and Stefani, M. (2002). Inherent toxicity of aggregates implies a common mechanism for protein misfolding diseases. *Nature* **416**, 507–511.

Calamai, M., Taddei, N., Stefani, M., Ramponi, G., and Chiti, F. (2003). Relative influence of hydrophobicity and net charge in the aggregation of two homologous proteins. *Biochemistry* **42**, 15078–15083.

DuBay, K. F., Pawar, A. P., Chiti, F., Zurdo, J., Dobson, C. M., and Vendruscolo, M. (2004). Prediction of the absolute aggregation rates of amyloidogenic polypeptide chains. *J. Mol. Biol.* **341**, 1317–1326.

Fezoui, Y., and Teplow, D. B. (2002). Kinetic studies of amyloid beta-protein fibril assembly. Differential effects of alpha-helix stabilization. *J. Biol. Chem.* **277**, 36948–36954.

Gutfreund, H. (1995). Kinetics for the Life Sciences. Cambridge University Press, UK.

Kelly, J. W. (1998). The alternative conformations of amyloidogenic proteins and their multi-step assembly pathways. *Curr. Opin. Struct. Biol.* **8**, 101–106.

Laurine, E., Gregoire, C., Fandrich, M., Engemann, S., Marchal, S., Thion, L., Mohr, M., Monsarrat, B., Michel, B., Dobson, C. M., Wanker, E., Erard, M., and Verdier, J. M. (2003). Lithostathine quadruple helical filaments form proteinase-K resistant deposits in Creutzfeldt-Jakob disease. *J. Biol. Chem.* **278**, 51770–51778.

Marcon, G., Plakoutsi, G., Canale, C., Relini, A., Taddei, N., Dobson, C. M., Ramponi, G., and Chiti, F. (2005). Amyloid formation from HypF-N under conditions in which the protein is initially in its native state. *J. Mol. Biol.* **347,** 323–335.

Pedersen, J. S., Christensen, G., and Otzen, D. E. (2004). Modulation of S6 fibrillation by unfolding rates and gatekeeper residues. *J. Mol. Biol.* **341,** 575–588.

Plakoutsi, G., Taddei, N., Stefani, M., and Chiti, F. (2004). Aggregation of the acylphosphatase from *Sulfolobus solfataricus. J. Biol. Chem.* **279,** 14111–14119.

Plakoutsi, G., Bemporad, F., Calamai, M., Taddei, N., Dobson, C. M., and Chiti, F. (2005). Evidence for a mechanism of amyloid formation involving molecular reorganisation within native-like precursor aggregates. *J. Mol. Biol.* **351,** 910–922.

Stefani, M., and Dobson, C. M. (2003). Protein aggregation and aggregate toxicity: New insights into protein folding, misfolding diseases and biological evolution. *J. Mol. Med.* **81,** 678–699.

Uversky, V. N., and Fink, A. L. (2004). Conformational constraints for amyloid fibrillation: The importance of being unfolded. *Biochem. Biophys. Acta* **1698,** 131–153.

Welker, E., Maki, K., Shastry, M. C., Juminaga, D., Bhat, R., Scheraga, H. A., and Roder, H. (2004). Ultrarapid mixing experiments shed new light on the characteristics of the initial conformational ensemble during the folding of ribonuclease A. *Proc. Natl. Acad. Sci. USA* **101,** 17681–17686.

[5] Direct Observation of Amyloid Growth Monitored by Total Internal Reflection Fluorescence Microscopy

By TADATO BAN and YUJI GOTO

Abstract

Most morphological investigations of amyloid fibrils have been performed with various microscopic methods. Among them, direct observation of fibril growth is possible using atomic force microscopy and fluorescence microscopy. Direct observation provides information about the rate and direction of growth at the single fibril level, which cannot be obtained from averaged ensemble measurements. In this chapter, we describe a new technique for the direct observation of amyloid fibril growth using total internal reflection fluorescence microscopy (TIRFM) combined with amyloid-specific thioflavin T (ThT) fluorescence. TIRFM has been developed to monitor single molecules by effectively reducing the background fluorescence in an evanescent field. One of the advantages of TIRFM is that one can selectively monitor fibrils lying along a glass slide, so that one can obtain the exact length of fibrils. This method was used to follow the kinetics of seed-dependent fibril growth of amyloid β (1–40). The fibril growth was a highly cooperative process, with the fibril ends extending at a constant rate. Because ThT binding is common to all

METHODS IN ENZYMOLOGY, VOL. 413   0076-6879/06 $35.00
DOI: 10.1016/S0076-6879(06)13005-0

amyloid fibrils, the present method will have general applicability to the real-time analysis of amyloid fibrils.

Introduction

The formation of amyloid fibrils is considered to be a nucleation-dependent process in which nonnative precursor proteins slowly associate to form nuclei (Dobson, 2003; O'Nuallain et al., 2004; Naiki et al., 1997; Uversky and Fink, 2004; see Chapter 3 by O'Nuallain and colleagues in this volume). This process is followed by an extension reaction, where the nucleus grows by sequential incorporation of more precursor protein molecules. This model has been validated by the observation that fibril-extension kinetics are accelerated by the addition of preformed fibrils (i.e., by a seeding effect). However, the mechanism of fibril formation by individual poly-peptide chains is not completely understood, and there are several variations of the nucleation-dependent model (Depace and Weissman, 2002; Goldsburg et al., 2000; Kad et al., 2003; Scheibel et al., 2001; Serio et al., 2000). To address the mechanism of amyloid fibril formation, it is important to observe the process at the single-fibril level. Recently, epifluorescence with a newly introduced fluorescent dye (Inoue et al., 2001) and atomic force microscopy (AFM) (Goldsbury et al., 1999; Hoyer et al., 2004; Ionescu-Zanetti et al., 1999; Khurana et al., 2003) have been utilized for the direct observation of individual amyloid fibrils. Although these techniques are quite useful in providing information on the mode of fibril growth, the need to introduce the fluorescence probe prevents their general application. Moreover, the strong interaction of amyloid proteins with the mica surface used in AFM measurements results in the formation of fibrils morphologically different from the intact amyloid fibrils (Goldsbury et al., 1999; Green et al., 2004).

Total reflection fluorescence microscopy (TIRFM) has been developed to monitor single molecules (Funatsu et al., 1995; Yamasaki et al., 1999) by effectively reducing the background fluorescence under the evanescent field formed on the surface of a quartz slide. When a laser is incident on the interface between a quartz slide (high reflection index) and the aqueous solution (low reflection index) at the critical angle for total internal reflec-tion, the evanescent field is produced beyond the interface in the solution (Fig. 1). Because the evanescent field is produced with a penetration depth of about 150 nm, the illumination is restricted to fluorophores either bound to the quartz slide surface or located close by, resulting in highly reduced background fluorescence. Furthermore, with the careful selection of optical elements, the background fluorescence can be reduced by 2000-fold compared with ordinary epifluorescence microscopy (Ishijima and Yanagida, 2001).

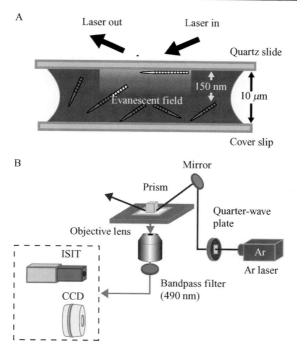

Fig. 1. Schematic representation of amyloid fibrils revealed by total reflection fluorescence microscopy (TIRFM). (A) Penetration depth of the evanescent field formed by the
total internal reflection of laser light is ≈150 nm for a laser light at 455 nm, so that only
amyloid fibrils lying in parallel with the quartz slide glass surface were observed. (B)
Schematic diagram of a prism-type TIRFM system on inverted microscopy. ISIT, image-
intensifier-coupled silicone-intensified target camera; CCD, charge-coupled device camera.

Additionally, the evanescent field turns out to be very useful in the
analysis of amyloid fibrils for the following reason. To obtain the exact
length of fibrils by conventional epifluorescence microscopy, one has to
analyze the image by three-dimensional reconstruction, because the orientation of fibrils relative to the slide glass is not always parallel to the quartz
surface. In contrast, because the penetration depth of the evanescent field
formed by the total internal reflection of laser light is quite shallow (≈150
nm for laser light at 455 nm), TIRFM selectively monitors long fibrils lying
along the slide glass (Fig. 1A). On the other hand, thioflavin T (ThT) is
known to bind rapidly to amyloid fibrils, accompanied by a dramatic
increase of fluorescence at around 485 nm when excited at 455 nm (Naiki
and Gejyo, 1999; Naiki et al., 1989). This makes ThT one of the most useful
probes to detect the formation of amyloid fibrils. Fluorescence at around

485 nm becomes useful in fluorescence microscopic studies that make use of lasers for the incident beam of excitation. By combining amyloid fibril-specific ThT fluorescence and TIRFM, it would be possible to observe the amyloid fibrils and their formation process without introducing any fluorescence reagent covalently bound to the protein molecule (Ban *et al.*, 2003, 2004).

TIRFM

The prism-type TIRFM system used to observe individual amyloid fibrils was developed based on an inverted microscope (IX70; Olympus, Tokyo, Japan) as described (Funatsu *et al.*, 1995; Wazawa and Ueda, 2005; Yamasaki *et al.*, 1999). The prism-type TIRFM is useful to obtain high-quality images with high signal-to-noise (S/N) ratios, although the thickness of water medium is limited to less than 50 μm (Wazawa and Ueda, 2005). In our case, the thickness was estimated to be about 10 μm from the fine focus stroke between the quartz slide and coverslip. Figure 1B shows a schematic diagram of the prism-type TIRFM system on inverted microscopy. The ThT molecule was excited using an Ar laser (Model 185F02-ADM; Spectra Physics, Mountain View, CA). The 460-nm line of the Ar laser was depolarized by passing through a quarter-wave plate. After passing through a quartz cubic prism, the laser was incident on a quartz slide at an angle of 68°. The gap between the quartz slide and cubic prism was filled with fluorescence-free glycerol. The fluorescence images were collected with an oil-immersion microscope objective lens (1.40 NA, 100×, PlanApo; Olympus). The fluorescence image was filtered with a bandpass filter (D490/30; Omega Optical, Brattleboro, VT) and visualized using an image intensifier (Model VS4-1845; Video scope International, Sterling, VA) coupled with a silicone-intensified target (SIT) camera (C2400-08; Hamamatsu Photonics, Shizuoka, Japan) or a charge-coupled device (CCD) camera (DP70; Olympus), depending on the fluorescence intensity of ThT.

Recently, TIRFM objective-type microscopes (Wazawa and Ueda, 2005) have become commercially available (Model IX71-ARCEVA; Olympus, Tokyo, Japan, and Model TIRF-C1; Nikon, Tokyo, Japan).

Dust contamination of the quartz surface usually causes high background noise, such that surface cleaning of quartz slides is critically important in TIRFM. Quartz substrates were first cleaned with 0.5% (v/v) Hellmanex (Hellma, Müllheim, Germany)/water and then treated with a solution mixture of NH_4OH (28% w/v)/H_2O_2 (30% w/v)/H_2O (0.05:1:5, v/v/v) for 15 min at 65°, followed by extensive rinsing with water. The quartz substrates were then dried in a vacuum oven at 110°.

Direct Observation of Amyloid β (1–40) Fibrils

Observation of Fibrils Prepared in Test Tube

Amyloid β (Aβ) (1–40) was purchased from Peptide Institute, Inc. (Osaka, Japan). The purity was >95% according to the elution pattern of high-performance liquid chromatography. Aβ(1–40) was dissolved in a 0.02% ammonia solution to 500 μM at 4°. Aβ(1–40) amyloid fibrils were prepared by the fibril extension method (Hasegawa *et al.*, 1999; Naiki and Nakakuki, 1996), in which the fragmented fibrils were extended with the monomeric peptides. Seed fibrils (i.e., fragmented fibrils) were prepared by sonication of 200-μl aliquots of a 10-μg/ml fibril solution using an UltraS homogenizer (VP-60S; TAITEC, Saitama, Japan) at output level 3, 15 × 2-sec pulses on ice. Seed fibrils were mixed with 100 μM monomeric Aβ (1–40) in polymerization buffer (50 mM Na phosphate, pH 7.5, and 100 mM NaCl) at a final concentration of 5 μg/ml of seed and 50 μM monomeric Aβ (1–40). Note that there was no agitation of solution during fibril formation.

Under our quiescent conditions, fibril formation of Aβ(1–40) depends critically on seeding (Fig. 2A) (Hasegawa *et al.*, 1999; Naiki and Nakakuki, 1996). After a 2-h incubation, the sample solution was diluted 20-fold with polymerization buffer and 100 μM ThT was added at a final concentration of 5 μM. An aliquot (14 μl) of sample solution was deposited on clean quartz substrate, and an image was obtained with TIRFM.

TIRFM revealed fluorescent fibrillar structures 1–2 μm in length (Fig. 2B). No such fibrillar structures were observed in the absence of ThT or fibrils. Although the diameters of fibrils cannot be determined from

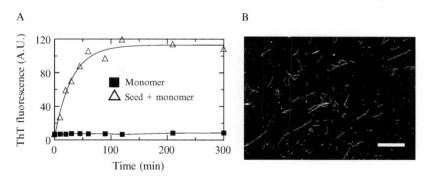

FIG. 2. Amyloid fibril formation of amyloid β (Aβ)(1-40) in a test tube. (A) Time course of fibril formation at pH 7.5 and 37° monitored using thioflavin T (ThT) fluorescence. Concentrations of Aβ(1–40) monomer and seed were 50 μM and 5 μg/ml, respectively. (B) Total reflection fluorescence microscopy (TIRFM) images of Aβ(1–40) fibrils prepared by the extension reaction. The scale bar represents 10 μm.

fluorescence images, the fibrillar images suggest relatively uniform diameters. Some fibrils were observed to be oriented in a similar direction (Fig. 2B). Such alignment might occur because of the fluid flow during the preparation of microscopic slides.

Real-Time Observation of Fibril Growth

Real-time observation of the growth of individual fibrils following seed-dependent extension was carried out on the surface of quartz slides (Fig. 3A–I). For the real-time observation, a sample mixture of 19 μl in a test tube contained 5 μg/ml of seed fibrils and 50 μM monomeric Aβ(1–40) in polymerization buffer (50 mM Na phosphate, pH 7.5, and 100 mM NaCl). After 100 μM ThT solution was added at a final concentration of

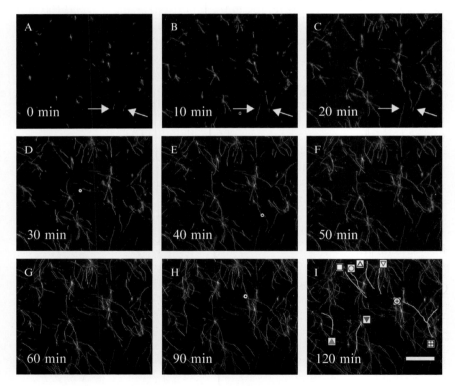

Fig. 3. Direct observation of amyloid β (Aβ)(1–40) fibril growth by total reflection fluorescence microscopy (TIRFM). (A–I) Real-time monitoring of fibril growth on glass slides. Arrows indicate the unidirectional growth of Aβ from a single seed fibril. In (I), the fibrils used for the kinetic analysis shown in Fig. 5 are indicated. The scale bar represents 10 μm. Modified and reproduced with permission from Ban *et al.* (2004).

5 μM, the sample mixture was deposited on the surface of a quartz slide and fibrils were observed every 2 min under TIRFM at 37°. The images were captured on a personal computer, and the lengths of the fibrils were calculated using Image-pro Plus (Media Cybernetics, Silver Spring, MD).

First, the ThT fluorescent spots of the seeds were observed (Fig. 3A). The growth of fibrils occurred concomitantly at many seeds. Although several fibrils often developed from apparently one seed, it is likely that the clustered seeds produced such a radial pattern. On the other hand, such a radial pattern was rarely observed for the fibrils prepared in a test tube (Fig. 2B), suggesting that the effects of the quartz surface might be involved. Once started, unidirectional growth continued, producing remarkably long fibrils more than 15 μm in length. We also observed unidirectional growth from obviously single seeds (Fig. 3A–C). Although we cannot exclude the possibility that interaction with the quartz surface prevented growth from the opposite end, it is likely that unidirectional growth holds for the fibril formation of Aβ. The diameter of fibrils that formed on the quartz surface was similar to that of fibrils prepared in solution (Fig. 2B), suggesting that similar fibrils were formed. Considering that TIRFM selectively monitors fibrils lying along the slide within 150 nm, the interaction of fibrils with the quartz surface caused the lateral growth. In addition, the combination of relatively rapid fibril growth and less aggregation of fibrils weakly fixed on the quartz surface enabled the formation of long fibrils.

Intriguingly, we occasionally observed growth to be aligned in a similar direction (e.g., vertically aligned growth in Fig. 4A), implying that the interaction with an ordered quartz surface significantly affected the direction of growth. Most of the growth is a homogeneous reaction extending a fibril of the same width as measured with ThT fluorescence (Fig. 4B, C). We sometimes observed a swinging motion of the growing head, resulting in a shift in the direction of growth, and consequently producing rugged fibrils (Fig. 4C). Notably, a dramatic swinging motion was occasionally observed (Fig. 4D), suggesting that the fibrils are flexible, bending in both horizontal and vertical directions to the plane of the quartz surface. We sometimes observed vanishing of fibril growing end (image at 70 min, arrow). Such vanishing may be caused by vertical bending of a fibril or growth in the vertical direction by which the fibril becomes located too far vertically compared with the depth of evanescent field. Re-emergence of vanished fibrils further demonstrates the flexibility of amyloid fibrils. It would be intriguing to quantify the flexibility of amyloid fibrils on the basis of the observed bending motions.

In addition, real-time observation revealed important images suggesting a transient loss of cooperativity in fibril growth (Fig. 4B). After the

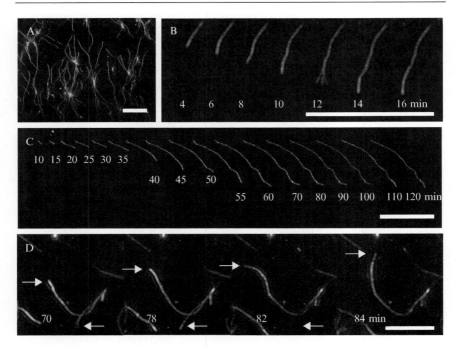

FIG. 4. Characteristic images of amyloid β (Aβ)(1–40) fibril growth revealed by total reflection fluorescence microscopy (TIRFM). (A) Vertically aligned image of fibrils. (B) Growth with transient fraying of the growing end at 12 min. (C) Growth with a swinging head producing a rugged fibril. The scale bars (A–C) are 10 μm. (D) Growth with flexible bending motion. Top and bottom arrows indicate horizontal and vertical bending, respectively, to the plane of the quartz surface. The scale bar represents 5 μm. Images in (A–C) modified and reproduced with permission from Ban *et al.* (2004).

cooperative growth with a blunt end (images from 6–10 min), the end frayed into three thinner filaments (image at 12 min). In the next step, braiding of the three filaments recovered the blunt end (image at 14 min), implying that the mature fibril is made up of three protofilaments.

Single Fibrillar Analysis of Aβ(1–40) Fibril Growth

The remarkable length of the fibrils enabled an exact analysis of the rate of growth of individual fibrils, in which the fibril length is plotted against the incubation time (Fig. 5A, B). The growth at the early and middle stages seems to occur in an all-or-none manner: when the fibril extends, the rate is almost constant (≈ 0.3 μm/min) independent of fibril species. There were cases where the growth paused briefly, possibly because of physical obstacles or local depletion of monomers. When the

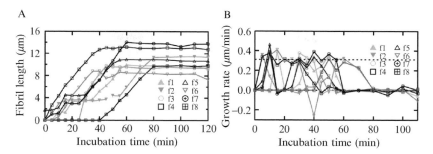

FIG. 5. Single fibrillar analysis of amyloid β (Aβ)(1–40) fibril growth. Fibril length (A) and growth rate (B) are plotted against incubation time. The fibrils used for the analysis shown are indicated in Fig. 3I. Modified and reproduced with permission from Ban *et al.* (2004).

growth restarted, however, a similar rate of 0.3 μm/min was regained. At around 40 min, although the growth rate decreased for some fibrils, there were still many fibrils extending at a rate of 0.3 μm/min. After 80 min, the entire extension ended, probably because of the depletion of monomeric Aβ(1–40). Similar discontinuous growth, termed the *stop-and-run mechanism*, was also observed in the growth process of α-synuclein protofibrils monitored by *in situ* AFM, which is capable of providing high-resolution images (Hoyer *et al.*, 2004).

The relatively constant rate of fibril growth at the early and middle stages implies similarity to the steady-state kinetics observed for enzymatic reactions. The enzyme and substrate correspond to the preformed fibril (or seed) and amyloidogenic precursor, respectively. The amyloidogenic precursors first interact with the growing ends of the fibrils, forming a precursor-fibril complex corresponding to the enzyme-substrate complex. The precursor-fibril complex is then converted to the amyloid conformation, corresponding to the catalytic process with the first-order rate constant of the chemical reaction (k_{cat}). The constant growth rate at the early and intermediate stages implies that the fibril growth is rate-limited by the transformation of the bound precursor to the fibrillar form, because the enzyme reaction is rate-limited by k_{cat}. Examining whether the constant growth rate of the fibril formation is common to other amyloidogenic proteins will be important. The similar multistep growth, termed the *dock-and-lock mechanism*, was also suggested in the Aβ or polyglutamine aggregation growth (Esler *et al.*, 2000; Bhattacharyya *et al.*, 2005). A dock-and-lock mechanism includes an initial reversible binding step (dock) and subsequent rate-limiting rearrangement (lock) of the peptide at the fibril end. The constant rate of fibril growth as observed here may be consistent with a dock-and-lock mechanism.

Possible Effects of Surface on the Growth

To interpret the TIRFM results, it is important to consider the effects of a glass surface on the fibril growth. In the case of *in situ* AFM observation, it has been argued that the interactions with the surface affect the morphology and elongation pattern of the fibrils. In the direct observation of amylin fibrils with *in situ* AFM, the diameter of amylin fibrils grown on the mica surface was different from that of fibrils grown in the test tube because of the constraint caused by adsorption to the mica surface (Goldsbury *et al.*, 1999). Furthermore, several investigations using *in situ* AFM reported that the morphology, size, and growth rate are greatly affected by characteristic features of the mica surface, such as charge, hydrophilicity, or hydrophobicity (Green *et al.*, 2004; Hoyer *et al.*, 2004; Kowalewski and Holtzman, 1999). The $A\beta(1-40)$ amyloid fibrils extended on the glass surface seem to be similar to those prepared in solution, except for a marked difference in the length of fibrils, probably because of the weaker interactions of fibrils with a quartz surface than with a mica surface.

Intriguingly, we observed aligned fibril growth both formed in a test tube (Fig. 2B) and extended on the quartz surface (Fig. 4A). Although the mechanisms might be different between the two cases, the flow of the solution and effects of the ordered surface are two important factors of alignment. Recently, because of the well-ordered nanostructures, amyloid fibrils have been focused on in the engineering fields (Hamada *et al.*, 2004; Reches and Gazit, 2003; Scheibel *et al.*, 2003). Here, aligned fibril growth on the quartz slide suggests the possibility of manipulating fibril growth on the slide surface.

Conclusion

One of the advantages of TIRFM is that one can selectively monitor fibrils lying along the slide glass, so that one can obtain the exact length of fibrils. Consequently, kinetic analysis is straightforward. Importantly, the remarkable length and clear image of the $A\beta(1-40)$ fibrils enabled analysis of the growth rate. The results revealed that the growth is a highly cooperative process extending the blunt ends at a constant rate.

Moreover, detailed kinetic analysis of individual fibril growth with clear real-time images will be possible with any amyloidogenic proteins, taking advantage of amyloid-specific ThT binding. We confirmed this with β_2-microglobulin amyloid fibrils responsible for dialysis-related amyloidosis and the octapeptide corresponding to the C-terminus derived from human medin (Ban *et al.*, 2003). Real-time observation of single fibril growth by the present method will provide critical insight into the mechanism of fibril formation.

On the other hand, it is important to keep in mind that the property of a glass surface and, additionally, the fluid flow might affect the fibril growth significantly. Because the interactions with membranes play an important role in fibril formation *in vivo*, analysis of the effects of the surface with the present method will be important to obtain further insight into the *in vivo* mechanism of amyloid fibril formation.

Acknowledgments

The authors acknowledge Hironobu Naiki (Fukui University), Tetsuichi Wazawa (Osaka University), and Daizo Hamada (Research Institute and Osaka Medical Center for Maternal and Child Health) for their support and encouragement. This work was supported by grants-in-aid from the Japanese Ministry of Education, Culture, Sports, Science, and Technology and by the Japan Society for Promotion of Science Research Fellowships for Young Scientists to T. Ban.

References

Ban, T., Hamada, D., Hasegawa, H., Naiki, H., and Goto, Y. (2003). Direct observation of amyloid fibril growth monitored by thioflavin T fluorescence. *J. Biol. Chem.* **278,** 16462–16465.

Ban, T., Hoshino, M., Takahashi, S., Hamada, D., Hasegawa, H., Naiki, H., and Goto, Y. (2004). Direct observation of Aβ amyloid fibril growth and inhibition. *J. Mol. Biol.* **344,** 757–767.

Bhattacharyya, A. M., Thakur, A. K., and Wetzel, R. (2005). Polyglutamine aggregation nucleation: Thermodynamics of a highly unfavorable protein folding reaction. *Proc. Natl. Acad. Sci. USA* **102,** 15400–15405.

Depace, A. H., and Weissman, J. S. (2002). Origins and kinetic consequences of diversity in Sup35 yeast prion fibrils. *Nat. Struct. Biol.* **9,** 389–396.

Dobson, C. M. (2003). Protein folding and misfolding. *Nature* **42,** 884–889.

Esler, W. P., Stimson, E. R., Jennings, J. M., Vinters, H. V., Ghilardi, J. R., Lee, J. P., Mantyh, P. W., and Maggio, J. E. (2000). Alzheimer's disease amyloid propagation by a template-dependent dock-lock mechanism. *Biochemistry* **39,** 6288–6295.

Funatsu, T., Harada, Y., Tokunaga, M., Saito, K., and Yanagida, T. (1995). Imaging of single fluorescent molecules and individual ATP turnovers by single myosin molecules in aqueous solution. *Nature* **374,** 555–559.

Goldsbury, C., Kistler, J., Aebi, U., Arvinte, T., and Cooper, G. J. (1999). Watching amyloid fibrils grow by time-lapse atomic force microscopy. *J. Mol. Biol.* **285,** 33–39.

Goldsbury, C. S., Wirtz, S., Müller, S. A., Sunderji, S., Wicki, P., Aebi, U., and Frey, P. (2000). Studies on the *in vitro* assembly of Aβ–40: Implications for the search for Aβ fibril formation inhibitors. *J. Struct. Biol.* **130,** 217–231.

Green, J. D., Goldsbury, C., Cooper, G. J. S., Kistler, J., and Aebi, U. (2004). Human amylin oligomer growth and fibril elongation define two distinct phases in amyloid formation. *J. Biol. Chem.* **279,** 12206–12212.

Hamada, D., Yanagihara, I., and Tsumoto, K. (2004). Engineering amyloidogenicity towards the development of nanofibrillar materials. *Trends Biotechnol.* **22,** 93–97.

Hasegawa, K., Yamaguchi, I., Omata, S., Gejyo, F., and Naiki, H. (1999). Interaction between Aβ(1-42) and Aβ(1-40) in Alzheimer's β-amyloid fibril formation *in vitro*. *Biochemistry* **38,** 15514–15521.

Hoyer, W., Cherny, D., Subramaniam, V., and Jovin, T. M. (2004). Rapid self-assembly of a-synuclein observed by *in situ* atomic force microscopy. *J. Mol. Biol.* **340,** 127–139.

Inoue, Y., Kishimoto, A., Hirao, J., Yoshida, M., and Taguchi, H. (2001). Strong growth polarity of yeast prion fiber revealed by single fiber imaging. *J. Biol. Chem.* **276**, 35227–35230.

Ionescu-Zanetti, C., Khurana, R., Gillespie, J. R., Petrick, J. S., Trabachino, L. C., Minert, S. A., Carter, S. A., and Fink, A. L. (1999). Monitoring the assembly of Ig light-chain amyloid fibrils by atomic force microscopy. *Proc. Natl. Acad. Sci. USA* **96**, 13175–13179.

Ishijima, A., and Yanagida, T. (2001). Single molecule nanobioscience. *Trends Biochem. Sci.* **26**, 438–444.

Kad, N. M., Myers, S. L., Smith, D. P., Smith, D. A., Radford, S. E., and Thomson, N. H. (2003). Hierarchical assembly of β_2-microglobulin amyloid *in vitro* revealed by atomic force microscopy. *J. Mol. Biol.* **330**, 785–797.

Khurana, R., Ionescu-Zanetti, C., Pope, M., Li, J., Nielson, L., Ramirez-Alvarado, M., Regan, A. L., Fink, A. L., and Carter, S. A. (2003). A general model for amyloid fibril assembly based on morphological studies using atomic force microscopy. *Biophys. J.* **85**, 1135–1144.

Kowalewski, T., and Holtzman, D. M. (1999). *In situ* atomic force microscopy study of Alzheimer's β-amyloid peptide on different substrates: New insights into mechanism of β-sheet formation. *Proc. Natl. Acad. Sci. USA* **96**, 3688–3693.

Naiki, H., and Nakakuki, K. (1996). First–order kinetic model of Alzheimer's β-amyloid fibril extension *in vitro. Lab. Invest.* **74**, 374–383.

Naiki, H., and Gejyo, F. (1999). Kinetic analysis of amyloid fibril formation. *Methods Enzymol.* **309**, 305–318.

Naiki, H., Higuchi, K., Hosokawa, M., and Takeda, T. (1989). Fluorometric determination of amyloid fibrils *in vitro* using the fluorescent dye, thioflavin T. *Anal. Biochem.* **177**, 244–249.

Naiki, H., Hashimoto, N., Suzuki, S., Kimura, H., Nakakuki, K., and Gejyo, F. (1997). Establishment of a kinetic model of dialysis–related amyloid fibril extension *in vitro. Amyloid* **4**, 223–232.

O'Nuallain, B., Williams, A. D., Westermark, P., and Wetzel, R. (2004). Seeding specificity in amyloid growth induced by heterologous fibrils. *J. Biol. Chem.* **279**, 17490–17499.

Reches, M., and Gazit, E. (2003). Casting metal nanowires within discrete self-assembled peptide nanotubes. *Science* **300**, 625–627.

Scheibel, T., Kowal, A. S., Bloom, J. D., and Lindquist, S. L. (2001). Bidirectional amyloid fiber growth for a yeast prion determinant. *Curr. Biol.* **11**, 366–369.

Serio, T. R., Cashikar, A. G., Kowal, A. S., Sawicki, G. J., Moslehi, J. J., Serpell, L., Arnsdorf, M. F., and Lindquist, S. L. (2000). Nucleated conformational conversion and the replication of conformational information by a prion determinant. *Science* **289**, 1317–1321.

Uversky, V. N., and Fink, A. L. (2004). Conformational constraints for amyloid fibrillation: The importance of being unfolded. *Biochim. Biophys. Acta* **1698**, 131–153.

Yamasaki, R., Hoshino, M., Wazawa, T., Ishii, Y., Yanagida, T., Kawata, Y., Higurashi, T., Sakai, J., Nagai, J., and Goto, Y. (1999). Single molecular observation of the interaction of GroEL with substrate proteins. *J. Mol. Biol.* **292**, 965–972.

Wazawa, T., and Ueda, M. (2005). Total internal reflection fluorescence microscopy in single molecule nanobioscience. *Adv. Biochem. Eng. Biotechnol.* **95**, 77–106.

[6]   Characterization of Amyloid Structures at the Molecular Level by Solid State Nuclear Magnetic Resonance Spectroscopy

By ROBERT TYCKO

Abstract

Solid state nuclear magnetic resonance (NMR) spectroscopy is particularly useful in structural studies of amyloid fibrils because solid state NMR techniques have unique capabilities as site-specific, molecular-level structural probes of noncrystalline materials. These techniques provide experimental data that strongly constrain the secondary, tertiary, and quaternary structures of amyloid fibrils, permitting the development of experimentally based structural models. Examples of techniques that are applicable to amyloid samples prepared with isotopic labeling of specific sites and to samples prepared with uniform isotopic labeling of selected residues are presented, illustrating the utility of the various techniques and labeling schemes. Information regarding the preparation of amyloid samples for solid state NMR measurements is also included.

Introduction

The phrase "solid state NMR" simply means nuclear magnetic resonance (NMR) techniques that are applicable to samples that are solids (crystalline or noncrystalline) or solid-like (e.g., phospholipid bilayer membranes, precipitated protein aggregates). As demonstrated by recent studies in our laboratory (Antzutkin *et al.*, 2000, 2002, 2003; Balbach 2000, 2002; Chan *et al.*, 2005; Gordon *et al.*, 2004; Oyler and Tycko, 2004; Petkova *et al.*, 2002, 2004, 2005, 2006; Tycko and Ishii, 2003) and in other laboratories (Benzinger *et al.*, 1998, 2000; Burkoth *et al.*, 2000; Gregory *et al.*, 1998; Heller *et al.*, 1996; Jaroniec *et al.*, 2002b, 2004; Kammerer *et al.*, 2004; Lansbury *et al.*, 1995; Laws *et al.*, 2001; Naito *et al.*, 2004; Ritter *et al.*, 2005; Siemer *et al.*, 2005), amyloid fibrils are excellent systems for solid state NMR investigations. This is because (1) amyloid fibrils, although noncrystalline, do have well-defined molecular structures, and therefore yield solid state NMR data that are of high quality and have clear interpretations; (2) amyloid fibrils can be prepared with selective or uniform isotopic labeling in the multimilligram quantities required for most solid state NMR measurements (see Chapter 2 by Teplow in this volume); (3) amyloid fibrils can be prepared in high concentrations by lyophilization

METHODS IN ENZYMOLOGY, VOL. 413   0076-6879/06 $35.00
                                                           DOI: 10.1016/S0076-6879(06)13006-2

or centrifugation, leading to good signal-to-noise ratios in the solid state NMR data; (4) the structural information available from solid state NMR measurements is arguably more direct and specific than information available from other measurements; and (5) the molecular structures of amyloid fibrils are of great current interest in the biomedical, biophysical, and biochemical research communities.

Based on X-ray fiber diffraction data, we know that the principal structural motifs in amyloid fibrils are cross-β motifs (i.e., ribbon-like β-sheets extending over the length of the fibrils) composed of β-strand peptide segments oriented approximately perpendicular to the long fibril axis and linked by backbone hydrogen bonds oriented approximately parallel to the long axis (Sunde and Blake, 1998; Tycko, 2004). The reality of the cross-β motif in fibrils formed by the β-amyloid peptide associated with Alzheimer's disease (Aβ) has been confirmed by electron microscopy (EM) (Serpell and Smith, 2000) and by recent solid state NMR measurements on oriented fibrils (Oyler and Tycko, 2004). Given that amyloid fibrils are constructed from β-sheets, the molecular structures of amyloid fibrils can be discussed at the levels of primary, secondary, tertiary, and quaternary structure, in analogy to the classification of structure in globular proteins and protein complexes. Primary structure refers to the amino acid sequence. For amyloid fibrils, secondary structure refers to the locations of β-strand and non-β-strand segments and of ordered and disordered segments. Tertiary structure refers to the arrangement of β-strands into parallel or antiparallel β-sheets. Quaternary structure refers to the relative orientation of and contacts between β-sheets. As explained below, solid state NMR data can be used to determine or place strong constraints on secondary, tertiary, and quaternary structures in amyloid fibrils. These constraints permit the development of complete structural models based entirely on experimental data (Burkoth et al., 2000; Jaroniec et al., 2004; Kammerer et al., 2004; Lansbury et al., 1995; Petkova et al., 2006; Ritter et al., 2005).

Determination of Secondary Structure

Constraints on secondary structure come from at least two types of solid state NMR measurements. One approach is simply to measure isotropic ^{13}C NMR chemical shifts in fibrils that are prepared with ^{13}C labels at specific carbonyl, α-carbon, or β-carbon sites (Balbach et al., 2002; Chan et al., 2005; Heller et al., 1996; Laws et al., 2001; Naito et al., 2004) or with uniformly ^{13}C-labeled residues (Balbach et al., 2000; Jaroniec et al., 2002b, 2004; Petkova et al., 2002, 2004, 2005, 2006; Ritter et al., 2005; Siemer et al., 2005). In β-strand segments, carbonyl and α-carbon lines are shifted upfield relative to "random coil" chemical shift values, whereas β-carbon lines are shifted

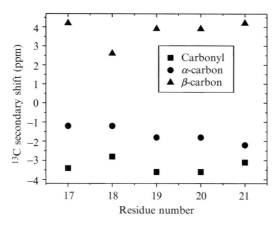

FIG. 1. ^{13}C nuclear magnetic resonance (NMR) secondary shifts (i.e., differences between observed chemical shifts and random coil shifts) for amyloid fibrils formed by Aβ_{16-22} (N-acetyl-KLVFFAE-amide). Negative secondary shifts for α-carbons and carbonyl carbons and positive secondary shifts for β-carbons indicate a β-strand containing residues 17–21 (Balbach *et al.*, 2000).

downfield. Observation of β-strand-like ^{13}C chemical shifts for several sequential residues provides strong evidence for a β-strand encompassing these residues. As an example, ^{13}C chemical shift data for Aβ_{16-22} fibrils (where Aβ_{n-m} means residues n through m of the Aβ peptide) are plotted in Fig. 1 as "secondary shifts," or the differences between observed chemical shifts and the corresponding random coil shifts.

The chemical shift measurements are carried out with magic-angle spinning (MAS), a standard solid state NMR technique in which samples are placed in cylindrical capsules (called MAS rotors) and spun rapidly about an axis that makes an angle, $\theta_m \equiv \cos^{-1}(\frac{1}{\sqrt{3}})$, with the magnetic field of the NMR spectrometer, using a pneumatic turbine system. MAS at 5–25-kHz frequencies averages out both chemical shift anisotropies (CSAs) and nuclear magnetic dipole-dipole couplings (e.g., between ^{13}C-^{13}C and ^{15}N-^{13}C pairs), resulting in relatively sharp NMR lines at the isotropic chemical shift positions. Dipole-dipole couplings to protons are not removed efficiently by MAS at readily achievable spinning frequencies, and therefore must be removed by high-power proton decoupling. In the case of samples with uniformly labeled residues, two-dimensional (2D) spectroscopy (or higher dimensional spectroscopy) is usually required for resolution and assignment of the ^{13}C MAS NMR lines (Balbach *et al.*, 2000; Jaroniec *et al.*, 2002b, 2004; Petkova *et al.*, 2002, 2004, 2005, 2006; Ritter *et al.*, 2005; Siemer *et al.*, 2005). Figure 2 shows examples of 2D solid state ^{13}C-^{13}C NMR spectra of Aβ_{1-40} fibrils synthesized with uniform ^{15}N and ^{13}C labeling of selected residues.

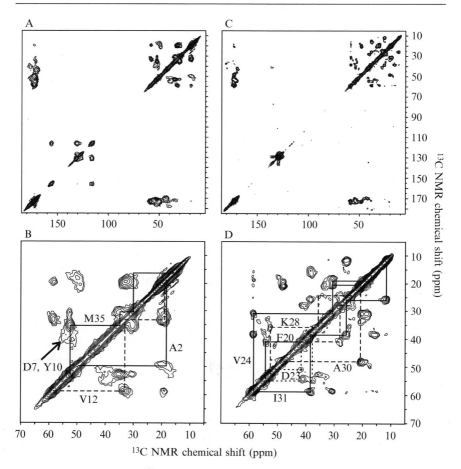

FIG. 2. Two-dimensional ^{13}C nuclear magnetic resonance (NMR) spectra of Aβ_{1-40} fibrils in lyophilized form, obtained with magic-angle spinning (MAS) under conditions that produce strong cross-peaks for directly bonded ^{13}C pairs (Ishii, 2001; Petkova *et al.*, 2002). Full spectra (A, C) and expansions of the aliphatic regions (B, D) are shown for fibrils that are uniformly ^{15}N,^{13}C-labeled at A2, D7, G9, Y10, V12, and M35 (A, B) or F10, D23, V24, K28, G29, A30, and I31 (C, D). Solid and dashed lines indicate cross-peak connectivities used for resonance assignment. Note the broad cross-peaks for A2 and D7, which support the existence of a disordered N-terminal segment in Aβ_{1-40} fibrils. Spectra were recorded in a 9.4-T field with MAS at approximately 23 kHz.

A more sophisticated and potentially more powerful approach is to use a class of solid state NMR techniques that are collectively called "tensor correlation methods" (Blanco and Tycko, 2001; Chan and Tycko, 2003; Costa *et al.*, 1997; Feng *et al.*, 1997; Ishii *et al.*, 1996; Schmidt-Rohr, 1996;

Weliky and Tycko, 1996). These techniques provide quantitative constraints on the relative orientations of pairs of chemical bonds (e.g., the relative orientation of a ^{15}N—H bond and a ^{13}C$_\alpha$—H bond within one residue, which depends on the backbone ϕ torsion angle for this residue) or of pairs of functional groups (e.g., the relative orientation of sequential ^{13}C-labeled backbone carbonyl sites, which depends on the backbone ϕ and ψ torsion angles between the labeled sites). Tensor correlation methods are particularly useful for determining sidechain conformations or backbone conformations at residues where the chemical shift data suggest a non-β-strand conformation (Antzutkin et al., 2003).

Structurally ordered segments in amyloid fibrils can be distinguished from disordered segments by measurements of ^{13}C MAS NMR linewidths. In lyophilized samples, ordered segments exhibit linewidths of roughly 2 ppm or less for backbone sites, whereas disordered segments exhibit linewidths greater than 3 ppm (Balbach et al., 2000; Petkova et al., 2002, 2004). For comparison, solid state ^{13}C MAS NMR linewidths for peptides that are tightly bound to antibodies, and are therefore structurally ordered, are 1.5–2.5 ppm in noncrystalline frozen solutions (Weliky et al., 1999; Sharpe et al., 2004). Fully hydrated amyloid samples may exhibit narrower lines for ordered segments (although not necessarily) and may exhibit missing signals for disordered segments that undergo large-amplitude motions when hydrated (Jaroniec et al., 2002b, 2004; Petkova et al., 2004; Ritter et al., 2005; Siemer et al., 2005). Figure 3 shows ^{13}C NMR linewidth data extracted from 2D spectra of Aβ_{1-40} fibrils.

To date, helical secondary structure has not been observed in amyloid fibrils by solid state NMR, although helical segments in peptides and proteins can be identified by either ^{13}C NMR chemical shift measurements (Havlin and Tycko, 2005) or tensor correlation methods (Blanco and Tycko, 2001; Chan and Tycko, 2003). Solid state NMR can also be used to characterize β-turn conformations (Sharpe et al., 2004; Weliky et al., 1999). Recent models for amyloid fibrils derived from solid state NMR and other data (Guo et al., 2004; Jimenez et al., 2002; Petkova et al., 2006; Ritter et al., 2005; Williams et al., 2004) assign the observed non-β-strand segments to "bend" or "loop" structures rather than to the standard β-turns invoked in some earlier models (George and Howlett, 1999; Lazo and Downing, 1998; Li et al., 1999).

Determination of Tertiary Structure

As reflected in structural models for Aβ fibrils published before 2000 (Chaney et al., 1998; George and Howlett, 1999; Lazo and Downing, 1998; Li et al., 1999), a consensus once existed that the cross-β motifs in amyloid

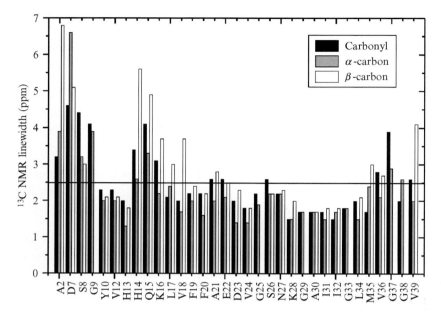

FIG. 3. ^{13}C nuclear magnetic resonance (NMR) linewidths (full width at half maximum) extracted from two-dimensional (2D) spectra of Aβ_{1-40} as in Fig. 2. These data indicate conformational disorder in the N-terminal segment, in the vicinity of Q15, and (to a lesser extent) in residues 35–39. Other residues are well ordered.

fibrils were composed of antiparallel β-sheets. This belief was overturned by solid state NMR studies of Aβ_{10-35} fibrils by Lynn, Meredith, Botto, and co-workers (Benzinger *et al.*, 1998, 2000; Burkoth *et al.*, 2000; Gregory *et al.*, 1998), which showed that Aβ_{10-35} fibrils contain only in-register parallel β-sheets (i.e., β-sheets in which residue k in one β-strand forms backbone hydrogen bonds with residues k−1 and k+1 of a neighboring β-strand). Subsequently, in-register parallel β-sheets have been discovered in other amyloid fibrils, including full-length Aβ fibrils (Antzutkin *et al.*, 2000, 2002; Balbach *et al.*, 2002; Petkova *et al.*, 2005; Torok *et al.*, 2002), by solid state NMR (Antzutkin *et al.*, 2000, 2002; Balbach *et al.*, 2002; Gordon *et al.*, 2004) and other techniques (Der-Sarkissian *et al.*, 2003; Jayasinghe and Langen, 2004; Torok *et al.*, 2002) (see Chapter 7 by Margittai and Langen in this volume). However, certain amyloid fibrils have been shown to contain antiparallel β-sheets, particularly fibrils formed by relatively short peptides (Balbach *et al.*, 2000; Gordon *et al.*, 2004; Kammerer *et al.*, 2004; Lansbury *et al.*, 1995; Petkova *et al.*, 2004), implying that there is no truly universal molecular structure for amyloid fibrils. Results obtained to date suggest that parallel β-sheets are favored in amyloid fibrils, because parallel alignment of

β-strand segments maximizes hydrophobic contacts (Antzutkin *et al.*, 2000; Petkova *et al.*, 2002) and permits favorable interactions among polar sidechains, dubbed "polar zippers" by Perutz (Perutz *et al.*, 1994), within a single β-sheet. Experimental evidence for polar zippers in amyloid fibrils has been obtained from solid state NMR (Chan *et al.*, 2005). Antiparallel β-sheets can be favored by electrostatic interactions but have been found in amyloid fibrils composed of peptides with hydrophobic segments only when the requirement for optimal hydrophobic contacts can also be met with antiparallel alignment of β-strand segments (Balbach *et al.*, 2000; Gordon *et al.*, 2004; Lansbury *et al.*, 1995; Petkova *et al.*, 2004).

Constraints on tertiary structure are obtained from measurements of nuclear magnetic dipole-dipole couplings, whose strength is proportional to $\gamma_I \gamma_S / R_{IS}^3$, where γ_I and γ_S are the gyromagnetic ratios of coupled nuclei I and S and R_{IS} is the internuclear distance. For a ^{13}C-^{13}C pair at $R_{IS} = 4.8$ Å (the interstrand distance in a β-sheet), the coupling strength is 69 Hz, implying a 15-ms time scale for evolution of NMR signals from ^{13}C nuclei with 4.8-Å separations. For a ^{15}N-^{13}C pair at $R_{IS} = 4.1$ Å (the approximate distance between an amide nitrogen and a carbonyl carbon in an interstrand hydrogen bond), the coupling strength is 45 Hz, implying a 22-ms time scale.

In-register parallel β-sheets can be identified by synthesizing a series of peptide samples with single ^{13}C labels at backbone or sidechain sites (Antzutkin *et al.*, 2000; Balbach *et al.*, 2002; Benzinger *et al.*, 1998; Gregory *et al.*, 1998). Because carbonyl and methyl carbon sites have relatively weak dipole-dipole couplings to proteins, and hence exhibit relatively long transverse spin relaxation times (T_2 values) under typical experimental conditions, it has proven most useful to label backbone carbonyl and alanine methyl sites. Measurements of intermolecular ^{13}C-^{13}C dipole-dipole couplings, as in Fig. 4 for Aβ_{1-40} fibrils, permit intermolecular distances up to at least 6 Å to be measured with approximately $\pm 10\%$ accuracy (limited by effects of T_2 relaxation and pulse sequence imperfections). Measurements of intermolecular distances between backbone carbonyl sites alone may be incapable of distinguishing in-register alignment of β-strands from a one-residue shift in alignment, because examination of molecular models for parallel β-sheets shows that a one-residue shift only increases the nearest-neighbor intermolecular distances for backbone carbonyl labels from 4.8 to approximately 5.2 Å. However, nearest-neighbor intermolecular distances for alanine methyl carbon labels increase from 4.8 to approximately 6.7 Å with a one-residue shift. Therefore, data like those in Fig. 4 can only be explained by an in-register parallel β-sheet structure.

Data in Fig. 4 were obtained under MAS, using a pulse sequence technique called "constant-time finite-pulse radiofrequency-driven recoupling" (fpRFDR-CT) (Balbach *et al.*, 2002; Ishii *et al.*, 2001). This technique

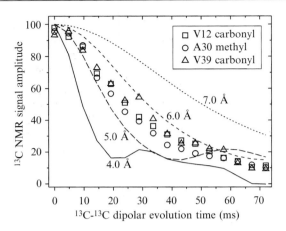

FIG. 4. Experimental constraints on intermolecular distances in $A\beta_{1-40}$ fibrils, using the constant-time finite-pulse radiofrequency-driven recoupling (fpRFDR-CT) technique (Balbach *et al.*, 2002; Ishii *et al.*, 2001) to measure ^{13}C-^{13}C nuclear magnetic dipole-dipole couplings in samples that are ^{13}C-labeled at single sites in V12, A30, or V39. Simulations for linear chains of ^{13}C nuclei with nearest-neighbor spacings from 4.0 to 7.0 Å are shown. These data indicate that the labeled residues participate in in-register parallel β-sheets. Data were obtained in a 9.4-T field with magic-angle spinning (MAS) at 20.0 kHz.

is one example of a class of solid state NMR methods called "recoupling" techniques, in which radiofrequency pulses are applied in synchrony with sample rotation to restore dipole-dipole couplings or other nuclear spin interactions that would otherwise be averaged out by MAS (Bennett *et al.*, 1998; Gullion and Schaefer, 1989; Hohwy *et al.*, 1998; Ishii, 2001; Mehta *et al.*, 1996). The fpRFDR-CT technique has the advantages of being relatively insensitive to T_2 effects, compatible with high-speed MAS, and effective even in the presence of the large ^{13}C CSA characteristic of carbonyl carbons.

NMR signal decay curves recorded under fpRFDR-CT or other dipolar recoupling techniques are primarily determined by the shortest internuclear distances (i.e., the strongest dipole-dipole couplings). Thus, these data do not readily distinguish a propagating in-register parallel β-sheet structure from, for example, a structure that alternates between in-register parallel and antiparallel (or shifted) alignment of neighboring β-strands. Multiple quantum ^{13}C NMR measurements, which are sensitive to the entire network of dipole-dipole couplings among many ^{13}C spins, can distinguish among these and other structures (Antzutkin *et al.*, 2000). Multiple quantum ^{13}C NMR data show that the in-register parallel alignment of peptide chains in $A\beta_{1-40}$ fibrils extends over at least four successive peptide chains in the β-sheets (and probably hundreds of β-strands).

Antiparallel β-sheets have also been demonstrated by solid state NMR to exist in amyloid fibrils. Measurements of either ^{13}C-^{13}C (Lansbury *et al.*, 1995) or ^{15}N-^{13}C (Balbach *et al.*, 2000; Gordon *et al.*, 2004; Kammerer *et al.*, 2004; Petkova *et al.*, 2004) intermolecular dipole-dipole couplings can be used to identify antiparallel β-sheets in samples with selective isotopic labeling. Measurements of intermolecular couplings between ^{15}N-labeled backbone amide sites and ^{13}C-labeled backbone carbonyl sites, using the rotational echo double resonance (REDOR) recoupling technique (Anderson *et al.*, 1995; Gullion and Schaefer, 1989), are particularly powerful, because the 4.1-Å internuclear distance for hydrogen-bonded backbone amide and carbonyl sites is significantly shorter than other intermolecular distances in β-sheets. REDOR data, such as those in Fig. 5, establish the registry of hydrogen bonds unambiguously.

Constraints on secondary and tertiary structures from fpRFDR-CT, multiple quantum NMR, REDOR, and related solid state NMR measurements

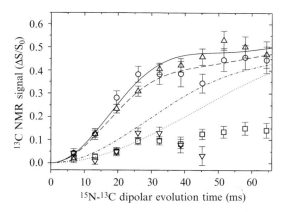

FIG. 5. Measurements of intermolecular dipole-dipole couplings between ^{15}N labels at backbone amide sites and ^{13}C labels at backbone carbonyl sites in amyloid fibrils formed by $A\beta_{16-22}$ with either acetyl (circles, downward triangles) or octanoyl (upward triangles, squares) groups at the N-terminus. These measurements use the rotational echo double resonance (REDOR) solid state nuclear magnetic resonance (NMR) technique (Anderson *et al.*, 1995; Gullion and Schaefer, 1989). Fibrils are formed from mixtures of molecules that are ^{15}N-labeled at A21 and molecules that are ^{13}C-labeled at L17 (circles, squares) or F20 (triangles), with an approximate 1.6:1.0 ratio of ^{15}N-labeled to ^{13}C-labeled molecules. Curves are simulations in which the labeled residues are hydrogen bonded in parallel (solid line) or antiparallel (dashed line) β-sheets or are shifted from hydrogen-bonded positions by one residue in parallel (dash-dotted line) or antiparallel (dotted line) β-sheets. These data indicate an antiparallel β-sheet structure with hydrogen bonds between residues 17+k and 21−k (k = 0, 1, 2, 3) for the acetyl peptide and an in-register parallel β-sheet structure for the octanoyl peptide (Gordon *et al.*, 2004). Data were obtained in a 9.4-T field with magic-angle spinning (MAS) at 5.00 kHz.

can be analyzed quantitatively by comparison of experimental data with accurate numerical simulations of the solid state NMR measurements. Such simulations contain no adjustable parameters other than the geometric parameters that describe the molecular structure, such as internuclear distances.

In samples with multiple uniformly labeled residues (or multiple residues ^{13}C-labeled at α-carbons), the hydrogen bond registry in antiparallel β-sheets can also be established from 2D solid state ^{13}C NMR spectra in which transfer of nuclear spin polarization between ^{13}C-labeled sites occurs in three steps: (1) transfer from ^{13}C nuclei to directly bonded protons, (2) transfer among protons that are within approximately 3 Å of one another, and (3) transfer from protons to directly bonded ^{13}C nuclei. Such "proton-mediated" transfers are particularly efficient for α-carbons that are directly opposite one another in antiparallel β-sheets, because the corresponding interstrand α-proton distances are approximately 2.1 Å. Proton-mediated 2D [13]C-^{13}C exchange spectra of antiparallel β-sheet structures therefore show strong, nonsequential cross-peaks connecting NMR lines of α-carbons that align with one another (Petkova *et al.*, 2004; Tycko and Ishii, 2003). Figure 6 shows examples of proton-mediated 2D spectra

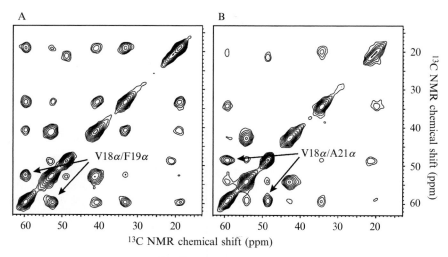

FIG. 6. Proton-mediated 2D [13]C-^{13}C exchange spectra of Aβ_{11-25} fibrils grown at pH 7.4 (A) or pH 2.4 (B), with uniform ^{15}N and ^{13}C labeling of V18, F19, F20, and A21. Strong cross-peaks between α-carbon chemical shifts of V18 and α-carbon chemical shifts of either F19 (A) or A21 (B) indicate the hydrogen bond registry in antiparallel β-sheets (Petkova *et al.*, 2004; Tycko and Ishii, 2003). At pH 7.4, residue 17+k is hydrogen-bonded to residue 20−k. At pH 2.4, residue 17+k is hydrogen-bonded to residue 22−k. Spectra were recorded in a 14.1-T field with magic-angle spinning (MAS) at 21.40 kHz.

for $A\beta_{11-25}$ fibrils prepared at pH 2.4 and pH 7.4. These spectra reveal that the hydrogen bond registry is pH-dependent, presumably because of variations in electrostatic interactions.

Determination of Quaternary Structure

Experimental constraints on quaternary structure from solid state NMR take the form of measurements of approximate distances, or "contacts," between sidechains that project above or below the β-sheets or between sidechain and backbone sites (see Chapter 10 by Shivaprasad and Wetzel in this volume for alternative approaches to quaternary structure determination). Because the shortest distances between backbone atoms in different β-sheets are roughly 8 Å or more, corresponding to ^{13}C-^{13}C dipole-dipole couplings of 15 Hz or less, constraints on quaternary structure cannot be obtained readily from backbone-backbone couplings.

Semiquantitative information about sidechain-sidechain distances can be obtained from 2D ^{13}C-^{13}C exchange spectra of amyloid fibril samples with multiple uniformly labeled residues, which are obtained under MAS and with exchange periods greater than 100 ms (Petkova et al., 2006). An example is shown in Fig. 7. Under these conditions, all possible intraresidue cross-peaks are observed (rather than only the one-bond or two-bond cross-peaks that appear with shorter exchange periods, as in Fig. 2), and many interresidue cross-peaks are observed for sequential pairs of labeled residues. In addition, nonsequential cross-peaks that can only arise from quaternary contacts may be observed. Because of the complexity of the dipole-dipole coupling networks in such measurements, internuclear distances cannot be determined precisely from the cross-peak intensities. However, basic principles of nuclear spin interactions and measurements on model systems imply that detectable cross-peaks in 2D ^{13}C exchange spectra with exchange periods less than 1 s can only occur if the internuclear distances are roughly 6–8 Å or less (Petkova et al., 2006).

Constraints on quaternary structure in amyloid fibrils can also be obtained from measurements of ^{15}N-^{13}C dipole-dipole couplings. Two-dimensional ^{15}N-^{13}C transferred echo double resonance (TEDOR) techniques (Hing et al., 1992; Jaroniec et al., 2002a; Michal and Jelinski, 1997) are particularly useful for detecting the proximity of specific sidechain carbon sites to specific backbone nitrogen sites in samples with multiple uniformly labeled residues (Petkova et al., 2006). An example is shown in Fig. 8. By comparing 2D ^{15}N-^{13}C TEDOR spectra of samples in which all molecules are isotopically labeled with spectra of samples in which labeled molecules are diluted in unlabeled molecules (prior to fibril formation), one can distinguish intermolecular ^{15}N-^{13}C contacts from intramolecular contacts.

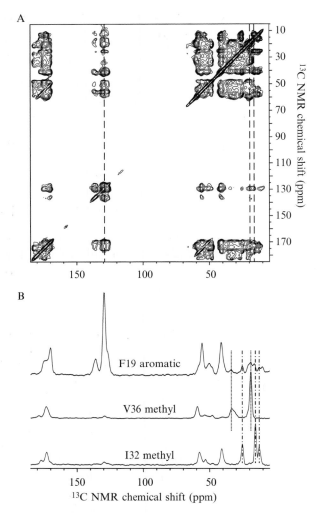

Fig. 7. (A) Two-dimensional (2D) ^{13}C-^{13}C exchange spectra of Aβ_{1-40} fibrils, prepared with uniform ^{15}N and ^{13}C labeling of K16, F19, A21, E22, I32, and V36, obtained with a 500-ms exchange period (Petkova *et al.*, 2006). (B) One-dimensional slices taken at positions indicated by the dashed lines in the 2D spectrum. The slice at the chemical shift of F19 aromatic carbons shows cross-peaks at the chemical shifts of V36 aliphatic carbons (dotted lines) and I32 aliphatic carbons (dot-dashed lines). These cross-peaks indicate contacts between aromatic and aliphatic sidechains, placing constraints on quaternary structure. The spectrum was recorded in a 14.1-T field with magic-angle spinning (MAS) at 18.00 kHz.

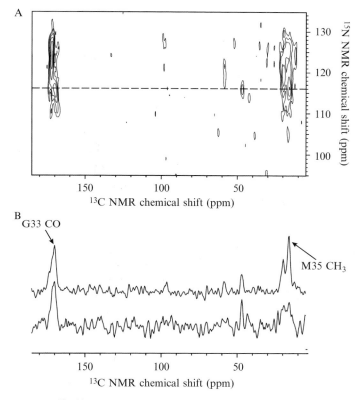

FIG. 8. (A) 2D ^{15}N-^{13}C transferred echo double resonance (TEDOR) spectrum of Aβ_{1-40} fibrils, prepared with uniform ^{15}N and ^{13}C labeling of I31, G33, M35, G37, and V39, obtained with 5.745-ms ^{15}N-^{13}C recoupling periods (Petkova *et al.*, 2006). Cross-peaks indicate the proximity of particular ^{13}C-labeled backbone and sidechain sites to ^{15}N-labeled backbone amide sites. (B) One-dimensional slices taken at the ^{15}N chemical shift of G33, indicated by the dashed line in the two-dimensional (2D) spectrum, for samples grown from labeled Aβ_{1-40} (top) and from a 2:1 mixture of unlabeled and labeled Aβ_{1-40} (bottom). The reduced amplitude of the M35 methyl cross-peak relative to the G33 carbonyl cross-peak in the isotopically diluted sample indicates that the M35 methyl/G33 amide contact is primarily intermolecular, and therefore represents a quaternary constraint. Spectra were recorded in a 14.1-T field with magic-angle spinning (MAS) at 11.140 kHz.

Construction of Molecular Models Based on Solid State NMR Data

In principle, the molecular structures of amyloid fibrils can be determined completely from solid state NMR data, with no assistance from other experimental techniques, in analogy to approaches commonly employed to determine complete structures of soluble proteins from solution

NMR data. However, this approach would require solid state NMR constraints on many torsion angles and many internuclear distances (both intramolecular and intermolecular) for each residue in the amyloid-forming peptide or protein. The limited resolution and signal-to-noise ratio in solid state NMR spectra and the difficulty of preparing fibril samples with arbitrary isotopic labeling patterns make this approach impractical, except possibly in the case of fibrils formed by relatively short peptides (Jaroniec et al., 2004).

Fortunately, information from other experimental techniques greatly simplifies the task of constructing realistic models for amyloid fibril structures. The fact that amyloid fibrils contain cross-β motifs implies that the number of possible quaternary structures is not large, once solid state NMR data that constrain the secondary and tertiary structures are available. This is because the cross-β motif requires that all β-strand segments be approximately perpendicular to a single axis (the long axis of the fibrils), with their backbone carbonyl and amide groups orientated to permit hydrogen bonds that are approximately parallel to the same axis. The number of possible quaternary structures is also strongly restricted by measurements of fibril dimensions in EM or atomic force microscopy (AFM) images and by measurements of the mass-per-length of amyloid fibrils by scanning transmission EM (Antzutkin et al., 2002; Goldsbury et al., 2000; Petkova et al., 2005). In addition, the structural model should make sense from the standpoint of physical chemistry (e.g., exposure of hydrophobic groups to aqueous solvent should be minimized, burial of unpaired charges in the fibril core should be avoided). Detection of a small number of unambiguous contacts between β-sheets by solid state NMR may then be sufficient to determine the correct quaternary structure, and hence the principal features of the full molecular structure. The precise details of sidechain conformations and backbone conformations at certain sites may remain undetermined, but these can be constrained by subsequent solid state NMR measurements as required.

These remarks apply to the determination of structural models for the cores of amyloid fibrils. As demonstrated definitively by biochemical experiments on yeast prion protein fibrils (Baxa et al., 2002), amyloid fibrils formed by large proteins can contain highly structured globular domains outside the fibril core. Determination of the structures of these domains in the context of amyloid fibrils would depend on solid state NMR strategies demonstrated recently in studies of microcrystalline globular proteins (Castellani et al., 2003; Lange et al., 2005).

Atomic coordinates for molecular models derived from solid state NMR constraints can be generated with molecular modeling and molecular dynamics (MD) software (Chan et al., 2005; Petkova et al., 2006).

A relatively simple approach to the generation of atomic coordinates that reflect the experimental information consists of three steps: (1) construction of a peptide in a conformation that is approximately consistent with available data by manual assignment of backbone torsion angles, using molecular modeling software; (2) construction of an initial model for a segment of the amyloid structure by manual combination of multiple copies of this peptide with appropriate translations and rotations; and (3) refinement of the model by alternating cycles of energy minimization and MD simulations, using appropriate software. In the energy minimization and MD simulations, experimental constraints on internuclear distances and on backbone and sidechain torsion angles are enforced by artificial harmonic potential functions. If the simulations are performed without explicit solvent, electrostatic interactions are turned off or attenuated by setting the dielectric constant to an artificially large value. To identify aspects of the structural model that are not restricted by experimental constraints, the MD simulations can be run at elevated temperatures and the results of multiple independent refinement attempts can be compared. Portions of the structure that are not constrained by experiments will appear to be disordered.

Sample Preparation for Solid State NMR

Amyloid fibrils are typically prepared by incubation of peptides or proteins in aqueous or mixed solvents under controlled conditions of concentration, pH, ionic strength, and temperature. Formation of fibrils can be monitored by EM or AFM imaging of aliquots of the incubating solution. However, it is important to recognize that EM and AFM images do not necessarily show all aggregated structures (nonfibrillar as well as fibrillar) in the solution. Ultimately, only the absence of signals attributable to nonfibrillar components in the solid state NMR spectra themselves is proof that the sample is fully fibrillized. Complete conversion to fibrils is essential, because all immobilized molecules in the sample will contribute approximately equally to the solid state NMR signals. Unlike EM and AFM, solid state NMR sees the entire sample.

Structural studies of amyloid fibrils, using the techniques discussed above, require roughly 1 micromole of isotopically labeled molecules to achieve an adequate signal-to-noise ratio in the data. With 1-micromole samples, 2D spectra are typically obtained in 0.5–10 days (depending on the details of the measurements) when the experiments are performed at room temperature and in 9.4- or 14.1-T magnetic fields. Solid state NMR measurements can be performed either on centrifuged pellets of fibrils or on lyophilized samples. We find that pelleted and lyophilized samples have identical chemical shifts in ^{13}C MAS NMR spectra (Petkova *et al.*, 2004), indicating that lyophilization does not perturb the structures of amyloid fibrils. Certain sidechain ^{13}C NMR linewidths are reduced in hydrated samples (either pelleted or lyophilized and subsequently rehydrated) relative to the linewidths in dry lyophilized samples, most likely reflecting sidechain motions in hydrated samples that are quenched in dry samples. Lyophilization permits a denser packing of fibrils into the MAS rotors, increasing the signal-to-noise ratio and facilitating high-speed MAS. Lyophilization also minimizes sample heating by radiofrequency pulses during NMR experiments.

EM images often show several distinct morphologies for fibrils that are formed by a single peptide or protein (Goldsbury *et al.*, 2000; Jimenez *et al.*, 2002). The distinct morphologies may differ in diameter, twist periodicity, and apparent propensity for lateral association. The sensitivity of ^{13}C NMR chemical shifts to subtle variations in molecular structure has allowed us to demonstrate that distinct fibril morphologies have different underlying molecular structures (Petkova *et al.*, 2005). The molecular structure propagates with the morphology when sonicated fibril fragments are used to seed the growth of new fibrils. Existing data suggest that fibrils with different morphologies have different quaternary structures and different conformations in non-β-strand segments. These findings imply that careful attention must be paid to fibril growth conditions, which can affect the relative abundance of various fibril polymorphs, and to the detailed appearance of fibrils in EM images before comparisons of structural data from different research groups are made. The most definitive proof that two different groups are studying the same fibril structure would be that the two groups observe the same sets of NMR chemical shifts in their samples.

Acknowledgments

This work was supported by the Intramural Research Program of the National Institute of Diabetes and Digestive and Kidney Diseases and by the Intramural AIDS Targeted Antiviral Program of the National Institutes of Health. Data presented in the figures were obtained by Drs. Yoshitaka Ishii, Aneta T. Petkova, and John J. Balbach. Additional contributions to this work by Drs. Oleg N. Antzutkin, Richard D. Leapman, Stephen C. Meredith, David J. Gordon, Nathan A. Oyler, Jerry C. C. Chan, Gerd Buntkowsky, Anant K. Paravastu, Simon J. Sharpe, and Wai-Ming Yau are gratefully acknowledged.

References

Anderson, R. C., Gullion, T., Joers, J. M., Shapiro, M., Villhauer, E. B., and Weber, H. P. (1995). Conformation of 1-^{13}C,^{15}N-acetyl-L-carnitine: Rotational echo double resonance nuclear magnetic resonance spectroscopy. *J. Am. Chem. Soc.* **117**, 10546–10550.

Antzutkin, O. N., Balbach, J. J., and Tycko, R. (2003). Site-specific identification of non-β-strand conformations in Alzheimer's β-amyloid fibrils by solid state NMR. *Biophys. J.* **84,** 3326–3335.

Antzutkin, O. N., Leapman, R. D., Balbach, J. J., and Tycko, R. (2002). Supramolecular structural constraints on Alzheimer's β-amyloid fibrils from electron microscopy and solid state nuclear magnetic resonance. *Biochemistry* **41,** 15436–15450.

Antzutkin, O. N., Balbach, J. J., Leapman, R. D., Rizzo, N. W., Reed, J., and Tycko, R. (2000). Multiple quantum solid state NMR indicates a parallel, not antiparallel, organization of β-sheets in Alzheimer's β-amyloid fibrils. *Proc. Natl. Acad. Sci. USA* **97,** 13045–13050.

Balbach, J. J., Petkova, A. T., Oyler, N. A., Antzutkin, O. N., Gordon, D. J., Meredith, S. C., and Tycko, R. (2002). Supramolecular structure in full-length Alzheimer's β-amyloid fibrils: Evidence for a parallel β-sheet organization from solid state nuclear magnetic resonance. *Biophys. J.* **83,** 1205–1216.

Balbach, J. J., Ishii, Y., Antzutkin, O. N., Leapman, R. D., Rizzo, N. W., Dyda, F., Reed, J., and Tycko, R. (2000). Amyloid fibril formation by $A\beta_{16-22}$, a seven-residue fragment of the Alzheimer's β-amyloid peptide, and structural characterization by solid state NMR. *Biochemistry* **39,** 13748–13759.

Baxa, U., Speransky, V., Steven, A. C., and Wickner, R. B. (2002). Mechanism of inactivation on prion conversion of the *Saccharomyces cerevisiae* Ure2 protein. *Proc. Natl. Acad. Sci. USA* **99,** 5253–5260.

Bennett, A. E., Rienstra, C. M., Griffiths, J. M., Zhen, W. G., Lansbury, P. T., and Griffin, R. G. (1998). Homonuclear radio frequency driven recoupling in rotating solids. *J. Chem. Phys.* **108,** 9463–9479.

Benzinger, T. L. S., Gregory, D. M., Burkoth, T. S., Miller-Auer, H., Lynn, D. G., Botto, R. E., and Meredith, S. C. (1998). Propagating structure of Alzheimer's β-amyloid$_{(10-35)}$ is parallel β-sheet with residues in exact register. *Proc. Natl. Acad. Sci. USA* **95,** 13407–13412.

Benzinger, T. L. S., Gregory, D. M., Burkoth, T. S., Miller-Auer, H., Lynn, D. G., Botto, R. E., and Meredith, S. C. (2000). Two-dimensional structure of β-amyloid$_{(10-35)}$ fibrils. *Biochemistry* **39,** 3491–3499.

Blanco, F. J., and Tycko, R. (2001). Determination of polypeptide backbone dihedral angles in solid state NMR by double quantum ^{13}C chemical shift anisotropy measurements. *J. Magn. Reson.* **149,** 131–138.

Burkoth, T. S., Benzinger, T. L. S., Urban, V., Morgan, D. M., Gregory, D. M., Thiyagarajan, P., Botto, R. E., Meredith, S. C., and Lynn, D. G. (2000). Structure of the β-amyloid$_{(10-35)}$ fibril. *J. Am. Chem. Soc.* **122,** 7883–7889.

Castellani, F., van Rossum, B. J., Diehl, A., Rehbein, K., and Oschkinat, H. (2003). Determination of solid state NMR structures of proteins by means of three-dimensional ^{15}N-^{13}C-^{13}C dipolar correlation spectroscopy and chemical shift analysis. *Biochemistry* **42,** 11476–11483.

Chan, J. C. C., and Tycko, R. (2003). Solid state NMR spectroscopy method for determination of the backbone torsion angle ψ in peptides with isolated uniformly labeled residues. *J. Am. Chem. Soc.* **125,** 11828–11829.

Chan, J. C. C., Oyler, N. A., Yau, W. M., and Tycko, R. (2005). Parallel β-sheets and polar zippers in amyloid fibrils formed by residues 10–39 of the yeast prion protein Ure2p. *Biochemistry* **44,** 10669–10680.

Chaney, M. O., Webster, S. D., Kuo, Y. M., and Roher, A. E. (1998). Molecular modeling of the $A\beta_{1-42}$ peptide from Alzheimer's disease. *Protein Eng.* **11,** 761–767.

Costa, P. R., Kocisko, D. A., Sun, B. Q., Lansbury, P. T., and Griffin, R. G. (1997). Determination of peptide amide configuration in a model amyloid fibril by solid state NMR. *J. Am. Chem. Soc.* **119,** 10487–10493.

Der-Sarkissian, A., Jao, C. C., Chen, J., and Langen, R. (2003). Structural organization of α-synuclein fibrils studied by site-directed spin labeling. *J. Biol. Chem.* **278,** 37530–37535.

Feng, X., Eden, M., Brinkmann, A., Luthman, H., Eriksson, L., Graslund, A., Antzutkin, O. N., and Levitt, M. H. (1997). Direct determination of a peptide torsional angle ψ by double-quantum solid state NMR. *J. Am. Chem. Soc.* **119,** 12006–12007.

George, A. R., and Howlett, D. R. (1999). Computationally derived structural models of the β-amyloid found in Alzheimer's disease plaques and the interaction with possible aggregation inhibitors. *Biopolymers* **50,** 733–741.

Goldsbury, C. S., Wirtz, S., Muller, S. A., Sunderji, S., Wicki, P., Aebi, U., and Frey, P. (2000). Studies on the *in vitro* assembly of $A\beta_{1-40}$: Implications for the search for $A\beta$ fibril formation inhibitors. *J. Struct. Biol.* **130,** 217–231.

Gordon, D. J., Balbach, J. J., Tycko, R., and Meredith, S. C. (2004). Increasing the amphiphilicity of an amyloidogenic peptide changes the β-sheet structure in the fibrils from antiparallel to parallel. *Biophys. J.* **86,** 428–434.

Gregory, D. M., Benzinger, T. L. S., Burkoth, T. S., Miller-Auer, H., Lynn, D. G., Meredith, S. C., and Botto, R. E. (1998). Dipolar recoupling NMR of biomolecular self-assemblies: Determining inter- and intrastrand distances in fibrilized Alzheimer's β-amyloid peptide. *Solid State Nucl. Magn. Reson.* **13,** 149–166.

Gullion, T., and Schaefer, J. (1989). Rotational echo double resonance NMR. *J. Magn. Reson.* **81,** 196–200.

Guo, J. T., Wetzel, R., and Ying, X. (2004). Molecular modeling of the core of $A\beta$ amyloid fibrils. *Proteins* **57,** 357–364.

Havlin, R. H., and Tycko, R. (2005). Probing site-specific conformational distributions in protein folding with solid state NMR. *Proc. Natl. Acad. Sci. USA* **102,** 3284–3289.

Heller, J., Kolbert, A. C., Larsen, R., Ernst, M., Bekker, T., Baldwin, M., Prusiner, S. B., Pines, A., and Wemmer, D. E. (1996). Solid state NMR studies of the prion protein H1 fragment. *Protein Sci.* **5,** 1655–1661.

Hing, A. W., Vega, S., and Schaefer, J. (1992). Transferred echo double resonance NMR. *J. Magn. Reson.* **96,** 205–209.

Hohwy, M., Jakobsen, H. J., Eden, M., Levitt, M. H., and Nielsen, N. C. (1998). Broadband dipolar recoupling in the nuclear magnetic resonance of rotating solids: A compensated C7 pulse sequence. *J. Chem. Phys.* **108,** 2686–2694.

Ishii, Y. (2001). ^{13}C-^{13}C dipolar recoupling under very fast magic angle spinning in solid state nuclear magnetic resonance: Applications to distance measurements, spectral assignments, and high-throughput secondary-structure determination. *J. Chem. Phys.* **114,** 8473–8483.

Ishii, Y., Terao, T., and Kainosho, M. (1996). Relayed anisotropy correlation NMR: Determination of dihedral angles in solids. *Chem. Phys. Lett.* **256,** 133–140.

Ishii, Y., Balbach, J. J., and Tycko, R. (2001). Measurement of dipole-coupled lineshapes in a many-spin system by constant-time two-dimensional solid state NMR with high-speed magic angle spinning. *Chem. Phys.* **266,** 231–236.

Jaroniec, C. P., Filip, C., and Griffin, R. G. (2002a). 3D TEDOR NMR experiments for the simultaneous measurement of multiple carbon-nitrogen distances in uniformly $^{13}C,^{15}N$-labeled solids. *J. Am. Chem. Soc.* **124,** 10728–10742.

Jaroniec, C. P., MacPhee, C. E., Astrof, N. S., Dobson, C. M., and Griffin, R. G. (2002b). Molecular conformation of a peptide fragment of transthyretin in an amyloid fibril. *Proc. Natl. Acad. Sci. USA* **99,** 16748–16753.

Jaroniec, C. P., MacPhee, C. E., Bajaj, V. S., McMahon, M. T., Dobson, C. M., and Griffin, R. G. (2004). High-resolution molecular structure of a peptide in an amyloid fibril determined by magic angle spinning NMR spectroscopy. *Proc. Natl. Acad. Sci. USA* **101,** 711–716.

Jayasinghe, S. A., and Langen, R. (2004). Identifying structural features of fibrillar islet amyloid polypeptide using site-directed spin labeling. *J. Biol. Chem.* **279,** 48420–48425.

Jimenez, J. L., Nettleton, E. J., Bouchard, M., Robinson, C. V., Dobson, C. M., and Saibil, H. R. (2002). The protofilament structure of insulin amyloid fibrils. *Proc. Natl. Acad. Sci. USA* **99,** 9196–9201.

Kammerer, R. A., Kostrewa, D., Zurdo, J., Detken, A., Garcia-Echeverria, C., Green, J. D., Muller, S. A., Meier, B. H., Winkler, F. K., Dobson, C. M., and Steinmetz, M. O. (2004). Exploring amyloid formation by a *de novo* design. *Proc. Natl. Acad. Sci. USA* **101,** 4435–4440.

Lange, A., Becker, S., Seidel, K., Giller, K., Pongs, O., and Baldus, M. (2005). A concept for rapid protein structure determination by solid state NMR spectroscopy. *Angew. Chem. Int. Ed. Engl.* **44,** 2089–2092.

Lansbury, P. T., Costa, P. R., Griffiths, J. M., Simon, E. J., Auger, M., Halverson, K. J., Kocisko, D. A., Hendsch, Z. S., Ashburn, T. T., Spencer, R. G. S., Tidor, B., and Griffin, R. G. (1995). Structural model for the β-amyloid fibril based on interstrand alignment of an antiparallel-sheet comprising a C-terminal peptide. *Nat. Struct. Biol.* **2,** 990–998.

Laws, D. D., Bitter, H. M. L., Liu, K., Ball, H. L., Kaneko, K., Wille, H., Cohen, F. E., Prusiner, S. B., Pines, A., and Wemmer, D. E. (2001). Solid state NMR studies of the secondary structure of a mutant prion protein fragment of 55 residues that induces neurodegeneration. *Proc. Natl. Acad. Sci. USA* **98,** 11686–11690.

Lazo, N. D., and Downing, D. T. (1998). Amyloid fibrils may be assembled from β-helical protofibrils. *Biochemistry* **37,** 1731–1735.

Li, L. P., Darden, T. A., Bartolotti, L., Kominos, D., and Pedersen, L. G. (1999). An atomic model for the pleated β-sheet structure of Aβ amyloid protofilaments. *Biophys. J.* **76,** 2871–2878.

Mehta, M. A., Gregory, D. M., Kiihne, S., Mitchell, D. J., Hatcher, M. E., Shiels, J. C., and Drobny, G. P. (1996). Distance measurements in nucleic acids using windowless dipolar recoupling solid state NMR. *Solid State Nucl. Magn. Reson.* **7,** 211–228.

Michal, C. A., and Jelinski, L. W. (1997). Redor 3D: Heteronuclear distance measurements in uniformly labeled and natural abundance solids. *J. Am. Chem. Soc.* **119,** 9059–9060.

Naito, A., Kamihira, M., Inoue, R., and Saito, H. (2004). Structural diversity of amyloid fibril formed in human calcitonin as revealed by site-directed ^{13}C solid state NMR spectroscopy. *Magn. Reson. Chem.* **42,** 247–257.

Oyler, N. A., and Tycko, R. (2004). Absolute structural constraints on amyloid fibrils from solid state NMR spectroscopy of partially oriented samples. *J. Am. Chem. Soc.* **126,** 4478–4479.

Perutz, M. F., Johnson, T., Suzuki, M., and Finch, J. T. (1994). Glutamine repeats as polar zippers: Their possible role in inherited neurodegenerative diseases. *Proc. Natl. Acad. Sci. USA* **91,** 5355–5358.

Petkova, A. T., Yau, W. M., and Tycko, R. (2006). Experimental constraints on quaternary structure in Alzheimer's β-amyloid fibrils. *Biochemistry* **45,** 498–512.

Petkova, A. T., Buntkowsky, G., Dyda, F., Leapman, R. D., Yau, W. M., and Tycko, R. (2004). Solid state NMR reveals a pH-dependent antiparallel β-sheet registry in fibrils formed by a β-amyloid peptide. *J. Mol. Biol.* **335,** 247–260.

Petkova, A. T., Leapman, R. D., Guo, Z. H., Yau, W. M., Mattson, M. P., and Tycko, R. (2005). Self-propagating, molecular-level polymorphism in Alzheimer's β-amyloid fibrils. *Science* **307,** 262–265.

Petkova, A. T., Ishii, Y., Balbach, J. J., Antzutkin, O. N., Leapman, R. D., Delaglio, F., and Tycko, R. (2002). A structural model for Alzheimer's β-amyloid fibrils based on experimental constraints from solid state NMR. *Proc. Natl. Acad. Sci. USA* **99,** 16742–16747.

Ritter, C., Maddelein, M. L., Siemer, A. B., Luhrs, T., Ernst, M., Meier, B. H., Saupe, S. J., and Riek, R. (2005). Correlation of structural elements and infectivity of the HET-s prion. *Nature* **435,** 844–848.

Schmidt-Rohr, K. (1996). Torsion angle determination in solid ^{13}C-labeled amino acids and peptides by separated-local-field double-quantum NMR. *J. Am. Chem. Soc.* **118,** 7601–7603.

Serpell, L. C., and Smith, J. M. (2000). Direct visualisation of the β-sheet structure of synthetic Alzheimer's amyloid. *J. Mol. Biol.* **299,** 225–231.

Sharpe, S., Kessler, N., Anglister, J. A., Yau, W. M., and Tycko, R. (2004). Solid state NMR yields structural constraints on the V3 loop from HIV-1 gp120 bound to the 447-52D antibody Fv fragment. *J. Am. Chem. Soc.* **126,** 4979–4990.

Siemer, A. B., Ritter, C., Ernst, M., Riek, R., and Meier, B. H. (2005). High-resolution solid state NMR spectroscopy of the prion protein HET-s in its amyloid conformation. *Angew. Chem. Int. Ed. Engl.* **44,** 2441–2444.

Sunde, M., and Blake, C. C. F. (1998). From the globular to the fibrous state: Protein structure and structural conversion in amyloid formation. *Q. Rev. Biophys.* **31,** 1–39.

Torok, M., Milton, S., Kayed, R., Wu, P., McIntire, T., Glabe, C. G., and Langen, R. (2002). Structural and dynamic features of Alzheimer's Aβ peptide in amyloid fibrils studied by site-directed spin labeling. *J. Biol. Chem.* **277,** 40810–40815.

Tycko, R. (2004). Progress towards a molecular-level structural understanding of amyloid fibrils. *Curr. Opin. Struct. Biol.* **14,** 96–103.

Tycko, R., and Ishii, Y. (2003). Constraints on supramolecular structure in amyloid fibrils from two-dimensional solid state NMR spectroscopy with uniform isotopic labeling. *J. Am. Chem. Soc.* **125,** 6606–6607.

Weliky, D. P., and Tycko, R. (1996). Determination of peptide conformations by two-dimensional magic angle spinning NMR exchange spectroscopy with rotor synchronization. *J. Am. Chem. Soc.* **118,** 8487–8488.

Weliky, D. P., Bennett, A. E., Zvi, A., Anglister, J., Steinbach, P. J., and Tycko, R. (1999). Solid state NMR evidence for an antibody-dependent conformation of the V3 loop of HIV-1 gp120. *Nat. Struct. Biol.* **6,** 141–145.

Williams, A. D., Portelius, E., Kheterpal, I., Guo, J. T., Cook, K. D., Xu, Y., and Wetzel, R. (2004). Mapping Aβ amyloid fibril secondary structure using scanning proline mutagenesis. *J. Mol. Biol.* **335,** 833–842.

[7] Spin Labeling Analysis of Amyloids and Other Protein Aggregates

By MARTIN MARGITTAI and RALF LANGEN

Abstract

Because of the enormous size of amyloid fibrils and their low tendency to form crystal lattices, it has been difficult to obtain high-resolution structural information on these aggregates. Magnetic resonance methods, such as solid-state nuclear magnetic resonance spectroscopy and electron paramagnetic resonance (EPR) spectroscopy, are promising new technologies

METHODS IN ENZYMOLOGY, VOL. 413
0076-6879/06 $35.00
DOI: 10.1016/S0076-6879(06)13007-4

by which to obtain molecular models. This chapter will focus on the application of EPR spectroscopy to amyloids and other protein aggregates. Site-directed spin labeling (SDSL), in combination with EPR spectroscopy, has been successfully used to study protein structure and the dynamics of soluble as well as membrane proteins. Recent studies indicate that this strategy is also well suited for studying amyloid fibrils. For example, an important outcome of the SDSL studies performed in our laboratory is that fibrils of amyloid β, islet amyloid polypeptide, α-synuclein, and tau have their β-strands aligned in an in-register, parallel fashion. Future studies promise to yield molecular information about fibril topography and protofilament arrangement and can be extended to include oligomeric structures.

Introduction

Protein misfolding is thought to be a key event in the etiology of numerous diseases; yet, it is poorly understood on a molecular level. Misfolding can result in aggregates ranging in size from a few kilodaltons to many megadaltons (Caughey and Lansbury, 2003; Dobson, 2003). These aggregates may include fibrils and prefibrillar intermediates as well as products that are off-pathway with respect to fibrillization. Site-directed spin labeling (SDSL), in conjunction with electron paramagnetic resonance (EPR) spectroscopy, can be used to study protein assemblies of any molecular weight; thus, in principle, all forms of protein aggregates should be amenable to this methodology. SDSL was originally developed for structural analysis of soluble and membrane proteins (for review, see Columbus and Hubbell, 2002; Feix and Klug, 1998; Hubbell et al., 1996, 1998, 2000). Its application to amyloidogenic proteins is still new and, thus far, has focused on fibrils, the pathological hallmarks of the diseases.

The basic strategy of SDSL entails the introduction of a small paramagnetic spin label that contains a single unpaired electron. Typically, the spin label is introduced by attachment to a cysteine. Although many spin labels have been used, by far the best characterized and most commonly employed label is the nitroxide label shown in Fig. 1 (Berliner et al., 1982), resulting in the side chain R1. The key advantages of this label as compared with other labels are its minimal structural perturbation and high structural sensitivity (Mchaourab et al., 1996).

EPR measurements may be used to determine spin label mobility and accessibility and to map distances within a protein. These quantities have been used successfully in determining local and global backbone folds and in defining structural transitions, local motions, domain movements, and subunit assembly (Columbus and Hubbell, 2002; Feix and Klug, 1998; Hubbell and Altenbach, 1994; Hubbell et al., 1996, 1998, 2000). Furthermore,

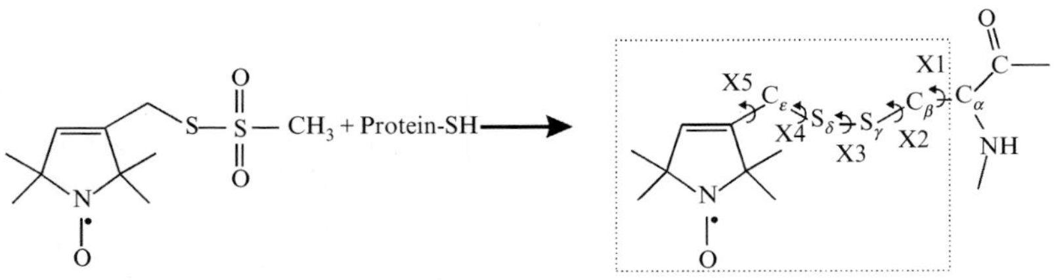

Side chain R1

FIG. 1. Reaction of the spin label [1-oxy-2,2,5,5-tetramethyl-pyrroline-3-methyl]-methane-thiosulfonate with a sulfhydryl group in a protein results in the side chain R1 (enclosed by a dotted square). Dihedral angles X1–X5 are indicated.

structural changes can be monitored in real time (in favorable cases, with millisecond resolution (e.g., Farahbakhsh et al., 1993; Langen et al., 1998; Steinhoff et al., 1994). Collectively, these findings represent an important foundation for similar work involving amyloid structures. The following sections provide a brief general overview of SDSL, along with a more detailed methodological account of its application to amyloid fibrils.

Mobility

R1 mobility is probably the most frequently examined quantity in SDSL and is a direct reflection of local structure. Mobility-related information is encoded in the line shape of the continuous-wave X-band EPR spectrum, which, for an improved signal-to-noise ratio, is recorded as a first derivative of the absorption spectrum. The single unpaired electron of R1 has a magnetic moment that can be oriented with or against the external magnetic field present in the EPR spectrometer. Transitions from the former orientation to the latter occur with the absorption of microwave energy when the resonance condition is met. In principle, this should result in a single-line EPR spectrum. However, because the nitrogen nucleus of the nitroxide moiety can take up three different spin states (with different magnetic moments), a hyperfine splitting occurs and a three-line EPR spectrum is observed. Mobility affects the shape of each of these lines in a characteristic way. High mobility of the label gives sharp, narrowly spaced lines, whereas low mobility results in broad and distant peaks.

The spectra in Fig. 2 depict examples of high and low mobilities. Both spectra were taken from annexin B12, a soluble protein whose structure is mainly α-helical (Luecke et al., 1995). Figure 2A shows an EPR spectrum of annexin B12 (labeled at position 147) after guanidine hydrochloride (GdnHCl)–induced unfolding. The sharp and narrowly spaced lines indicate

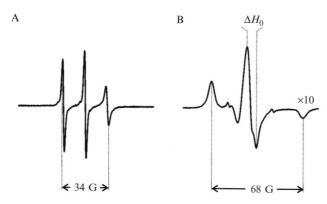

FIG. 2. Mobility of R1 determines electron paramagnetic resonance spectral line shape. Annexin B12 is spin-derivatized at position 147. (A) Three sharp peaks indicate high mobility in the GdnHCl denatured state. (B) Broad, highly separated peaks and small amplitudes reflect low mobility of the buried spin label in the native fold (Isas *et al.*, 2004). For a comparison of line shapes, the spectrum in B is amplified 10-fold. The scan width for both spectra is 100 G. Dotted lines mark outer peaks. Separation between outer peaks is indicated for both spectra. The peak-to-peak distance of the central resonance, defined as the central line width (ΔH_0), is indicated in B. Commonly, the inverse of the central line width (ΔH_0^{-1}) is determined (for comparison purposes, sometimes depicted as scaled mobility [Hubbell *et al.*, 2000]).

high mobility, as is typically seen for unfolded regions (Isas *et al.*, 2002; Margittai *et al.*, 2001; Mchaourab *et al.*, 1996). In contrast, the EPR spectrum of the folded form (Fig. 2B) exhibits broad lines and an increased separation between outer peaks (68 G). This spectrum is characteristic of a completely immobilized label (Isas *et al.*, 2002; Margittai *et al.*, 2001; Mchaourab *et al.*, 1996) and agrees well with this residue being buried in the protein interior (Isas *et al.*, 2004; Luecke *et al.*, 1995). Apart from these two extreme cases, spectra reflecting intermediate mobilities are oftentimes observed. Such spectra are typically observed for sites that are either exposed on a surface of an ordered structure or have partial tertiary contacts.

Two frequently used measures of mobility are (1) the breadth of the spectrum (typically quantified by the second moment $<H^2>$) and (2) the inverse of the central line width, ΔH_0^{-1} (Fig. 2B). Systematic studies on a number of helical proteins have indicated that these two mobility parameters can be used to distinguish between labels that are in loop, helix surface, tertiary contact, or buried positions (Isas *et al.*, 2002; Margittai *et al.*, 2001; Mchaourab *et al.*, 1996). An asymmetrically solvated secondary structure can be identified by scanning through consecutive sites of singly labeled mutants and identifying the periodicity in the mobility of the side chain R1 (mobility scan). This approach is based on different inherent periodicities of secondary structural elements (Hubbell *et al.*, 1996). For example, in the case of an α-helix, a mobility scan will give rise to periodic oscillations between high (surface exposed sites) and low (buried sites) mobilities, with a periodicity of ≈ 3.6 amino acids.

Numerous studies using X-ray crystallography (Langen *et al.*, 2000), spectral simulations (Columbus *et al.*, 2001), and mutagenesis (Mchaourab *et al.*, 1996) have yielded detailed molecular models that explain how local structure modulates the R1 mobility in α-helical proteins. In the case of helix surface sites, the Sδ atom of the disulfide bond has been shown to interact with the C_a hydrogen atom in the backbone (Langen *et al.*, 2000). This interaction, together with the high barrier of rotation around the S—S bond, is thought to restrict the R1 motion around all but the terminal two bonds of the linker (X4 and X5 in Fig. 1). This compact structure is likely to make the label sensitive to backbone dynamics and helps to reduce nearest neighbor interactions. Although β-sheet proteins are not as well characterized, here, nearest neighbor interactions are clearly more important, especially in the case of center strands (Lietzow and Hubbell, 2004). A first systematic study revealed that exposed sites in four edge strands had spectra similar to those obtained from helix surface sites (Lietzow and Hubbell, 2004). In contrast, surface residues in interior strands often had interactions with residues on neighboring strands. These interactions appeared to be modulated by strand twists.

Accessibility

EPR experiments are based on monitoring transitions between spin states. These transitions can be influenced by the presence of paramagnetic colliders, which affect the relaxation properties of R1. The accessibility data that can be obtained from this effect provide important structural information.

The most commonly used paramagnetic colliders are O_2 and Ni(II) ehylene diamine-N, N'-diacetic acid NiEDDA. In the case of soluble proteins, buried residues tend to be inaccessible to both colliders, whereas surface-exposed sites give rise to higher accessibilities. Thus, the degree of surface exposure governs the overall accessibility to both colliders (Hubbell *et al.*, 1998).

The systematic scanning of consecutive cysteines through an asymmetrically solvated secondary structure allows one to determine its periodicity and, thus, the type of secondary structure. Such scans have been successfully applied using accessibility data; for soluble proteins, the conclusions reached have been similar to those from mobility scans (Hubbell *et al.*, 1996).

For membrane proteins, additional information can be obtained, because O_2 is hydrophobic and preferentially partitions into membranes, whereas the hydrophilic collider NiEDDA preferentially partitions into the aqueous environment (Hubbell *et al.*, 1998). As a consequence, lipid-exposed sites are primarily O_2 accessible, whereas solvent-exposed sites are mainly NiEDDA accessible. Because an increased immersion depth causes an increase in the O_2 accessibility and a decrease in the NiEDDA accessibility, the immersion depth of a given spin-labeled site can be determined from the ratio of the O_2 and NiEDDA accessibilities (Altenbach *et al.*, 1994). A scan of residues thus allows a determination of backbone orientation with respect to the bilayer plane.

Distances

When two spin labels are within close proximity of each other, additional characteristic changes can be detected in the EPR spectra. Quantitative analysis of these changes can be used to determine inter- and intramolecular distances up to 25 Å.

Dipolar interactions between labels (8–25 Å) are usually measured under rigid lattice conditions, such as in frozen solution (Rabenstein and Shin, 1995). A recent study showed, however, that slowly tumbling proteins at room temperature can be treated as "frozen" in the analysis with fairly accurate distance results (Altenbach *et al.*, 2001). In the special case of fast protein tumbling (molecular weight (MW) = 15 kDa), measurements can also be made at room temperature and distances can be derived from line-broadening effects (Mchaourab *et al.*, 1997). In the frozen state, of course, distances can also be obtained for these smaller proteins or peptides from the aforementioned dipolar interactions.

When spin labels are in direct contact (and their orbitals overlap), an exchange interaction occurs. In the special case wherein multiple spin labels are in van der Waals contact, spin exchange results in a narrowing of the spectrum (Molin, 1980). Such a result has been observed at high spin label concentrations in solution, in membranes and micelles (Molin, 1980), and in ordered crystals (Lajzerowicz-Bonneteau, 1976).

Recent developments in pulsed EPR spectroscopy have further increased the range of distances that can be measured between labels. Using the double electron-electron resonance (DEER) method, distances between spin labels ranging from 15 to 80 Å can be determined (Jeschke, 2001).

Application to Amyloids and Other Protein Aggregates

Mobility, accessibility, and distance are the primary quantities obtained using EPR of R1. Mostly applied to fibrils have been mobility and distance analyses. For example, mobility analysis has been used in fibrils of amyloid β (Aβ), α-synuclein, and islet amyloid polypeptide (IAPP) to define "domain organization" by determining which regions of those proteins become ordered in the fibril (Der-Sarkissian *et al.*, 2003; Jayasinghe and Langen, 2004; Torok *et al.*, 2002). In the yeast prion Sup35, mobility data supported the existence of different fibril types (Tanaka *et al.*, 2004).

Distance information has proven very useful in obtaining structural information. For Aβ (Torok *et al.*, 2002), α-synuclein (Der-Sarkissian *et al.*, 2003), IAPP (Jayasinghe and Langen, 2004), and tau (Margittai and Langen, 2004), we found parallel arrangement of β-strands. Another EPR study on transthyretin revealed antiparallel arrangement of β-strands (Serag *et al.*, 2002), emphasizing that both strand arrangements can exist in amyloid fibrils (also see Gordon *et al.*, 2004)).

Accessibility has not yet been used extensively; its use has been limited to the study of membrane-bound amyloidogenic proteins, such as α-synuclein (see below).

The following section discusses in more detail how experiments are performed and which strategies are chosen. We discuss spin-labeling studies of the large isoform of tau (441 amino acids) as our primary example (Margittai and Langen, 2004).

Sample Preparation

Mutagenesis/Peptide Synthesis. To ensure specific labeling, all native cysteines (two in the case of tau) are replaced by appropriate amino acids (typically Ser or Ala) and a single cysteine is introduced at the site of interest. For this, we choose standard polymerase chain reaction (PCR) techniques, such as the QuikChange method from Stratagene. In the case of peptides, the cysteine is introduced during solid-phase synthesis. In general, it is important to ensure that the cysteine substitutions do not cause major structural perturbations of the native protein or otherwise interfere with fibril formation. Proteins with many cysteines are naturally less apt for SDSL.

Purification. In order to minimize oxidation of the cysteines, care must be taken to ensure that every purification step involves appropriate amounts of reducing agents and metal chelators. For unambiguous data interpretation, proteins should be highly pure (>95%) and free of contamination from other cysteine-containing proteins.

A typical purification protocol for tau (Margittai and Langen, 2004) is described here. Bacterial pellets containing overexpressed tau are solubilized in extraction buffer (20 mM Pipes, pH 6.5, 500 mM NaCl, 5 mM ethylenediaminetetraacetic acid (EDTA), 50 mM β-mercaptoethanol) and stored at −80°. Frozen pellets are thawed and incubated for 20 min at 80°.

After 5 min on ice, samples are sonicated for 60 sec at power setting 10 in an ultrasonic cell disruptor (Microson; Misonic, Farmingdale, NY). Cell debris is pelleted for 30 min at 13,000g. Heat-stable tau in the supernatant is precipitated with 60% ammonium sulfate for 1 h and subsequently centrifuged for 10 min at 13,000g. The tau pellet is solubilized in H_2O, 5 mM dithiothreitol (DTT), passed through an Acrodisc syringe filter (Pall) with a pore size of 0.45 μm, and immediately loaded onto a UnoS column (BioRad). Protein is eluted with a linear NaCl gradient (buffer A: 50 mM NaCl, 20 mM Pipes, pH 6.5, 0.5 mM EDTA, 2 mM DTT; buffer B: 1000 mM NaCl, 20 mM Pipes, pH 6.5, 0.5 mM EDTA, 2 mM DTT). Protein fractions are analyzed by sodium dodecyl sulfate-polyacrylamide gel electrophoresis (SDS-PAGE) and pooled according to purity. DTT is added (5 mM), and samples are subjected to size exclusion chromatography using a Superdex 200 column from Pharmacia (elution buffer: 100 mM NaCl, 20 mM Pipes, pH 6.5, 1 mM EDTA, 2 mM DTT). Fractions are again analyzed by SDS-PAGE. Fractions containing pure tau protein are pooled, adjusted to 7 mM DTT, and precipitated with an equal volume of MeOH. Precipitation is allowed to take place overnight on ice. Pellets are spun down for 10 min at 10,500g and washed with 1 ml MeOH and 2 mM DTT. Pellets are stored until further use at $-80°$.

Labeling. Protein pellets (30–120 nM) are resuspended in 200 μl 6-M GdnHCl to ensure that the protein is highly accessible to the spin label and that fibril formation starts from a homogeneous pool of monomers. Use of the denaturant is a general precaution used specifically for tau, which is mainly a random coil in its monomeric form but becomes ordered upon oligomerization. It should be noted that in soluble and membrane proteins, even interior sites could be labeled; thus, in many cases, it might be preferable to leave out the denaturant.

In order to avoid air oxidation of the cysteines, the labeling reaction should be started quickly after solubilizing the protein. For labeling, we generally employ a 10-fold molar access of spin label (stock solution: 40 mg nitroxide label (Toronto Research Chemicals) in 1 ml dimethyl sulfoxide (DMSO)) and incubate for approximately 1 h at room temperature. Excess label and GdnHCl are removed by elution over PD-10 gel filtration columns from Amersham (elution buffer: 10 mM Hepes, pH 7.4, 100 mM NaCl). Under these conditions, the protein usually is completely labeled. It is possible to test for complete labeling using mass spectrometry. Because the disulfide-linked spin label R1 can be released by the addition of reducing agents, all further steps are carried out in their absence. Protein concentrations are determined by absorbance measurements at 276 nm in the presence of 6 M GdnHCl. Final protein concentrations are adjusted for each experiment. Under the given conditions, aggregation of full-length tau is very slow, to the point that no aggregation is observed during this period of sample preparation.

Aggregation

A myriad of protocols exist for the formation of fibrils. Even in the case of tau, various protocols are in use (see references in Barghorn and Mandelkow, 2002). In the work of our group, tau fibril (typically referred to as filament) formation is induced by adding heparin (Goedert *et al.*, 1996) at a protein/heparin molar ratio of 4:1. Filaments are formed under agitation at room temperature for at least 8 days.

To ensure the formation of fibrils and not random aggregates, it is always important to analyze the samples by either atomic force microscopy or electron microscopy. Filaments are separated from nonpolymerized tau molecules by centrifugation at 160,000g for 30 min. The transparent pellets are washed, recentrifuged, and taken up in 12-μl buffer.

Instrumentation

EPR Spectrometer and Resonators. In our laboratory, we use X-band EPR spectrometers from Bruker (Karlsruhe, Germany), which include magnets (types ER070 and ER073), microwave bridges (ER041X at X-band), and controller units EMX080 and EMX081 (composed of a microwave bridge controller, modulation amplifier, and field controller).

Depending on the nature of the experiments, different resonators can be used. For accessibility measurements, we use either a dielectric resonator (ER4123D) or a two-loop one-gap resonator (XP-0201). For line shape analysis, we use either an HS resonator with a cylindrical cavity (ER4119HS) or a two-loop one-gap resonator (XP-0201; Jagmar, Krakow, Poland). The loop gap resonator requires a very small sample volume ($\approx$2 μl). It is ideally suited to measure fibrils that have been pelleted in the capillary. The HS resonator can accommodate larger sample volumes (typically 10–20 μl) and can be used to measure more diluted samples. Despite different concentrations, similar amounts (as low as $\approx$50 pM) are required for detection under favorable conditions. Larger amounts of spin-labeled proteins might be required in the event the signal amplitude is significantly reduced because of immobilization of R1 or the presence of strong spin-spin interactions (see below).

Sample Holder (Capillaries). Fibrils are inserted into round quartz capillaries (0.60-mm inner diameter × 0.84-mm outer diameter × 100-mm length, two open ends) from Vitro Com Inc. (NJ) by capillary flow or suction. One end of the capillary is sealed with Critoseal. Alternatively, one end of the capillary may be sealed in a flame and the fibrils are transferred with a glass pipette. In this case, fibrils are subsequently centrifuged for 3 min at 1000g in a tabletop centrifuge. The use of proper capillaries is essential. Manufacturing defects may result in background signals. For accessibility measurements, samples are transferred into gas-permeable TPX capillaries (L & M EPR Supplies), which allow diffusion of gases into and out of the sample.

Measurement

Experiment. In a typical continuous-wave X-band EPR experiment, the microwave frequency is kept constant ($\approx$9.75 GHz), although the magnetic field is swept. For the resonators mentioned above, the microwave power is usually set at a value between 2 and 12 mW in order to avoid saturation effects. The width of the field modulation at 100 kHz is generally optimized for the best signal-to-noise ratio without introducing distortion caused by overmodulation. The time and number of scans depend upon signal intensity. The scan width is typically 150 G. However, in the case of spectral broadening, as occurs in the case of strong spin-spin interactions, the scan width may have to be increased to 200 G or greater in order to collect the full breadth of the spectrum. For comparison, spectra are normalized by double integration to the same amount of spin label. The first integration restores the absorption spectrum from the experimentally determined first-derivative spectrum. The second integration defines the area beneath the absorption line and is directly proportional to the number of spins.

Strategy for Distance and Mobility Measurements. Spin-spin interactions and mobility are known to influence spectral line shape. In the case of monomeric proteins, it is easy to distinguish between these effects, because spectral broadening caused by spin-spin interactions occurs only when two (or more) labels are introduced into the same protein. In this case, distances can be obtained from the dipolar broadening function (for slow tumbling) (Rabenstein and Shin, 1995) or from line broadening (for fast tumbling) (Mchaourab *et al.*, 1997), as stated in the previous section. A fundamental requirement for both approaches is the need to compare spectra from interacting and noninteracting spin labels. The most convenient way to obtain a spectrum for the noninteracting state is to record an EPR spectrum of a sample that has been labeled with a mixture of paramagnetic and diamagnetic labels (see below). The diamagnetic label we use in our laboratory is a structural analogue (Gross *et al.*, 1999) of the paramagnetic label and has no unpaired electron (Fig. 3); thus, this label gives no signal in the EPR

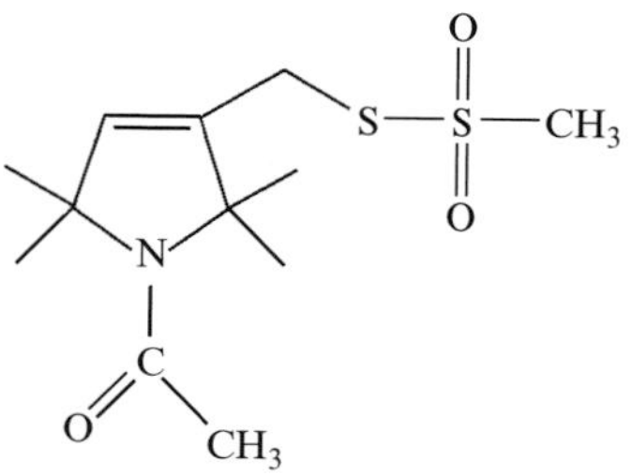

FIG. 3. Structure of the diamagnetic label [1-acetyl-2,2,5,5-tetramethyl-pyrroline-3-methyl]-methanethiosulfonate. This label is a structural analogue of the paramagnetic label depicted in Fig. 1.

experiment. Alternatively, one can record the EPR spectra of singly labeled proteins (one for each site in the double mutants). This approach, however, requires additional mutagenesis and is more labor-intensive.

Protein oligomers may result in spectral line broadening even though only single sites are labeled. This happens when spin labels in different monomers are in close proximity to each other (25 Å or less). Such intermolecular interactions are very common in fibrils (Der-Sarkissian et al., 2003; Jayasinghe and Langen, 2004; Margittai and Langen, 2004; Torok et al., 2002). Again, in order to determine distances, one needs to record a spectrum without spin-spin interactions. This can be achieved either by labeling the protein with a mixture of paramagnetic label and diamagnetic analogue, as described above, or by diluting the spin-derivatized cysteine mutants with wild-type protein (Cys-less) prior to fibril formation.

Although intramolecular distance measurements in amyloid fibrils have not yet been reported, this should, in principle, be possible. An important requirement will be to exclude intermolecular interactions, and this could be achieved by mixing a small fraction of a completely labeled double mutant with a large access of unlabeled protein.

Thus far, all of the spin-spin interactions reported for fibrils have come from singly labeled proteins and, therefore, must have arisen from intermolecular interactions. In our study of tau, we came upon a highly unique case (Margittai and Langen, 2004). Here, EPR spectra from core residues in the filament resulted in single lines (without the typical hyperfine splitting). Such spectra originate from spin exchange between neighboring spin labels. The absence of hyperfine splitting can be explained by the delocalization of the unpaired electron, which is possible only if multiple labels are in van der Waals contact. Because consecutive sites in the filament core exhibit spin exchange, we concluded that strands in different tau molecules are aligned in parallel and that the same residues are stacked perfectly along the fiber axis (Margittai and Langen, 2004). A dilution series clearly differentiates between dipolar and exchange interactions. As the percentage of spin-labeled monomers in the fibril increases, the amplitude of the central resonance first decreases because of line broadening (Fig. 4A, B). At higher percentages of spin-labeled monomers, however, the amplitude increases again, because a greater number of labels are in van der Waals contact, resulting in exchange narrowing. If only dipolar interactions are present, the amplitude continues to decrease (dotted line in Fig. 4A) (e.g., Berengian et al., 1999; Langen et al., 1998).

As mentioned above, mobility and spin-spin interactions can have significant effects on the line shapes of R1 EPR spectra. Thus, in order to obtain reliable mobility parameters, it is essential to deconvolute the effects of mobility and spin-spin interaction. Based upon our experience, it is very likely that the X-band EPR spectra obtained from fibrillar proteins labeled at a single site will contain components of spin-spin interaction; for this reason, the spin dilution methods described above must be employed.

Oftentimes, line shape analysis clearly enables a distinction between monomeric and fibrillar states. Soluble monomers of tau (α-synuclein, IAPP, and Aβ) are, for the most part, unstructured. Spin-labeled mutants in the soluble form give rise to EPR spectra with three narrow lines and large amplitudes (exemplified for positions 308 and 403 in Fig. 4C left). Upon filament formation, spin labels in the filament core (represented by position 308) become immobilized, as observed by the large separation between the outer peaks and the reduced amplitudes in the spectrum of the spin-diluted sample (Fig. 4C upper right). Regions outside the filament core may undergo less restructuring. A spin label at position 403, for example, remains mobile. The spectrum retains its three sharp lines (Fig. 4C lower right).

Perceptibly, the spectra for unfolded and filamentous tau at position 308 in Fig. 4C are very similar to those of unfolded and native annexin B12 at position 147 in Fig. 2. The mobility of the spin label is a reliable quantity that can be applied to any protein regardless of size and fold, thus allowing, as in the case of amyloid fibrils, the inference of structural information from unknown structures.

Accessibility. Thus far, accessibility studies have been performed primarily on amyloidogenic proteins bound to membranes. In many cases, membrane interactions have been implicated in the misfolding and toxicity of these proteins.

Using accessibility analysis allowed us to identify an elongated α-11/3 helical structure of α-synuclein upon binding of the monomer to the lipid bilayer (Jao et al., 2004). Here, we use this study as an example to describe briefly some basic aspects of accessibility measurements. Following the preparations of liposome-bound spin-labeled α-synuclein (see Jao et al., 2004 for additional details), a sample (4 μl) is loaded into gas-permeable

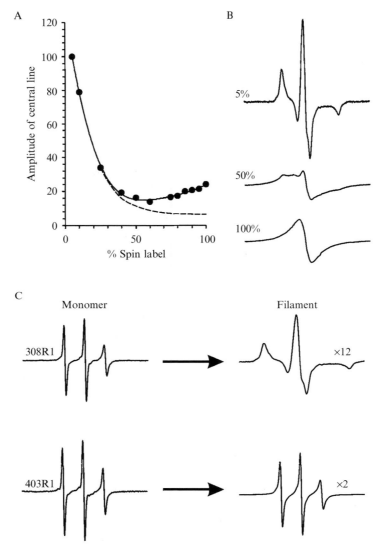

FIG. 4. Influence of spin-spin interactions and R1 mobility on line shape. In all examples, tau is spin-derivatized at position 308. Spectra are normalized by double integration. (A) Tau filaments containing different ratios of paramagnetic and diamagnetic labels result in varying amplitudes of central lines. Amplitudes are plotted as a function of percentage of para- magnetic label (solid line). Spin exchange is observed at high percentages of paramagnetic label and results in increased amplitude as compared with intermediate percentages (e.g., 50%). The dotted line schematically indicates a case wherein only dipolar interactions but not spin exchange are observed. Here, the amplitude continues to decrease even at high percentages of spin label (Margittai and Langen, 2004). (B) Electron paramagnetic resonance

TPX capillaries and subsequently positioned into a dielectric resonator. Accessibilities to O_2 and 3 mM NiEDDA are then determined using power saturation methods.

In a power saturation experiment, one can determine the effect of a given collider on the saturation properties of R1 by recording the EPR spectra at different microwave powers ($\approx$0.1–80 mW). At higher power settings, the EPR signals begin to saturate and the signal amplitudes begin to decrease. In the presence of paramagenetic colliders, additional relaxation mechanisms cause saturation to occur at higher microwave power levels. To quantify the effects of the colliders, power saturation analysis is done in their presence or absence. Measurements taken in the absence of any colliders require the purging of molecular oxygen (resulting from air exposure of a sample); this is accomplished by flushing the resonator with nitrogen gas prior to and during the recordings. Similarly, the NiEDDA power saturation curves must also be obtained in a nitrogen atmosphere. In contrast, O_2 power saturation measurements require equilibration with air. Precise accessibilities are then determined by fitting each power saturation curve (amplitude of the central resonance versus microwave power) for the various conditions as described in more detail elsewhere (Altenbach et al., 1994; Oh et al., 2000).

In principle, it should be possible to extend this approach to the study of the membrane-perturbing oligomeric form of α-synuclein and to apply this methodology to other amyloidogenic proteins and aggregates as well. If intermolecular spin-spin interactions are present, as in amyloid fibrils, samples must be spin-diluted before accessibility measurements are taken.

Experimental Considerations. Although the application of SDSL to amyloid fibrils has worked remarkably well, its success depends on a number of important requirements.

Sample purity is of paramount importance; that is, the sample must be completely and selectively labeled and be free of contamination. Although labeling efficiency is generally tested by mass spectrometry, labeling specificity and protein contaminations are tested by spin labeling a cysteine-free variant of the given protein. In the absence of background labeling, no

---

spectra of tau filaments at three different percentages of paramagnetic label taken at 150-G scan width. (C) Formation of tau filaments from monomeric protein results in major spectral changes. Spectra for monomeric tau (positions 308 and 403) are on the left. Spectra for filamentous tau are on the right. The 12-fold amplified spectrum is taken from B and represents filaments at 5% spin label. The spectrum is rescaled for better comparison with the spectrum from soluble tau. Spectra are depicted at 100 G. Adapted with permission from Margittai and Langen (2004).

signal should be observed. If a signal is observed, the purification protocol should be further refined.

The quality of all fibril preparations should be checked by atomic force microscopy or electron microscopy to verify, for each sample, that only fibrils (versus other aggregates) are formed. Even if pure fibril preparations are obtained from proteins that lack any background labeling, residual contamination can arise from free spin label. The fast-tumbling spin label gives rise to very sharp lines with high relative intensity. Thus, even minute amounts of label can cause significant contamination of the spectrum, resulting in spectra with three sharp lines. Such contamination may be prevented by washing the fibrils multiple times before transferring them into the capillary. Careful washing of the fiber sample also ensures the absence of contamination from monomers, which, too, may result in spectra with a significant component of three sharp lines.

For distance measurements, incomplete labeling of the protein poses a problem, because this results in increased separation between the spin labels. Incomplete labeling would most likely occur in amyloid proteins that are either not fully reduced or are already folded prior to fibrillization and wherein cysteines are buried in the protein interior. In the latter case, labeling efficiency may be improved by partial denaturing, higher spin label-to-protein ratios, and extended incubation times.

Outlook

SDSL, in conjunction with EPR spectroscopy, is a powerful new tool by which to study amyloid structure. For soluble and membrane proteins, EPR spectroscopy has produced three-dimensional (3-D) models (Dong et al., 2005; Perozo et al., 1998; Poirier et al., 1998) that were in good agreement with crystal structures that were solved at nearly the same time. Use of the full repertoire of methods described above should facilitate the construction of 3-D models for amyloid fibrils as well. Considering that amyloid fibrils from the same protein can adopt different structures depending on the conditions (e.g., Kihara et al., 2005; Krishnan and Lindquist, 2005; Petkova et al., 2005; Tanaka et al., 2004), multiple 3-D models would probably have to be generated. The combination of SDSL and EPR spectroscopy should be ideally suited to decipher these differences.

Studies on soluble and membrane proteins have shown that changes in mobility and distance can be followed in real time, sometimes allowing for millisecond time resolution. This strategy has been applied in monitoring protein folding, oligomerization, and rigid body movements as well as protein insertion into membranes (reviewed in Hubbell et al., 1996, 1998, 2000). A first kinetic analysis of fibril formation has been performed on a small peptide derived from the prion protein (Lundberg *et al.*, 1997). Because spin labels can be attached at any site in the protein, EPR might serve to resolve differences in folding rates within an amyloid fibril, especially if multiple domains are involved. Furthermore, structural changes that relate to fibril maturation, such as the formation of new contacts between protofilaments, could be monitored in real time.

Finally, EPR spectroscopy is a promising means by which to pursue the study of oligomers further, including their interconversions.

Acknowledgments

The authors thank Drs. Sajith Jayasinghe, Christian Altenbach, John Voss, and Jens Lagerstedt for helpful comments on the manuscript. This research was supported by grants to R. Langen from the Larry L. Hillblom Foundation, the Beckman Foundation, the John Douglas French Alzheimer's Foundation, and the National Institutes of Health (P50 AG05142) as well as by a fellowship grant to M. Margittai from the Larry L. Hillblom Foundation.

References

Altenbach, C., Greenhalgh, D. A., Khorana, H. G., and Hubbell, W. L. (1994). A collision gradient method to determine the immersion depth of nitroxides in lipid bilayers: Application to spin-labeled mutants of bacteriorhodopsin. *Proc. Natl. Acad. Sci. USA* **91,** 1667–1671.

Altenbach, C., Oh, K. J., Trabanino, R. J., Hideg, K., and Hubbell, W. L. (2001). Estimation of inter-residue distances in spin labeled proteins at physiological temperatures: Experimental strategies and practical limitations. *Biochemistry* **40,** 15471–15482.

Barghorn, S., and Mandelkow, E. (2002). Toward a unified scheme for the aggregation of tau into Alzheimer paired helical filaments. *Biochemistry* **41,** 14885–14896.

Berengian, A. R., Parfenova, M., and Mchaourab, H. S. (1999). Site-directed spin labeling study of subunit interactions in the alpha-crystallin domain of small heat-shock proteins. Comparison of the oligomer symmetry in alphaA-crystallin, HSP 27, and HSP 16.3. *J. Biol. Chem.* **274,** 6305–6314.

Berliner, L. J., Grunwald, J., Hankovszky, H. O., and Hideg, K. (1982). A novel reversible thiol-specific spin label: Papain active site labeling and inhibition. *Anal. Biochem.* **119,** 450–455.

Caughey, B., and Lansbury, P. T. (2003). Protofibrils, pores, fibrils, and neurodegeneration: Separating the responsible protein aggregates from the innocent bystanders. *Annu. Rev. Neurosci.* **26,** 267–298.

Columbus, L., and Hubbell, W. L. (2002). A new spin on protein dynamics. *Trends Biochem. Sci.* **27,** 288–295.

Columbus, L., Kalai, T., Jeko, J., Hideg, K., and Hubbell, W. L. (2001). Molecular motion of spin labeled side chains in alpha-helices: Analysis by variation of side chain structure. *Biochemistry* **40,** 3828–3846.

Der-Sarkissian, A., Jao, C. C., Chen, J., and Langen, R. (2003). Structural organization of alpha-synuclein fibrils studied by site-directed spin labeling. *J. Biol. Chem.* **278,** 37530–37535.

Dobson, C. M. (2003). Protein folding and misfolding. *Nature* **426,** 884–890.

Dong, J., Yang, G., and Mchaourab, H. S. (2005). Structural basis of energy transduction in the transport cycle of MsbA. *Science* **308,** 1023–1028.

Farahbakhsh, Z. T., Hideg, K., and Hubbell, W. L. (1993). Photoactivated conformational changes in rhodopsin: A time-resolved spin label study. *Science* **262,** 1416–1419.

Feix, J. B., and Klug, C. S. (1998). Spin labeling: The next millenium. *In* "Biological Magnetic Resonance" (C. S. Berliner, ed.), Vol. 14, pp. 251–281. Plenum Press, New York.

Goedert, M., Jakes, R., Spillantini, M. G., Hasegawa, M., Smith, M. J., and Crowther, R. A. (1996). Assembly of microtubule-associated protein tau into Alzheimer-like filaments induced by sulphated glycosaminoglycans. *Nature* **383,** 550–553.

Gordon, D. J., Balbach, J. J., Tycko, R., and Meredith, S. C. (2004). Increasing the amphiphilicity of an amyloidogenic peptide changes the beta-sheet structure in the fibrils from antiparallel to parallel. *Biophys. J.* **86,** 428–434.

Gross, A., Columbus, L., Hideg, K., Altenbach, C., and Hubbell, W. L. (1999). Structure of the KcsA potassium channel from Streptomyces lividans: A site-directed spin labeling study of the second transmembrane segment. *Biochemistry* **38,** 10324–10335.

Hubbell, W. L., and Altenbach, C. (1994). Investigation of structure and dynamic in membrane proteins using site-directed spin labeling. *Curr. Opin. Struct. Biol.* **4,** 566–573.

Hubbell, W. L., Cafiso, D. S., and Altenbach, C. (2000). Identifying conformational changes with site-directed spin labeling. *Nat. Struct. Biol.* **7,** 735–739.

Hubbell, W. L., Gross, A., Langen, R., and Lietzow, M. A. (1998). Recent advances in site-directed spin labeling of proteins. *Curr. Opin. Struct. Biol.* **8,** 649–656.

Hubbell, W. L., Mchaourab, H. S., Altenbach, C., and Lietzow, M. A. (1996). Watching proteins move using site-directed spin labeling. *Structure* **4,** 779–783.

Isas, J. M., Langen, R., Haigler, H. T., and Hubbell, W. L. (2002). Structure and dynamics of a helical hairpin and loop region in annexin 12: A site-directed spin labeling study. *Biochemistry* **41,** 1464–1473.

Isas, J. M., Langen, R., Hubbell, W. L., and Haigler, H. T. (2004). Structure and dynamics of a helical hairpin that mediates calcium-dependent membrane binding of annexin B12. *J. Biol. Chem.* **279,** 32492–32498.

Jao, C. C., Der-Sarkissian, A., Chen, J., and Langen, R. (2004). Structure of membrane-bound alpha-synuclein studied by site-directed spin labeling. *Proc. Natl. Acad. Sci. USA* **101,** 8331–8336.

Jayasinghe, S. A., and Langen, R. (2004). Identifying structural features of fibrillar islet amyloid polypeptide using site-directed spin labeling. *J. Biol. Chem.* **279,** 48420–48425.

Jeschke, G., Pannier, M., and Spiess, H. W. (2001). "Biological Magnetic Resonance." Kluwer, Amsterdam.

Kihara, M., Chatani, E., Sakai, M., Hasegawa, K., Naiki, H., and Goto, Y. (2005). Seeding-dependent maturation of beta2-microglobulin amyloid fibrils at neutral pH. *J. Biol. Chem.* **280,** 12012–12018.

Krishnan, R., and Lindquist, S. L. (2005). Structural insights into a yeast prion illuminate nucleation and strain diversity. *Nature* **435,** 765–772.

Lajzerowicz-Bonneteau, J. (1976). Molecular structures of nitroxides. *In* "Spin Labeling Theory and Applications" (J. Berliner, ed.). Academic Press, New York. 239–249.

Langen, R., Isas, J. M., Luecke, H., Haigler, H. T., and Hubbell, W. L. (1998). Membrane-mediated assembly of annexins studied by site–directed spin labeling. *J. Biol. Chem.* **273,** 22453–22457.

Langen, R., Oh, K. J., Cascio, D., and Hubbell, W. L. (2000). Crystal structures of spin labeled T4 lysozyme mutants: Implications for the interpretation of EPR spectra in terms of structure. *Biochemistry* **39,** 8396–8405.

Lietzow, M. A., and Hubbell, W. L. (2004). Motion of spin label side chains in cellular retinol-binding protein: Correlation with structure and nearest-neighbor interactions in an antiparallel beta-sheet. *Biochemistry* **43,** 3137–3151.

Luecke, H., Chang, B. T., Mailliard, W. S., Schlaepfer, D. D., and Haigler, H. T. (1995). Crystal structure of the annexin XII hexamer and implications for bilayer insertion. *Nature* **378,** 512–515.

Lundberg, K. M., Stenland, C. J., Cohen, F. E., Prusiner, S. B., and Millhauser, G. L. (1997). Kinetics and mechanism of amyloid formation by the prion protein H1 peptide as determined by time-dependent ESR. *Chem. Biol.* **4,** 345–355.

Margittai, M., Fasshauer, D., Pabst, S., Jahn, R., and Langen, R. (2001). Homo- and heterooligomeric SNARE complexes studied by site-directed spin labeling. *J. Biol. Chem.* **276,** 13169–13177.

Margittai, M., and Langen, R. (2004). Template-assisted filament growth by parallel stacking of tau. *Proc. Natl. Acad. Sci. USA* **101,** 10278–10283.

Mchaourab, H. S., Lietzow, M. A., Hideg, K., and Hubbell, W. L. (1996). Motion of spin-labeled side chains in T4 lysozyme. Correlation with protein structure and dynamics. *Biochemistry* **35,** 7692–7704.

Mchaourab, H. S., Oh, K. J., Fang, C. J., and Hubbell, W. L. (1997). Conformation of T4 lysozyme in solution. Hinge-bending motion and the substrate-induced conformational transition studied by site-directed spin labeling. *Biochemistry* **36,** 307–316.

Molin, Y. N., Salikhov, K. M., and Zamaraev, K. I. (1980). Spin exchange. *In* "Spin Exchange." Springer, Berlin.

Oh, K. J., Altenbach, C., Collier, R. J., and Hubbell, W. L. (2000). Site-directed spin labeling of proteins. Applications to diphtheria toxin. *Methods Mol. Biol.* **145,** 147–169.

Perozo, E., Cortes, D. M., and Cuello, L. G. (1998). Three-dimensional architecture and gating mechanism of a K+ channel studied by EPR spectroscopy. *Nat. Struct. Biol.* **5,** 459–469.

Petkova, A. T., Leapman, R. D., Guo, Z., Yau, W. M., Mattson, M. P., and Tycko, R. (2005). Self-propagating, molecular-level polymorphism in Alzheimer's beta-amyloid fibrils. *Science* **307,** 262–265.

Poirier, M. A., Xiao, W., Macosko, J. C., Chan, C., Shin, Y. K., and Bennett, M. K. (1998). The synaptic SNARE complex is a parallel four-stranded helical bundle. *Nat. Struct. Biol.* **5,** 765–769.

Rabenstein, M. D., and Shin, Y. K. (1995). Determination of the distance between two spin labels attached to a macromolecule. *Proc. Natl. Acad. Sci. USA* **92,** 8239–8243.

Serag, A. A., Altenbach, C., Gingery, M., Hubbell, W. L., and Yeates, T. O. (2002). Arrangement of subunits and ordering of beta-strands in an amyloid sheet. *Nat. Struct. Biol.* **9,** 734–739.

Steinhoff, H. J., Mollaaghababa, R., Altenbach, C., Hideg, K., Krebs, M., Khorana, H. G., and Hubbell, W. L. (1994). Time-resolved detection of structural changes during the photocycle of spin-labeled bacteriorhodopsin. *Science* **266,** 105–107.

Tanaka, M., Chien, P., Naber, N., Cooke, R., and Weissman, J. S. (2004). Conformational variations in an infectious protein determine prion strain differences. *Nature* **428,** 323–328.

Torok, M., Milton, S., Kayed, R., Wu, P., McIntire, T., Glabe, C. G., and Langen, R. (2002). Structural and dynamic features of Alzheimer's Abeta peptide in amyloid fibrils studied by site-directed spin labeling. *J. Biol. Chem.* **277,** 40810–40815.

[8] Hydrogen/Deuterium Exchange Mass Spectrometry Analysis of Protein Aggregates

By INDU KHETERPAL, KELSEY D. COOK, and RONALD WETZEL

Abstract

The elucidation of the structure of amyloid fibrils and related aggregates is an important step toward understanding the pathogenesis of diseases like Alzheimer's disease, which feature protein misfolding and/or aggregation. However, the large size, heterogeneous morphology, and poor solubility of amyloid-like fibrils make them resistant to high-resolution structure determination. Using amyloid fibrils and protofibrils of the Alzheimer's plaque peptide amyloid β as examples, we describe here the use of hydrogen/deuterium exchange methods in conjunction with electrospray ionization mass spectrometry to determine regions of the peptide involved in β-sheet network when it is incorporated into protein aggregates. The advantages of this method are low sample utilization and high speed. The basic methodology exploits the fact that protons either involved in H-bonded secondary structures or buried in a protein's core structure exchange more slowly with deuterium than do solvent-exposed and non-H-bonded protons. Details of all aspects of this methodology, including sample preparation, data acquisition, and data analysis, are described. These data provide insights into the structures of monomers, protofibrils, and fibrils and to the structural relations among these states.

Introduction

Hydrogen/deuterium (H/D) exchange methods have been used for several decades to examine aspects of secondary structure in globular proteins (Busenlehner and Armstrong, 2005; Engen and Smith, 2001; Englander, 2000; Englander and Kallenbach, 1983; Hamuro *et al.*, 2005; Hoofnagle *et al.*, 2003; Komives, 2005; Raschke and Marqusee, 1998; Teilum *et al.*, 2005; Wales and Engen, 2006; Woodward, 1993). This method exploits the fact that protons either involved in H-bonded secondary structures, such as α-helix and β-sheet, or buried in a protein's core structure exchange more slowly than protons in solvent-exposed and non-H-bonded regions. By assessing the susceptibility of different exchangeable protons in the structure to exchange in D_2O, regions of the protein involved in secondary structural elements and how they fluctuate in the native state

METHODS IN ENZYMOLOGY, VOL. 413   0076-6879/06 $35.00
DOI: 10.1016/S0076-6879(06)13008-6

and during folding can be assessed. This technique has been used to study the conformational dynamics of native proteins that are capable of making amyloid fibrils (Nettleton and Robinson, 1999). For application to amyloid fibrils and other aggregates, however, certain technical problems had to be solved. Fortunately, after the first description of the analysis of amyloid fibril structure by H/D exchange (Kheterpal *et al.*, 2000), progress has been rapid and has extended to various aggregated forms of amyloid β (Aβ) (Ippel *et al.*, 2002; Kheterpal *et al.*, 2003a,b; Kraus *et al.*, 2003; Luhrs *et al.*, 2005; Olofsson *et al.*, 2006; Wang *et al.*, 2003; Whittemore *et al.*, 2005; Williams *et al.*, 2004, 2005); β2-microglobulin (Hoshino *et al.*, 2002; Yamaguchi *et al.*, 2004); transthyretin (Olofsson *et al.*, 2004); α-synuclein (Del Mar *et al.*, 2005); prion proteins (Kuwata *et al.*, 2003; Nazabal *et al.*, 2003, 2005; Ritter *et al.*, 2005); and model systems, such as SH3 domain (Carulla *et al.*, 2005) and surfactant protein C (Hosia *et al.*, 2002). The main techniques applied have been mass spectrometry (MS) (Carulla *et al.*, 2005; Del Mar *et al.*, 2005; Hosia *et al.*, 2002; Kheterpal *et al.*, 2000, 2003a,b; Kraus *et al.*, 2003; Nazabal *et al.*, 2003, 2005; Wang *et al.*, 2003; Williams *et al.*, 2004, 2005) and nuclear magnetic resonance (NMR) (Carulla *et al.*, 2005; Hoshino *et al.*, 2002; Ippel *et al.*, 2002; Kuwata *et al.*, 2003; Luhrs *et al.*, 2005; Olofsson *et al.*, 2004, 2006; Ritter *et al.*, 2005; Whittemore *et al.*, 2005; Yamaguchi *et al.*, 2004). Some H/D exchange methods have the capability of providing spatial resolution down to a single residue level to determine whether a particular amino acid in a polypeptide is involved in the hydrogen-bonded core structure when this protein/peptide is incorporated into amyloid fibrils. The data obtained using H/D exchange methods are consistent with and complementary to other biophysical techniques and aid in building a detailed amyloid fibril structural model, which is necessary to understand disease pathogenesis fully and may also aid in development of therapeutics.

The technical challenges in applying H/D exchange methods to aggregates are significant. A typical H/D exchange experiment on globular proteins involves incubation of the protein in a deuterated solution with real-time monitoring of exchange. Because fibrils, however, are too large and too heterogeneous for direct MS or NMR detection, rapid methods are needed for dissolution of aggregates while preserving the H/D exchange information. An H/D exchange experiment on amyloid therefore consists of a number of steps, including (1) exposure of aggregates to D$_2$O, (2) quenching of exchange with rapid and efficient disaggregation and dissolution of aggregates, (3) determination of the number (and, preferably, location) of protons exchanged, and (4) application of any necessary corrections for loss of exchange information attributable to back/forward-exchange during analysis.

The different analytical methods capable of following an H/D exchange experiment have strengths and weaknesses. The ideal technique should have (1) high sensitivity, (2) rapid analysis time that minimizes the opportunities for postquenching exchange while maximizing the number of kinetic data points that can be collected during an exchange reaction, (3) resolution at the single amino acid residue level so that the role of each backbone amide hydrogen in aggregate structure can be mapped, and (4) the ability to identify and quantify different aggregate structures in the sample. NMR, which detects the loss of individual N-H peaks in the spectrum as the H is exchanged with D, can provide resolution at the single residue level but requires relatively long analysis times and high sample concentrations. Fourier transform infrared (FTIR) requires high concentrations of analyte and can only provide global exchange information but has the strength of allowing data collection in the fibril state, and thus continuous, real-time data collection without the need for quenching and dissolution (Baello et al., 2000; Vigano et al., 2004). MS, which determines the number of protons exchanged by an increase in the mass of the polypeptide, features high sensitivity and rapid analysis time but has more challenges to obtaining single residue resolution. In addition, on-line MS, as described here, is the only one of these methods that is capable of identifying multiple aggregate structures in a sample.

There are a variety of possible aggregates of the AD plaque peptide Aβ. This peptide exists in a variety of sequence lengths, with the predominant ones being 1–40 and 1–42. Aβ can grow into a number of conformationally distinct mature amyloid fibrils (Petkova et al., 2005). In addition, a variety of aggregation states, such as oligomers and protofibrils, are populated during in vitro growth of mature fibrils (Lashuel et al., 2003). In this chapter, our examples are Aβ(1-40) protofibrils as well as mature amyloid fibrils grown in phosphate-buffered saline (PBS) at 37° without shaking ("quiescent" fibrils). Using these aggregates, we describe our approach to obtaining H/D exchange information using electrospray ionization (ESI)-MS and an on-line sample preparation scheme that minimizes analysis time. Another popular and effective approach is the use of proteolysis and fragment isolation by high-performance liquid chromatography (HPLC) prior to analysis (Del Mar et al., 2005; Smith et al., 1997; Wang et al., 2003; Zhang and Smith, 1993).

Technological Considerations in H/D Exchange-MS

A schematic of the H/D exchange-MS analysis experiment is presented in Fig. 1. Fibrils are suspended in a deuterated buffer to initiate the exchange reaction, and the reaction mixture is incubated at room temperature.

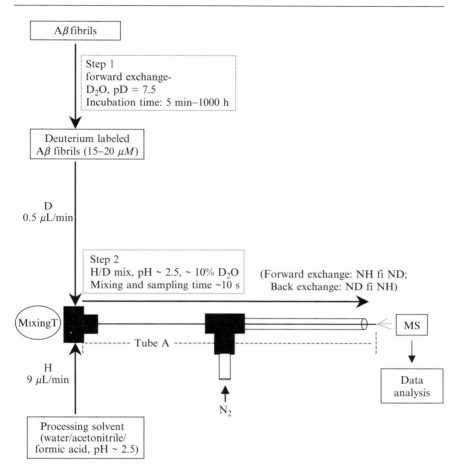

FIG. 1. Schematic of the experimental setup of the hydrogen/deuterium exchange-mass spectrometry (MS) experiment. Processing solvent is optimized to quench exchange, disaggregate and dissolve fibrils, and prepare sample for MS.

After a controlled time interval (ranging from a few seconds to many hours), an aliquot of this sample is infused into a mixing T, where it is mixed with an acidic aqueous solution that quenches exchange (exchange into amide protons is slowest at pH ≈2.5 [Wuthrich and Wagner, 1979]) and initiates fibril dissolution before streaming into the MS. Although the goal of the experiment is to assess the number of protons exchanging with deuterium in the amyloid fibril sample (step 1 in Fig. 1), in practice, some artifactual exchange will inevitably occur as the solutions mix and the fibrils dissolve (step 2 in Fig. 1), despite the slow rate of exchange at pH 2.5. This may be either forward- or back-exchange (addition or loss of D, respectively), compromising the desired amyloid exchange information. The solution to the problem of artifactual exchange, which is an issue for both MS and NMR approaches to H/D exchange studies of amyloid, is first to optimize the data collection protocols to minimize artifactual exchange and second to employ methods for correction of the data for the residual artifactual exchange that cannot be avoided. These methods are described below.

Experimental Setup

The aggregate suspension is infused into one arm of a 250-μm internal diameter Teflon T (Mixing T, Valco; Fig. 1) using a syringe pump (Harvard Biosciences model 11) and a 75-μm internal diameter fused silica capillary. Disaggregating solvent is pumped through a second capillary to another arm of the T using a second syringe pump. The third arm of the mixing T is connected to an ESI probe via an $\approx$25 cm long and $\approx$100-μm internal diameter stainless steel capillary (Tube A in Fig. 1). The length and diameter of tube A, and the flow rates, control the time from mixing to ionization. This represents a trade-off between adequate time for dissolution and minimum time for artifactual exchange; we have found the latter to be the more important consideration, and thus have kept this time short ($\approx$10 s).

Tube A passes through a larger tube, accommodating nitrogen nebulizing gas (to assist the spray process). We therefore refer to this as a "coaxial probe" (distinguishing it from a triaxial probe as discussed later). Although the experimental setup is simple, several factors noted below have to be carefully optimized, controlled, and monitored in order to obtain reproducible, accurate, and precise data.

Sample Preparation and Buffer Exchange of Monomers

Deuterated 2 mM Tris-DCl buffer is freshly prepared under nitrogen (to exclude atmospheric moisture) by the addition of 10 μl of 100 mM d-Tris-DCl (in D_2O) buffer to a 500-μl ampule of 99.9 % D_2O (Sigma). Dry monomeric peptide is rigorously disaggregated as described (see Chapter 3 by O'Nuallain and colleagues in this volume). For protonated monomeric samples, disaggregated peptide is then lyophilized and redissolved in 2 mM Tris-HCl (pH 7.5). Fully deuterated monomer samples are prepared by gently dissolving disaggregated monomer in freshly prepared 2 mM d-Tris-DCl, pD 7.5 (pD values were read directly from either pH paper or a pH meter and have not been corrected for any isotope effects). Vigorous mixing at this step can result in precipitation and must therefore be avoided. To ensure that no higher order structures are present that can seed aggregation reactions, solutions of monomeric peptide can be subjected to high-speed ultracentrifugation (see Chapter 3 by O'Nuallain and colleagues in this volume) and the supernatant carefully collected for further analysis. For H/D exchange kinetics experiments on monomer, however, samples were infused into the T immediately after dissolution into d-Tris-DCl without ultracentrifugation in order to capture early deuteration time points. The minimum time from addition of deuterated buffer to the collection of the first spectrum is ≈5 min. We have monitored exchange into Aβ(1–40) monomer for up to 800 h and found it does not increase after the initial 5 min, demonstrating complete exchange (Kheterpal et al., 2000). The concentration of monomer is determined by diluting an aliquot into 1% aqueous trifluoroacetic acid (TFA; Pierce), followed by HPLC analysis against a standard Aβ solution quantified by amino acid composition analysis (see Chapter 3 by O'Nuallain and colleagues in this volume). The final monomer concentrations in all our experiments were in the range of 10–30 μM. Protonated and deuterated monomeric solutions can be stored at −20° for several months.

Preparation, Isolation, and Buffer Exchange of Aggregates

Preparation. In general, Aβ(1-40) amyloid fibrils were prepared at 37° in PBS without agitation (Petkova et al., 2005; Williams et al., 2006) as described (see Chapter 3 by O'Nuallain and colleagues in this volume). To prepare fully deuterated aggregates (required for the artifactual exchange correction [see below]), disaggregated monomer is dissolved in D_2O at 0.25 mg/ml to obtain completely deuterated monomer and lyophilized overnight. Dry deuterated monomer is then redissolved in freshly prepared 2 mM NaOD in D_2O, followed by an equal volume of deuterated 2 × PBS (20 mM phosphate, 276 mM sodium chloride, 5.4 mM potassium chloride, pD 7.4). This solution is centrifuged at high speed to remove any aggregates, and the supernatant is incubated at 37° (see Chapter 3 by O'Nuallain and colleagues in this volume). Fibril growth is monitored using thioflavin T fluorescence (LeVine, 1999), and morphology is examined by electron microscopy.

Isolation and Buffer Exchange of Amyloid Fibrils. Because high ionic strength buffers compromise the sensitivity of MS detection, exchange reactions have to be carried out in low ionic strength deuterated buffer. Mature amyloid fibrils (e.g., 50 μl of 0.2 μg/μl) are isolated by centrifugation at ≈20,000g for 15–30 min in a bench-top centrifuge. The resulting fibril pellet is decanted and washed with 100 μl of 2 mM Tris-HCl, followed by centrifugation for 30 min. This longer centrifugation time is required because fibrils do not pellet well in low ionic strength buffer; centrifugation at very high g-forces is not a viable option because ultracentrifuged pellets are difficult to quickly resuspend for the H/D exchange measurement. At this stage, it is critical to remove as much of the protonated buffer as possible from the pellet. Gel loading tips are ideal for removing final traces of the protonated solvent without disturbing the pellet. Two hundred microliters of freshly prepared deuterated 2 mM Tris-DCl buffer (as described for buffer exchange of monomers) is added, under nitrogen, to the fibril pellet. This marks the start of the exchange reaction. The solution is vortexed vigorously to resuspend the fibrils from the pellet, after which aliquots from the exchanging suspension can be infused into one arm of the mixing T using a syringe pump (Fig. 1). Early exchange is monitored by continuous infusion and mass spectral data acquisition for the first hour. Afterward, an aliquot of the reaction mixture incubating at room temperature is examined once every few hours. We have followed kinetics of deuterium exchange into fibrils from 5 min to longer than 1000 h (Kheterpal et al., 2000). Exchange of deuterium into fibrils is nearly complete after 24 h; hence, this time point has been selected for the examples shown here. The concentration of these partially deuterated fibrils is measured by dissolving a 20-μl aliquot of the suspension in 100 μl 10% TFA, followed by HPLC analysis (see Chapter 3 by O'Nuallain and colleagues in this volume). The monomer-equivalent concentration in fibril samples used in our experiments is usually 10–20 μM.

Isolation and Buffer Exchange of Protofibrils. Small protofibrillar aggregates observed early in the growth of fibrils can be isolated using size exclusion chromatography (SEC) (Lashuel and Grillo-Bosch, 2005; Lashuel et al., 2003; Walsh et al., 1997). For H/D exchange-MS analysis, these small oligomeric aggregates need to be suspended in low ionic strength buffer, but buffer exchange of these intermediate species presents a special problem because they do not pellet at low-speed spins. Although collection at high speed is possible (see Chapter 11 by Mok and Howlett in this volume), the integrity of the sample may be compromised by the rigorous methods required for resuspension of the pellet. We therefore developed a filtration method to process these protofibrillar samples (Kheterpal et al., 2003a). To minimize dissolution and/or further aggregation of protofibrils, they are stored on ice and all centrifugations are performed at 3°. A 250–500-μl aliquot of protofibril fraction collected using SEC (Lashuel et al., 2003) is placed in an YM-10 microcon tube (Millipore) and centrifuged at 14,000g for 30–50 min (aggregates are retained above the filter); to avoid formation of a highly concentrated protofibril sample that can readily form fibrils, the centrifugation time is selected to prevent their concentration above the filter by more than fivefold. The amount of filtrate is measured, and an equivalent amount of deuterated Tris buffer is added, and this marks the start of the exchange reaction. This step is repeated two to three times until the residual PBS buffer concentration in the solution above the filter is equal to 2% of its initial value. The typical time to prepare a sample in this manner is ≈2.5 h, during which time, of course, H/D exchange has already begun. When buffer exchange is complete, the sample is immediately infused into the mixing T for MS analysis. To prepare protonated control samples, buffer exchange as described above is performed with protonated Tris buffer. The final concentrations of these samples are determined by either amino acid composition analysis or by HPLC. Hydrogen exchange data on protofibrils are not presented or discussed here but have been published (Kheterpal *et al.*, 2003a).

Dissolution of Aggregates

Fast H/D exchange kinetics measurements require identification of a processing solvent that quickly dissolves fibrils within the mixing T, quenches exchange, and is MS-compatible for efficient ionization. The typical flow rates used for infusing sample (0.5 μl/min) and solvent (9 μl/min) into the mixing T serve to dilute the sample by a factor of ≈20 and establish a ≈10-s dwell time from mixing to spray. An input mass concentration of aggregate, equivalent to 10–20 μM peptide concentration, resulting in a 0.5–1 μM final concentration of the peptide, is adequate for MS detection. This means that the chosen solvent must be capable of dissolving such concentrations of the fibrils. Because the dissolution solvent must also provide quenching, it is important to explore solvents near the pH for minimum exchange of backbone amide protons, pH ≈2.5 (Wuthrich and Wagner, 1979). Reduced dissolution temperatures can also be explored to minimize exchange further, but these may retard dissolution kinetics (see below). Finally, the processing solvent must be conducive to high-sensitivity detection of peptides in ESI-MS, for example, by containing a significant portion of acetonitrile. Using this set of constraints (some of which may be instrument dependent), we determined that a mixture of 50% water, 50% acetonitrile, and 0.5% formic acid, with pH ≈2.5, is optimal to dissolve the highest possible concentration of fibrils in the mixing T and in the capillary prior to ESI (Kheterpal *et al.*, 2003b). To quantitate the percentage of fibrils being dissolved on-line, signals derived from fibril samples are compared with a standard curve generated by injection of known concentrations of monomer into the arm of the mixing T usually employed for fibrils. For a series of fibril suspensions of increasing concentration, the signal intensity reaches a maximum beyond which there is no gain in signal with higher concentration. Thus, there is apparently a limit to the concentration of fibrils (≈15 μM) that can be completely dissolved in the ≈10 s available between mixing and spray tip. Working above this limit results in more frequent clogging of the mixing T and capillary and potentially unreliable data. Similar results for fibril dissolution were obtained with 1–42 fibrils and with a triaxial probe (described below) (Chen *et al.*, 2006).

Data Collection

All the data shown here have been collected using a Micromass Quattro II triple quadrupole mass spectrometer (Manchester, UK). Experiments were performed in the positive ion mode. All spectra were acquired in the multichannel accumulation mode with mass range from 650–1150 Da for intact peptide at a scan rate of 250 Da/s for 1 min.

To obtain reproducible mixing time between sample and solvent, and hence reproducible data, the syringe pumps are turned on and off simultaneously. Aggregates tend to settle in the syringe during long-term kinetics experiments, resulting in a loss of MS signal; it is therefore important to make sure that the sample in the syringe is either changed frequently (30–60 min in our hands) or is remixed periodically. It is also very important to clean the T and the capillaries used for sample and solvent delivery by flushing with water and solvent, respectively, between each run. Adequate cleanliness is validated by collecting data for a blank run between each new sample. Any clogs in the T can be removed by dismantling the T and sonicating it in the processing solvent.

Because the processing solvent is generally protonated and the sample exchange buffer is deuterated, there is always a percentage of D_2O present in the final solvent mixture. The expected percentage can be easily estimated from the ratio of the solvent and the sample flow rates and can be rigorously determined by comparing the ratio of the acetonitrile peak at mass-to-charge (m/z) 43 ($CH_3CN + H^+$) and at m/z 44 ($CH_3CN + D^+$). We monitor the ratio of this pair of peaks during data collection to ensure consistency in flow rates. Typical experiments with sample and solvent flow rates of 0.5 and 9 μl/min, respectively, result in 10% D_2O in the final solvent mixture. Thus, the ratio of acetonitrile peaks at 43 and 44 is 10:1. Any blockage in the experimental setup will affect the mixing rate, resulting in either greater or lower than 10% D_2O in the final mixture. This change is reflected in the ratio of the acetonitrile peaks at m/z 43 and 44.

The Artifactual Exchange Problem

The protocol and apparatus (Fig. 1) outlined above expose the sample to a mixture of H_2O and D_2O in the mixing T and beyond for a total of about 10 s. Because fibril dissolution generates monomers, and thus exposes previously protected sites, this gives rise to an opportunity for spurious incorporation of deuterium in protected sites (forward-exchange) as well as loss of deuteration at sites that were exposed during incubation (back-exchange) during analysis. Collectively, these two processes are designated as "artifactual exchange." Back-exchange could be eliminated by using D_2O instead of H_2O in the processing solvent, but this would simply result in an increase in an indeterminate amount of artifactual forward-exchange after dissolution of fibrils.

Even under optimal exchange quenching conditions at pH 2.5, the fast-exchanging protons at acidic and basic terminal and side chain sites completely equilibrate with deuterium during processing (Wuthrich and Wagner, 1979), so that deuterium incorporation reflects the final composition of the processing solvent, erasing the exchange protection information at these sites. This fast exchange in the side chain and terminal protons is impossible to stop but is also simple to correct for, so it does not compromise our ability to obtain quality data on backbone exchange protection. In contrast, not only is artifactual exchange at backbone amide protons more difficult to correct for but the accuracy and precision of the corrected data decrease as artifactual exchange levels increase. Therefore, it is critical to minimize the latter type of artifactual exchange.

Minimizing Artifactual Exchange. To minimize artifactual exchange into backbone amide positions, the ESI source parameters as well as the solvent composition have to be optimized empirically. $A\beta$ monomer at pH 2.5 contains 66 exchangeable protons (side chains and amide protons). When a fully deuterated monomer sample is mixed with deuterated processing solvent (50:50:0.2 (v/v/v) $D_2O/CH_3CN/HCOOH$), an increase in mass (relative to the corresponding experiment using protonated sample and solvents) of 65.8 ± 0.2, within the experimental error of the expected incorporation of 66 deuteriums, was obtained (Kheterpal *et al.*, 2000). This validates the fundamental approach of using ESI-MS to follow H/D exchange and suggests that the monomer has no sites immune to H/D exchange in 2 mM Tris DCl, pD 7.4 (Kheterpal *et al.*, 2000).

Another important outcome of this experiment is the demonstration that no artifactual back-exchange is occurring during the electrospray process. This is not a trivial point, because we have found that in some ESI sources, significant artifactual back-exchange can occur during the droplet phase prior to entering the inlet of the mass spectrometer. This back-exchange, which derives from moisture in the air, can be resolved by adding an extra N_2 line into the source.

Several source parameters, such as cone voltage, emitter voltage, and source temperature, can also have a significant effect on the ES performance and/or extent of artifactual isotope exchange (Kheterpal *et al.*,

2003b). The parameters to be optimized will vary depending on the design of the ESI source. To test the effect of these parameters and to optimize them for minimum artifactual exchange, we analyzed fully deuterated Aβ (1–40) monomer using protonated processing solvent under various conditions and compared the resulting levels of deuterium content. Fully deuterated monomer was infused at a rate of 1 μl/min and mixed in the T with the processing solvent (18 μl/min), yielding a final sample mixture in which 10% of the water came from D_2O (for convenience, we will subsequently refer to this composition as 10% D_2O, neglecting the acetonitrile). If there was no back-exchange occurring during analysis, a maximum of 39 amide protons should be measured. Deviation from this number indicates the level of back-exchange. For a Micromass Quattro II ESI source, we found the source temperature to have the largest effect on the amount of back-exchange (Kheterpal *et al.*, 2003b). A temperature change from 35° to 100° resulted in increase of back-exchange from 4 to 11 amide protons. In our studies, this exchange could be reduced at lower temperatures, but with an accompanying reduction in the signal-to-noise ratio (S/N). The loss of signal at low temperature could only partially be compensated for by increasing the cone voltage; above 40 V, peptide fragmentation became significant. Therefore, a temperature of 35° and cone voltage of 40 V were selected for this system. Similar studies were carried out to assess the effect of emitter voltage and flow rates of the N_2 nebulizing and drying gases; no significant effects on S/N or the amount of exchange were observed. The final emitter voltage used was 3.5 kV, and the flow rates of N_2 nebulizing and drying gases were 20 and 275 L/h, respectively.

In the apparatus shown in Fig. 1, the sample temperature could not be controlled. All experiments were carried out at room temperature. With modifications in this setup, artifactual exchange could probably be reduced if the sample temperature was at or below 0°. However, the lower temperature would likely make it more difficult to dissolve fibrils quickly for fast analysis. Because the amount of artifactual exchange was similar to that reported for globular proteins with cooled probes, no modifications were undertaken (Del Mar *et al.*, 2005; Ehring, 1999; Engen *et al.*, 2002; Smith *et al.*, 1997).

Correction for Artifactual Exchange. Correction for artifactual exchange into rapidly exchanging side chains and terminal protons is straightforward. The deuterium incorporation into these sites reflects the percentage of deuterium in the final solvent composition. For our experiments, this composition is 10%; therefore, 10% of all side chain and terminal protons will be deuterated. This can be corrected by simply subtracting 2.7 (10% of the 27 protons of this type found in Aβ[1–40]) from the experimentally measured molecular mass of the (partially deuterated) peptide.

The remaining level of exchange (D_{meas} = experimental mass − molecular weight − 2.7) is the experimentally measured number of backbone amide protons exchanging with deuteriums. Correction for artifactual exchange into these backbone positions is much more complicated than into rapidly exchanging positions. The methods to account for this exchange in globular proteins have been developed (Zhang and Smith, 1993) and have also been recently applied to H/D exchange studies on amyloid fibrils (Del Mar et al., 2005; Wang et al., 2003; see Chapter 9 by Nazabal and Schmitter). We have tested various correction schemes for their independence from the quenching and sample processing conditions in our experimental set-up (Kheterpal et al., 2003b). In other words, for each prospective correction method, we changed solvent and sample flow rates into the mixing T as a convenient means of varying the quenching conditions. The total flow rate determines the time available for artifactual exchange, whereas the relative flow rates of the sample and quenching solvent determine the final solvent composition. Because neither of these should affect the prequenching deuteration level in the fibrils (step 1 in Fig. 1), the constancy of the corrected value obtained at various flow rates provides a sensitive test for the robustness of the correction method. A description of the correction methods tested and validation plots showing robustness as described above are presented elsewhere (Kheterpal et al., 2003b). For routine measurements, we recommend using the following method. This method has been tested on measurements of both intact and proteolytic fragments of $A\beta(1-40)$ and for both coaxial and triaxial probes (proteolyzed fragments and triaxial probe are described below).

This method involves measuring forward-exchange (F.E.) and back-exchange (B.E.) for fully protonated and deuterated samples, respectively, using the same workup conditions as the partially exchanged sample of interest and calculating the corrected deuterium content (D_{corr}) in the backbone amides of the target sample according to

$$D_{corr} = D_{meas} + B.E. - F.E. \qquad (1)$$

where B.E. and F.E. are as described next. Forward-exchange is measured by infusing the protonated form of the appropriate sample (fibrils, protofibrils, or monomer depending on the sample of interest) in protonated buffer into one arm of the T and mixing it with a "quenching" solvent containing sufficient D_2O to give a final solvent composition identical to that obtained in the analysis of the sample of interest (10% D_2O in the protocols described here). Because the solvent mixture during sample processing is the sole source of deuterium, this amount of exchange is the forward-exchange taking place during sample workup, or:

$$\text{F.E.} = M_0 - M \tag{2}$$

where M_0 is the measured molecular mass of the forward-exchange (after correcting for fast exchanging protons as described above) sample described above and M is the molecular mass of the monomer in protonated fibrils. The back-exchange value is obtained by measuring the exchange of a fully deuterated sample treated with protonated processing solvent. Fully deuterated fibrils and protofibrils are grown in deuterated buffers as described in the fibril preparation section. If there were no back-exchange taking place during sample processing, 39 deuteriums would be measured for these samples. Any deviation from 39 represents the amount of back-exchange taking place during sample processing, or:

$$\text{B.E.} = N - (M_{100} - M) \tag{3}$$

where M_{100} is the measured molecular mass of the fully deuterated sample (after correction for exchange in fast exchanging protons as described above) and N is the total number of backbone amide protons (39 for $A\beta$ [1–40]).

H/D Exchange into $A\beta(1–40)$ Fibrils

Figure 2A presents representative partial mass spectra (normalized and smoothed) obtained from analysis, with protonated processing solvent, of fully protonated fibrils ("H-fibrils," blue), partially deuterated fibrils ("d-fibrils" incubated in deuterated buffer for 24 h, red), and fully deuterated fibrils ("D-fibrils," black, M_{100} sample) using a co-axial probe. Also included in Fig. 2 are data for protonated fibrils (green) run with a processing solvent providing 10% D_2O (the M_0 sample described above). Centroids of the unresolved isotopic envelopes were used to calculate average molecular masses for each sample. We generally collect all data in triplicate, but for demonstration purposes, Fig. 2 shows only one data set for the +6 charge state. Charge states +5 and +6 were consistently observed with good S/N; therefore, molecular masses obtained from these charge states (corrected by subtracting labile and ionizing protons as described below) were averaged for all calculations. For data obtained using instruments with higher mass resolution, centroids are calculated by considering all the individual isotopic peaks; this process can become difficult and ambiguous for deuterated samples featuring broadened, overlapping distributions.

The spectrum obtained from partially deuterated fibrils (Fig. 2) shows two $A\beta$ populations represented by two peaks, in which the centroid

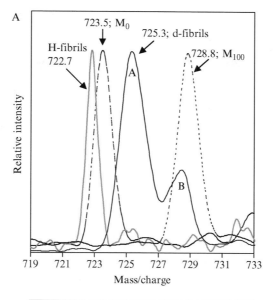

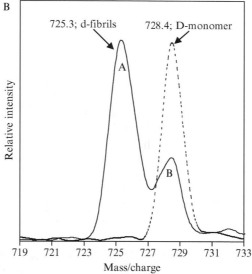

FIG. 2. (A) Electrospray ionization (ESI)-mass spectrometry (MS) of protonated fibrils (gray), M_0 and M_{100} (dot-dashed and dotted lines, respectively) control samples, and partially deuterated fibrils (black) after 24 h of exchange. (B) ESI-MS of partially deuterated fibrils (black; same as those shown in [A]) and fully deuterated monomer (dotted). The data shown here have been smoothed using the Savitzky Golay method implemented in the MassLynx program (Waters) and normalized in a spreadsheet program.

position of peak A varies with exchange time, whereas that of peak B remains constant at $\approx$m/z 728.4, just below the mass obtained from D-fibrils (M_{100}). A representative spectrum from a fully deuterated mono-mer sample is compared with the same d-fibril spectrum in Fig. 2B. From this figure, it can be seen that the B peak coincides almost exactly with that from the deuterated monomer. We observe less artifactual back-exchange in fully deuterated (D-) fibrils compared with fully deuterated monomer, so that the D-fibril peak falls at a higher m/z than the D-monomer peak. This is attributable to the finite time required for fibril dissolution, which reduces the time available for back-exchange, explaining why we grow fully deuterated fibrils for our back-exchange control samples.

The existence of a fully deuterated component (peak B) in the analysis of Aβ(1–40) fibrils and other amyloid systems has been reported (Carulla et al., 2005; Kheterpal et al., 2000, 2003a; Wang et al., 2003). In our experiments, fibrils are isolated by centrifugation immediately prior to subjecting them to deuterium exchange; hence, the completely deuterated Aβ component (peak B) cannot result from monomer present in the original fibril suspension. Although Aβ(1–40) fibrils are known to dissociate partially to give an equilibrium concentration of monomer (O'Nuallain et al., 2005), preliminary data show that the monomer concentrations present in the supernatant (determined by HPLC) are too low to account for all of the B peak. A mechanism involving monomer detachment, full exchange of the detached monomer, and reincorporation of the deuterated monomer into the fibril has been invoked to rationalize the existence of a fully exchanged peak in another amyloid system (Carulla et al., 2005) and may be active here. It is also possible that the peak B is derived from a population of peptide molecules that are bound to the aggregate but not via highly protective hydrogen bonding. Although the source of peak B requires further investigation, it normally does not interfere with analysis of the important exchange protection events.

The number of deuteriums exchanged into Aβ fibrils is calculated using the molecular mass obtained from the centroid of peak A, and the step-by-step procedure is presented below. Data on the protonated (M_0) and fully deuterated (M_{100}) forms of the sample, as required by the correction scheme, should be collected in triplicate at least once a day. We use the molecular mass measured for the protonated sample (M) instead of the theoretical molecular mass (4329.9 Da) to correct for the day-to-day variability in mass calibration of the instrument, although the latter two numbers generally agree to within 0.5 Da.

1. Calculate the average molecular mass for each sample. For protonated samples, such as in Fig. 2A (blue trace), this is calculated as $((m/z)*z) - z$, where z is the charge state. For samples in which a mixture

of H/D was used, such as in Fig. 2A (green, black, and red traces), the final ratio of H and D in the processing solvent is also reflected in the excess protons/deuterons gained during ionization (see data collection in the section on technological considerations in H/D exchange-MS). Thus, for the +6 charge state shown in Fig. 2, 10% of the six ionizing protons are deuterated. Therefore, the neutral molecular mass of these samples is calculated as $((m/z)*z) - z - (z*(\%D_2O/100))$; we must subtract 6.6 Da (instead of just 6) to account for the mass of the ionizing protons/deuterons. For example, the m/z obtained for the samples in Fig. 2 are 725.3, 723.48, and 728.81 for d-fibrils, M_0, and M_{100} samples, respectively. These represent a +6 charge state. Given the 10% D_2O in the final solvent mixture, the calculated molecular masses of the neutrals corresponding to these ions would be 4345.2, 4334.28, and 4366.26, respectively.

Comment: If an HPLC step is used prior to MS analysis, the bulk solvent is generally 100% protonated. In that case, protons gained during ionization and fast exchanging side chain and terminal protons are not deuterated, and no extra correction for these protons is required.

2. Correct for artifactual exchange into side chain and terminal protons. For wild-type (wt) Aβ(1–40), this involves subtracting 2.7 (10% of 27) from the experimentally measured molecular mass obtained after step 1 above. For the d-fibrils, M_0, and M_{100} samples in Fig. 2A, the resulting molecular masses will be 4342.5, 4331.58, and 4363.56, respectively. The difference between this molecular mass and the measured molecular mass of the monomer in protonated fibrils (4330.38 in Fig. 2A) provides the measured number of backbone amide protons exchanging with deuteriums (12.12 for partially deuterated fibrils in Fig. 2). This value can be used qualitatively to compare different fibril preparations, various mutant fibrils (Williams et al., 2004), and various structural states of aggregation (Kheterpal et al., 2003a; Williams et al., 2005). However, because full correction for artifactual exchange into backbone positions can modify results, we recommend full correction of data used for such comparisons.

3. Correct for artifactual exchange in amide protons. Forward-exchange is the difference between the side chain–corrected molecular mass of the M_0 sample and the mass of the protonated sample. For the data shown in Fig. 2A, the forward-exchange is 1.2. Back-exchange is the number of deuteriums lost during the sample processing. We measured 33.18 (4363.56–4330.38) deuteriums in the M_{100} sample; therefore, 5.82 (39–33.18) deuteriums have back-exchanged to protons during sample workup. The corrected number of deuteriums for the data shown in Fig. 2 is thus 16.74 (12.12 + 5.82 − 1.2).

The corrected number of deuteriums corresponds to the number of backbone amide protons in the aggregate exchanged with deuteriums during the exchange reaction, that is, when peptide was incorporated into the fibril structure. We have analyzed various fibril preparations of wt Aβ (1–40) and fibril preparations from Aβ(1–40) mutant peptides (Williams *et al.*, 2004) as well as protofibrils (Williams *et al.*, 2005) and used this corrected value quantitatively to compare the levels of protective secondary structure in these samples. Reproducibility for these experiments is generally within 1 Da. To obtain more detailed information on the structure of amyloid fibrils, we also measured kinetics of H/D exchange by monitoring deuteration as a function of incubation time. Results indicate that quiescent (Petkova *et al.*, 2005) Aβ(1–40) amyloid fibrils grown in PBS (Williams *et al.*, 2006) consist of a very rigid core structure, most likely an H-bonded β-sheet structure, which involves 50% of the 39 Aβ backbone amides (Kheterpal *et al.*, 2003b; Whittemore *et al.*, 2005). The H/D exchange kinetics of fibrils suggest a number of distinct exchange categories, and the rate of exchange of protons with deuteriums can be calculated (Anderegg, 1996). We find that 13 amide protons of Aβ(1–40) fibrils exchange with deuteriums at the same rate ($k_{ex} = 9$ h^{-1}) as that observed in the monomeric state. Perhaps these protons are exposed and not involved in β-sheet formation. A set of $\approx$5 protons exchange with rates ($k_{ex} = 0.01$ h^{-1} – 0.16 h^{-1}) comparable to those observed for protons involved in secondary structure in globular proteins. Another set of $\approx$20 protons is very resistant to exchange, indicating a very rigid core structure in the amyloid fibrils. Although this broad information is useful, more structural insight could be derived from discerning the locations of these various classes of protons within the sequence.

Achieving Higher Spatial Resolution by On-Line H/D Exchange-MS with Proteolysis

The ability to use MS to identify individual residues or regions undergoing exchange depends on either coupling the MS with proteolysis or using tandem MS. Our efforts to use tandem MS to obtain single residue resolution in H/D exchange-MS of Aβ(1–40) fibrils without proteolysis have generated ambiguous results (I. Kheterpal, B. DaGue, K. D. Cook, R. Wetzel, 2001, unpublished data), perhaps because of rearrangement of backbone amide hydrogens and deuteriums in the gas phase, as reported in other studies (Deng *et al.*, 1999; Hoerner *et al.*, 2004; Jorgensen *et al.*, 2005; Kweon and Hakansson, 2006). MS following proteolytic fragmentation, however, has proven informative.

The coupling of H/D exchange with proteolysis by pepsin followed by liquid chromatographic separation and detection by MS has been used for several years to study protein conformation (Busenlehner and Armstrong, 2005; Del Mar *et al.*, 2005; Ehring, 1999; Hoofnagle *et al.*, 2003; Smith *et al.*, 1997; Wales and Engen, 2006; Wang *et al.*, 2002, 2003). Pepsin is an ideal enzyme for coupling fragmentation with H/D exchange because its pH of maximum activity coincides with the pH of minimum exchange that is desirable in the processing solvent for amyloid analysis. We first incorporated proteolysis into our on-line H/D exchange-MS experiments by adding pepsin to the processing solvent in order to quench, disaggregate, dissolve, and proteolyze the sample in the mixing T and capillary simultaneously prior to ESI. However, we found that the enzyme activity is significantly reduced in the presence of acetonitrile, necessitating a high enzyme concentration (40:1 [w/w] relative to Aβ) (Chen *et al.*, 2006). As a result, the dominant peaks in the spectrum were those arising from autolysis of pepsin. A solution to this problem is to delay addition of the acetonitrile (which is necessary for efficient ESI) until immediately prior to spray formation. We describe here this direct on-line proteolysis approach prior to MS analysis featuring sequential processing of the analyte using a triaxial probe (Kheterpal *et al.*, 2006).

On-Line Proteolysis Using Triaxial Probe

A schematic of the triaxial probe is presented in Fig. 3. This probe consists of an innermost fused silica capillary (tube A, 50-μm internal diameter and 360-μm outer diameter) surrounded by an intermediate stainless steel tube with a 410-μm internal diameter and a 720-μm outer diameter for delivering acetonitrile (tube B), which, in turn, is surrounded by an outermost stainless steel tube with an 860-μm inner diameter and an 1100-μm outer diameter. This is in contrast to the format of two stainless steel tubes of the coaxial probe used in the experiments described above (Fig. 1). Aβ samples and solvent (0.1 μg/μl pepsin solution in 0.5% formic acid in H$_2$O, resulting in 5:1 weight ratio of pepsin/peptide after the mixing) are mixed into the first T coupling to promote aggregate dissolution and proteolysis in the innermost fused silica tube. Acetonitrile with 0.5% formic acid is infused through the middle stainless steel tube, mixing with the dissolved and proteolyzed sample only near the tip of the electrospray emitter. Nebulizing gas (nitrogen) is delivered to the region via the outermost stainless steel tube. The flow rates of the sample, pepsin solvent, acetonitrile, and nitrogen as well as the distance to the end of tube B and to the end of the fused silica capillary (tube A) are optimized to achieve good mixing with acetonitrile and to achieve good sensitivity. Typical flow rates for sample, pepsin in 0.5% formic acid, and acetonitrile with 0.5%

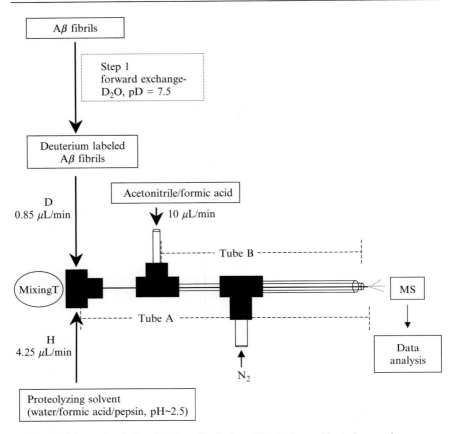

FIG. 3. Schematic of the triaxial probe designed for hydrogen/deuterium exchange-mass spectrometry experiments in conjunction with on-line proteolysis. Processing solvent is optimized to quench exchange, disaggregate and dissolve fibrils, and generate defined peptide fragments; the acetonitrile-formic acid stream prepares the same for the mass spectrometer.

formic acid were 0.85, 4.25, and 10 μl/min, respectively, resulting in a final solution composition in which 16% of the water was derived from D_2O. The flow rates of nebulizing and drying gas were 50 and 300 L/h, respectively. The distance between the ends of tube B and tube A is 5 mm, corresponding to a mixing time of 1.6 s when the total liquid flow rate is ≈15 μl/min. This approach provides effective and rapid on-line proteolysis while minimizing artifactual exchange.

As one test for comparing the performance of the coaxial and triaxial probes, the algorithm for correcting for artifactual exchange described above was applied to the triaxial probe. For these experiments, the probe was operated in a pseudo-co-axial mode; that is, our "standard" quenching solution (50:50:0.5 (v/v/v) $H_2O/CH_3CN/HCOOH$) was used instead of proteolyzing solvent and acetonitrile. Using these conditions, deuterium exchange into the intact fibrils was measured at 15 μl/min while maintaining 16% D_2O in the final solvent mixture (Fig. 4A). Spectra corresponding to the fully protonated and fully deuterated $A\beta(1-40)$ monomer obtained using the triaxial probe are also included for comparison. The spectra are very similar to those obtained using the coaxial probe (Fig. 2), including the presence of "peak B," which corresponds to the m/z obtained from fully deuterated monomer. The corrected mass for the A peak following 24-h exchange into $A\beta(1-40)$ fibrils using the triaxial probe (17.1 ± 0.6) agreed within experimental error with the value obtained using the coaxial probe (17 ± 1.1) (Kheterpal et al., 2006; Whittemore et al., 2005), consistent with an accurate assessment of H/D exchange protection within the fibril.

Proteolysis of $A\beta(1-40)$ Amyloid Fibrils

Fig. 4(B–D) presents representative mass spectra obtained for three prominent pepsin proteolysis fragments—20–34, 1–19, and 35–40—after 24 h of exchange. Spectra corresponding to the fully protonated and fully deuterated $A\beta(1-40)$ monomer are also included in each panel. Fragment identities based on mass were confirmed by MS/MS using a Q-Star XL quadrupole time-of-flight hybrid instrument (Applied Biosystems, Foster City, CA). Methods for determining the corrected number of backbone amide protons exchanging with deuteriums in each of these three fragments are described in detail next.

$[20–34]^{2+}$

This fragment clearly shows two peaks (partially protected A and fully deuterated B) similar to those obtained in the analysis of intact fibrils. The deuterium exchanged into this fragment was determined from the m/z of peak A in Fig. 4B. The charge state obtained for this peptide is +2, based on its mass and confirmation by MS/MS as described above. Forward- and back-exchange were calculated as described above in the sections on technological considerations in H/D exchange-MS and H/D exchange into $A\beta(1-40)$ fibrils for the coaxial probe. In this case, the total number of amide protons is 14 and total number of side chain and terminal protons is 10. (Note that even if residue 20 was protected in the fibril, it becomes the N-terminal site in the proteolytic fragment and will therefore be statistically deuterated.) The resulting corrected number of deuteriums exchanged into this peptide after 24 h was only 4.3, indicating that most of

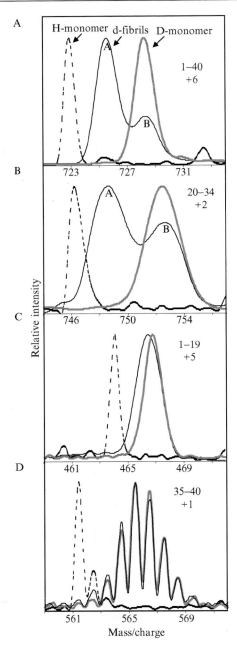

FIG. 4. Electrospray ionization-mass spectrometry of intact Aβ (1–40) fibrils (A) and proteolytic fragments 20–34 (B), 1–19 (C), and 35–40 (D) after 24 h of exchange. The corresponding spectra for protonated monomer (dashed) and fully deuterated monomer (gray) are overlaid for each fragment.

the 14 amino acids in this segment of the peptide were protected from exchange.

$[1-19]^{5+}$

If the B peaks in Figs. 2 and 4A derive from fully exchanged molecules as described above, proteolysis of these molecules would be expected to result in analogous B peaks for all product peptides. However, there is only one peak evident in the spectrum of the [1–19] fragment derived from partially deuterated fibrils (Fig. 4C, charge state of 5+ is shown), and that peak is at an m/z slightly lower than expected for the B peak. This can be explained by the relatively low resolution, which would cause coalescence of the A and B peaks to a single peak representing the weighted average. This is consistent with the low level of protection expected for this region (Petkova et al., 2005; Williams et al., 2004).

Assuming then that there are two peaks in Fig. 4C (as for the [1–40] and [20–34] spectra), but that, in the case of the 1–19 Fragment, they are not resolved, the m/z of the partially exchanged component (peak A) was obtained by deconvolution of this composite peak, as follows. First, the m/z of the peak derived from fully deuterated Aβ monomer was determined (466.8). The mass spectrum of [1–19] obtained from partially deuterated fibrils was then deconvoluted into A and B peaks using the Pearson VII equation in the PeakFit (Systat) software. The Pearson VII equation was determined to fit the mass spectra of monomer and fibrils best in comparison to other available functions (e.g., Gaussian). For the deconvolution, the number of peaks was fixed at two, the position of peak B was fixed at the value determined from fully exchanged monomer, and the relative abundance of peaks A and B was fixed at the value measured for the 20–34 fragment (A:B ≈1:0.6, Fig. 4B). This ratio was selected in preference to the (clearly larger) ratio for the intact monomer (Fig. 4A), because proteolysis was observed to attenuate the [1–40] B peak (data not shown) preferentially, consistent with longer exposure to the processing solvent (possibly attributable to more rapid detachment of B-type monomers from the fibrils). Data from all fragments are part of a single spectrum and should reflect the same relative contributions from materials derived from [1–40] A and B peaks (the ratio of A and B obtained from the fragment [20–34] comprises more nearly an "internal standard"). Nonlinear regression analysis was run iteratively until convergence of the single variable constituting the position of the [1–19] A peak, giving a value of 466.2 for the data shown in Fig. 4C. The corrected number of backbone amide protons exchanging after 24 h of incubation in deuterated buffer was thus 10.2, based on the centroid of the calculated peak A; roughly 8 (of a possible 18) amide protons are protected in this portion of the molecule.

$[35\text{–}40]^{1+}$

The C-terminal peptide 35–40 showed the least amount of protection as illustrated by almost complete overlap of the mass spectra derived from partially deuterated fibrils and fully deuterated monomer (Fig. 4D). As discussed above, one would still expect there to be distinct contributions from the [1–40] A and B peaks. These components can be deconvoluted as described below, but the change in the centroid after subtracting out the contribution from the fully deuterated species is minimal. For $A\beta$ (1–40), this difference (number of deuteriums calculated with and without deconvolution) is within the error bar of the measurement ($\approx$0.2 Da).

The mass spectra of the singly charged 35–40 fragment are isotopically resolved, which makes fitting using Peakfit (as described above) difficult. Instead, molecular masses were calculated assuming that the centroid of the isotopic distribution is a simple linear combination of the centroids of A and B distributions, with the latter estimated from the data for the deuterated monomer:

$$\text{Centroid}_{\text{d-fibril}} = 0.4 * \text{Centroid}_{\text{d-monomer}} + 0.6 * \text{Centroid}_{\text{x}} \qquad (4)$$

Here, $\text{Centroid}_{\text{d-fibril}}$ and $\text{Centroid}_{\text{d-monomer}}$ are the experimental values from the partially deuterated fibril and the fully deuterated monomer (564.2 and 564.4, respectively, from Fig. 4D), and $\text{Centroid}_{\text{x}}$ is the desired value for the unresolved A peak. The weighting factors (0.4 and 0.6) are those derived from Fig. 4B as described above for the [20–34] fragment. Solving for $\text{Centroid}_{\text{X}}$ in the above equation yields the deconvoluted m/z of 564.1. The corrected number of deuteriums exchanged into this fragment after 24 h of H/D exchange time was 3.9 (of a possible 5).

These data for exchange into $A\beta$ fragments in the fibril state are internally consistent with previous analyses on the intact peptide. The sum of the number of amide protons exchanging in fragments [1–19], [20–34], and [35–40] is 18.4 $\pm$ 0.6, which, within error, is equivalent to the 17.1 $\pm$ 0.6 backbone amide protons determined to be exchanging with deuterium for the intact peptide under similar conditions. Exchange or protection of two amide protons (from the 19–20 and 34–35 cleavage sites) would not be detectable in the analysis of proteolytic fragments. If these were unprotected (and therefore exchanged) in the fibril, the revised total for exchanging amide protons in the fragments would be 20.4, which is significantly higher than the full-length value of 17.1. This suggests that these two "invisible" protons are protected in the fibril, consistent with most models of $A\beta$ fibril structure (Williams *et al.*, 2006). Assuming that protection is an all-or-none phenomenon, these data suggest that 8, 10, and 1 backbone amide protons are protected in fragments 1–19, 20–34, and 35–40, respectively.

Conclusions

We have developed and applied methods using H/D exchange-MS to map the hydrogen bonding pattern and extent of the β-sheet network in amyloid fibrils and protofibrils. These methods can be used to obtain information on particular aggregates as well as to compare structures of different aggregates. To localize the hydrogen-bonding interaction sites within the peptide, we have coupled H/D exchange-MS with on-line proteolysis using pepsin. The methods have been validated in that the results obtained with wt Aβ(1–40) quiescent amyloid fibrils agree very well with other structural approaches applied to identical fibrils (Whittemore et al., 2005; Williams et al., 2006). The on-line H/D exchange-MS method of analysis of aggregate structure is rapid and reproducible.

A major limitation of the pepsin method described here is the limited number of overlapping fragments obtained. To obtain single residue resolution, future development and applications of these methods will include use of other enzymes tolerant of low pH and/or the use of multienzyme digestion, which should allow better sequence coverage. Another potential approach is MS/MS, in which fragmentation occurs in the gas phase (Deng et al., 1999; Hoerner et al., 2004; Jorgensen et al., 2005; Kweon and Hakansson, 2006) and deuterium incorporation at a single residue level can be directly obtained. However, use of this approach is questionable until issues related to scrambling of amide hydrogen atoms during gas phase are definitively resolved.

Acknowledgments

The authors thank John R. Engen, S. Douglass Gilman, and Ravi Kodali for helpful comments on the manuscript, and Erica D'Spain for her assistance with the preparation of the figures. This work was supported by the National Institutes of Health, under grant R01AG18927 (R. Wetzel) and F32 AG05869 (I. Kheterpal). I. Kheterpal also acknowledges support by the National Science Foundation grant EPS-0346411 and the Louisiana Board of Reagents.

References

Anderegg, R. J. (1996). Using Deuterium Exchange and Mass Spectrometry to Study Protein Structure. In "Mass Spectrometry in the Biological Sciences" (R. J. Burlingame and A. L. Carr, eds.), pp. 85–104. Humana Press, Totowa, NJ.

Baello, B. I., Pancoska, P., and Keiderling, T. A. (2000). Enhanced prediction accuracy of protein secondary structure using hydrogen exchange Fourier transform infrared spectroscopy. Anal. Biochem. 280, 46–57.

Busenlehner, L. S., and Armstrong, R. N. (2005). Insights into enzyme structure and dynamics elucidated by amide H/D exchange mass spectrometry. Arch. Biochem. Biophys. 433, 34–46.

Carulla, N., Caddy, G. L., Hall, D. R., Zurdo, J., Gairi, M., Feliz, M., Giralt, E., Robinson, C. V., and Dobson, C. M. (2005). Molecular recycling within amyloid fibrils. *Nature* **436,** 554–558.

Chen, M., Kheterpal, I., Wetzel, R., and Cook, K. D. (2006). A triaxial probe for on-line proteolysis coupled with hydrogen exchange-electrospray mass spectrometry. Manuscript submitted.

Del Mar, C., Greenbaum, E. A., Mayne, L., Englander, S. W., and Woods, V. L. (2005). Structure and properties of alpha-synuclein and other amyloids determined at the amino acid level. *Proc. Natl. Acad. Sci. USA* **102,** 15477–15482.

Deng, Y., Pan, H., and Smith, D. L. (1999). Selective isotope labeling demonstrates that hydrogen exchange at individual peptide amide linkages can be determined by collision-induced dissociation mass spectrometry. *J. Am. Chem. Soc.* **121,** 1966–1967.

Ehring, H. (1999). Hydrogen exchange/electrospray ionization mass spectrometry studies of structural features of proteins and protein/protein interactions. *Anal. Biochem.* **267,** 252–259.

Engen, J. R., Bradbury, E. M., and Chen, X. (2002). Using stable-isotope-labeled proteins for hydrogen exchange studies in complex mixtures. *Anal. Chem.* **74,** 1680–1686.

Engen, J. R., and Smith, D. L. (2001). Investigating protein structure and dynamics by hydrogen exchange MS. *Anal. Chem.* **73,** 256A–265A.

Englander, S. W. (2000). Protein folding intermediates and pathways studied by hydrogen exchange. *Ann. Rev. Biophys. Biomol. Struct.* **29,** 213–238.

Englander, S. W., and Kallenbach, N. R. (1983). Hydrogen exchange and structural dynamics of proteins and nucleic acids. *Q. Rev. Biophys.* **16,** 521–655.

Hamuro, Y., Weber, P. C., and Griffin, P. R. (2005). High-throughput analysis of protein structure by hydrogen/deuterium exchange mass spectrometry. *Methods Biochem. Anal.* **45,** 131–157 (2 plates).

Hoerner, J. K., Xiao, H., Dobo, A., and Kaltashov, I. A. (2004). Is there hydrogen scrambling in the gas phase? Energetic and structural determinants of proton mobility within protein ions. *J. Am. Chem. Soc.* **126,** 7709–7717.

Hoofnagle, A. N., Resing, K. A., and Ahn, N. G. (2003). Protein analysis by hydrogen exchange mass spectrometry. *Ann. Rev. Biophys. Biomol. Struct.* **32,** 1–25.

Hoshino, M., Katou, H., Hagihara, Y., Hasegawa, K., Naiki, H., and Goto, Y. (2002). Mapping the core of the beta(2)-microglobulin amyloid fibril by H/D exchange. *Nat. Struct. Biol.* **9,** 332–336.

Hosia, W., Johansson, J., and Griffiths, W. J. (2002). Hydrogen/deuterium exchange and aggregation of a polyvaline and a polyleucine alpha-helix investigated by matrix-assisted laser desorption ionization mass spectrometry. *Mol. Cell. Proteomics* **1,** 592–597.

Ippel, J. H., Olofsson, A., Schleucher, J., Lundgren, E., and Wijmenga, S. S. (2002). Probing solvent accessibility of amyloid fibrils by solution NMR spectroscopy. *Proc. Natl. Acad. Sci. USA* **99,** 8648–8653.

Jorgensen, T. J., Gardsvoll, H., Ploug, M., and Roepstorff, P. (2005). Intramolecular migration of amide hydrogens in protonated peptides upon collisional activation. *J. Am. Chem. Soc.* **127,** 2785–2793.

Kheterpal, I., Chen, M., Cook, K. D., and Wetzel, R. (2006). Structural differences in Aβ amyloid protofibrils and fibrils mapped by hydrogen exchange-mass spectrometry with on-line proteolytic fragmentation. *J. Mol. Biol.,* in press.

Kheterpal, I., Lashuel, H. A., Hartley, D. M., Walz, T., Lansbury, P. T., Jr., and Wetzel, R. (2003a). Aβ protofibrils possess a stable core structure resistant to hydrogen exchange. *Biochemistry* **42,** 14092–14098.

Kheterpal, I., Wetzel, R., and Cook, K. D. (2003b). Enhanced correction methods for hydrogen exchange-mass spectrometric studies of amyloid fibrils. *Protein Sci.* **12,** 635–643.

Kheterpal, I., Zhou, S., Cook, K. D., and Wetzel, R. (2000). Aβ amyloid fibrils possess a core structure highly resistant to hydrogen exchange. *Proc. Natl. Acad. Sci. USA* **97,** 13597–13601.

Komives, E. A. (2005). Protein-protein interaction dynamics by amide H/2H exchange mass spectrometry. *Int. J. Mass Spectrom.* **240,** 285–290.

Kraus, M., Bienert, M., and Krause, E. (2003). Hydrogen exchange studies on Alzheimer's amyloid-β peptides by mass spectrometry using matrix-assisted laser desorption/ionization and electrospray ionization. *Rapid Commun. Mass Spectrom.* **17,** 222–228.

Kuwata, K., Matumoto, T., Cheng, H., Nagayama, K., James, T. L., and Roder, H. (2003). NMR-detected hydrogen exchange and molecular dynamics simulations provide structural insight into fibril formation of prion protein fragment 106–126. *Proc. Natl. Acad. Sci. USA* **100,** 14790–14795.

Kweon, H. K., and Hakansson, K. (2006). Site-specific amide hydrogen exchange in melittin probed by electron capture dissociation Fourier transform ion cyclotron resonance mass spectrometry. *Analyst* **131,** 275–280.

Lashuel, H. A., and Grillo-Bosch, D. (2005). *In vitro* preparation of prefibrillar intermediates of amyloid-beta and alpha-synuclein. *Methods Mol. Biol.* **299,** 19–33.

Lashuel, H. A., Hartley, D. M., Petre, B. M., Wall, J. S., Simon, M. N., Walz, T., and Lansbury, P. T. (2003). Mixtures of wild-type and a pathogenic (E22G) form of Aβ 40 *in vitro* accumulate protofibrils, including amyloid pores. *J. Mol. Biol.* **332,** 795–808.

LeVine, H., III. (1999). Quantification of beta-sheet amyloid fibril structures with thioflavin T. *Methods Enzymol.* **309,** 274–284.

Luhrs, T., Ritter, C., Adrian, M., Riek-Loher, D., Bohrmann, B., Doeli, H., Schubert, D., and Riek, R. (2005). 3D structure of Alzheimer's amyloid-beta(1-42) fibrils. *Proc. Natl. Acad. Sci. USA* **102,** 17342–17347.

Nazabal, A., Dos Reis, S., Bonneu, M., Saupe Sven, J., and Schmitter, J.-M. (2003). Conformational transition occurring upon amyloid aggregation of the HET-s prion protein of Podospora anserina analyzed by hydrogen/deuterium exchange and mass spectrometry. *Biochemistry* **42,** 8852–8861.

Nazabal, A., Maddelein, M. L., Bonneu, M., Saupe, S. J., and Schmitter, J. M. (2005). Probing the structure of the infectious amyloid form of the prion-forming domain of HET-s using high resolution hydrogen/deuterium exchange monitored by mass spectrometry. *J. Biol. Chem.* **280,** 13220–13228.

Nettleton, E. J., and Robinson, C. V. (1999). Probing conformations of amyloidogenic proteins by hydrogen exchange and mass spectrometry. *Methods Enzymol.* **309,** 633–646.

O'Nuallain, B., Shivaprasad, S., Kheterpal, I., and Wetzel, R. (2005). Thermodynamics of A beta(1-40) amyloid fibril elongation. *Biochemistry* **44,** 12709–12718.

Olofsson, A., Ippel, J. H., Wijmenga, S. S., Lundgren, E., and Ohman, A. (2004). Probing solvent accessibility of transthyretin amyloid by solution NMR spectroscopy. *J. Biol. Chem.* **279,** 5699–5707.

Olofsson, A., Sauer-Eriksson, A. E., and Ohman, A. (2006). The solvent protection of Alzheimer amyloid-beta-(1-42) fibrils as determined by solution NMR spectroscopy. *J. Biol. Chem.* **281,** 477–483.

Petkova, A. T., Leapman, R. D., Guo, Z. H., Yau, W. M., Mattson, M. P., and Tycko, R. (2005). Self-propagating, molecular-level polymorphism in Alzheimer's beta-amyloid fibrils. *Science* **307,** 262–265.

Raschke, T. M., and Marqusee, S. (1998). Hydrogen exchange studies of protein structure. *Curr. Opin. Biotechnol.* **9**, 80–86.

Ritter, C., Maddelein, M.-L., Siemer Ansgar, B., Luhrs, T., Ernst, M., Meier Beat, H., SaupeSven, J., and Riek, R. (2005). Correlation of structural elements and infectivity of the HET-s prion. *Nature* **435**, 844–848.

Smith, D. L., Deng, Y., and Zhang, Z. (1997). Probing the non-covalent structure of proteins by amide hydrogen exchange and mass spectrometry. *J. Mass Spectrom.* **32**, 135–146.

Teilum, K., Kragelund, B. B., and Poulsen, F. M. (2005). Application of hydrogen exchange kinetics to studies of protein folding. *Protein Folding Handbook* **2**, 634–672.

Vigano, C., Smeyers, M., Raussens, V., Scheirlinckx, F., Ruysschaert, J. M., and Goormaghtigh, E. (2004). Hydrogen-deuterium exchange in membrane proteins monitored by IR spectroscopy: A new tool to resolve protein structure and dynamics. *Biopolymers* **74**, 19–26.

Wales, T. E., and Engen, J. R. (2006). Hydrogen exchange mass spectrometry for the analysis of protein dynamics. *Mass Spectrom. Rev.* **25**, 158–170.

Walsh, D. M., Lomakin, A., Benedek, G. B., Condron, M. M., and Teplow, D. B. (1997). Amyloid beta-protein fibrillogenesis. Detection of a protofibrillar intermediate. *J. Biol. Chem.* **272**, 22364–22372.

Wang, L., Pan, H., and Smith, D. L. (2002). Hydrogen exchange-mass spectrometry. Optimization of digestion conditions. *Mol. Cell. Proteomics* **1**, 132–138.

Wang, S. S. S., Tobler, S. A., Good, T. A., and Fernandez, E. J. (2003). Hydrogen exchange-mass spectrometry analysis of β-amyloid peptide structure. *Biochemistry* **42**, 9507–9514.

Whittemore, N. A., Mishra, R., Kheterpal, I., Williams, A. D., Wetzel, R., and Serpersu, E. H. (2005). Hydrogen-deuterium (H/D) exchange mapping of Aβ (1-40) amyloid fibril secondary structure using nuclear magnetic resonance spectroscopy. *Biochemistry* **44**, 4434–4441.

Williams, A. D., Portelius, E., Kheterpal, I., Guo, J.-T., Cook, K. D., Xu, Y., and Wetzel, R. (2004). Mapping Aβ amyloid fibril secondary structure using scanning proline mutagenesis. *J. Mol. Biol.* **335**, 833–842.

Williams, A. D., Sega, M., Chen, M., Kheterpal, I., Geva, M., Berthelier, V., Kaleta, D. T., Cook, K. D., and Wetzel, R. (2005). Structural properties of Aβ protofibrils stabilized by a small molecule. *Proc. Natl. Acad. Sci. USA* **102**, 7115–7120.

Williams, A. D., Shivaprasad, S., and Wetzel, R. (2006). Alanine scanning mutagenesis of Aβ (1-40) amyloid fibril stability. *J. Mol. Biol.* **357**, 1283–1294.

Woodward, C. (1993). Is the slow-exchange core the protein folding core? *Trends Biochem. Sci.* **18**, 359–360.

Wuthrich, K., and Wagner, G. (1979). Nuclear magnetic resonance of labile protons in the basic pancreatic trypsin inhibitor. *J. Mol. Biol.* **130**, 1–18.

Yamaguchi, K. I., Katou, H., Hoshino, M., Hasegawa, K., Naiki, H., and Goto, Y. (2004). Core and heterogeneity of beta(2)-microglobulin amyloid fibrils as revealed by H/D exchange. *J. Mol. Biol.* **338**, 559–571.

Zhang, Z., and Smith, D. L. (1993). Determination of amide hydrogen exchange by mass spectrometry: A new tool for protein structure elucidation. *Protein Sci.* **2**, 522–531.

[9] Hydrogen-Deuterium Exchange Analyzed by Matrix-Assisted Laser Desorption-Ionization Mass Spectrometry and the HET-s Prion Model

By Alexis Nazabal and Jean-Marie Schmitter

Abstract

Hydrogen/deuterium (H/D) exchange analyzed by mass spectrometry (HXMS) is a valuable tool for the investigation of protein conformation and dynamics. After exchange, the sample is generally submitted to electrospray ionization for mass analysis. Matrix-assisted laser desorption ionization (MALDI) has been used in a limited number of studies but has several significant advantages that include simplification of the spectra attributable to a predominance of singly charged ions, speed of analysis, sensitivity, and low H/D back-exchange level. MALDI-HXMS has been used to study amyloid aggregates from the HET-s prion protein. Our results underline the ability of this method to determine solvent accessibility within the amyloid aggregates, reaching a resolution of one to four amino acids. To achieve a complete peptide mass fingerprint of the protein, we have taken benefits of an ion trap operating in liquid chromatography-MS/MS mode. MALDI time-of-flight-MS was then used to determine deuterium incorporation within each peptide along the sequence of HET-s. The combined advantages of these two instruments yield a suitable solution for HXMS experiments that require highly resolved peptide mass fingerprints, high sensitivity, and speed of analysis for deuterium incorporation measurements.

Introduction

The exchange of protein amide protons with deuterons that occurs in deuterated solvents depends on solvent accessibility, hydrogen bonding, and spatial arrangement of amino acids into structure elements like helices and β-sheets. Thus, the solvent accessibility of a protein under different states, such as a soluble form and an aggregated form, may be monitored by hydrogen/deuterium (H/D) exchange analyzed by mass spectrometry (HXMS). This information can be refined when enzymatic cleavage is performed in order to reveal the distribution of deuterons within the sequence of a protein having a known primary structure. However, reliable information can only be obtained when proteolysis occurs under conditions of quenched

METHODS IN ENZYMOLOGY, VOL. 413   0076-6879/06 $35.00
DOI: 10.1016/S0076-6879(06)13009-8

exchange (i.e., at low temperature and pH). Indeed, at $0°$ and pH 2.5, the H/D exchange rate is slower by about five to six orders of magnitude than at room temperature and pH 7 (Bai *et al.*, 1993). In these conditions, the protease of choice is most often pepsin, in spite of its broad specificity.

Regarding MS, HXMS studies are often conducted in the electrospray ionization (ESI) mode. In that case, peptide mapping is achieved by means of on-line coupling with a liquid chromatograph providing the required separation of peptides. Because both sample preparation and separation steps must be conducted under conditions of quenched exchange, experimental constraints on the chromatographic setup are strong. Further, corrections for H/D back-exchange must be carefully applied.

Alternatively, matrix-assisted laser desorption ionization (MALDI) can be used instead of ESI for HXMS analysis, as demonstrated for the first time by Mandell *et al.* (1998). In MALDI mode, the sample is co-crystallized with an organic matrix and deposited on a metallic target plate. The matrix concentration (about 10 mM in aqueous acid solution) is usually 10,000 times higher than the sample concentration, and the sample-matrix mixing process leads to an efficient quench of the H/D exchange reaction, provided that the solvent evaporation is fast enough. Further, specific characteristics of MALDI, such as speed of spectral acquisition and sensitivity, render this method well suited for HXMS analysis. In the case of MALDI-HXMS, we always observed in- and back-exchange levels in the range of 12–16%. This is similar to the values that we and other authors have observed in the case of ESI-HXMS (Resing and Ahn, 1998).

One of the major differences between MALDI-HXMS and ESI-HXMS is the nature of the ions generated by these ionization methods. After MALDI ionization, most of the ions observed for peptides produced by proteolytic cleavage are singly charged, which strongly reduces the number of peaks observed on the spectra. In ESI mode, multiply charged ions are generated, leading to a higher complexity of mass spectra observed for the same peptides. Most of the time, a chromatographic separation is required before the ionization step to reduce overlaps of peaks of multiply charged ions. This overlapping phenomenon of isotopic clusters increases after H/D exchange and can lead to confusion in peak assignments. The use of MALDI ionization reduces the overlapping phenomenon, and thus greatly improves the quality of information obtained by HXMS experiments.

Limitations of MALDI-HXMS are mostly related to selective desorption-ionization effects encountered for complex peptide mixtures. Although these selective desorption-ionization effects depend on several factors, a general trend is the dominance of arginine-containing peptides in MALDI mass spectra. As a result of this phenomenon, complete sequence coverage of a protein of interest might be difficult to obtain by this methodology alone. Microfractionation by step-gradient elution of peptides from pipette tips packed with reversed-phase chromatographic support is an easy way to overcome this problem (Belghazi *et al.*, 2001).

The nature of interactions leading to the stable structures of amyloid aggregates is investigated by biochemical and biophysical methods (e.g., Chapter 5 by Ban and Goto, Chapter 8 by Kheterpal and colleagues, Chapter 7 by Margittai and Langen, Chapter 6 by Tycko, and Chapter 10 by Shivaprasad and Wetzel in this volume). Recent contributions to the structural study of amyloids have been made by biochemical methods like H/D exchange (Kheterpal *et al.*, 2000, 2003; Kraus *et al.*, 2003; Hoshino *et al.*, 2002) or limited proteolysis (Kheterpal *et al.*, 2001). Few studies have been aimed at the characterization of interactions at the molecular level in amyloid aggregates constituted by prion proteins (Nazabal *et al.*, 2003b, 2005a; Ritter *et al.*, 2005). The difficulties of such investigations are related to the heterogeneity of prion aggregate samples and the high molecular weight of the monomers. Furthermore, the infectious properties of aggregated mammalian prion proteins require secure experimental conditions. To get rid of the infectivity issue when investigating the structure of prion amyloids, we have taken benefit of the HET-s prion protein model.

The HET-s protein (289 amino acids) of the filamentous fungus *Podospora anserina* is a prion protein involved in a genetically programmed cell death reaction termed *heterokaryon incompatibility*. This cell death reaction is triggered when the prion form of HET-s interacts with a natural variant of HET-s called HET-S. The latter is devoid of the prion behavior and differs from HET-s by only 13 residues (Coustou *et al.*, 1997). HET-s aggregates specifically *in vivo* upon transition to the prion state (Coustou-Linares *et al.*, 2001). This fungal prion protein is a valuable model for the exploration of the mechanism of prion propagation, especially because infectious material can be generated *in vitro* from a recombinant protein. A recombinant full-length HET-s protein forms amyloid aggregates *in vitro*. Such fibers obtained *in vitro* are infectious, indicating that the HET-s prion can propagate as a self-perpetuating amyloid aggregate (Dos Reis *et al.*, 2002). The introduction of amyloid aggregates of recombinant HET-s into *P. anserina* cells induces the [Het-s] prion with a very high efficiency (Maddelein *et al.*, 2002). Thus, amyloids generated *in vitro* represent infectious material that can propagate the [Het-s] prion. The structure of the HET-s protein displays two distinct domains: a globular region starting from the N-terminus and ending approximately at residue 230 and a C-terminal domain that was first proposed to be poorly structured (residues 230 to 289) (Balguerie *et al.*, 2003). In fibrils, the C-terminal domain of HET-s forms a protease-resistant amyloid core, whereas the N-terminal globular domain remains accessible to proteolysis.

A C-terminal domain (residues 218–295) of HET-s was shown to be sufficient for [Het-s] propagation *in vivo* and amyloid formation *in vitro* (Balguerie *et al.*, 2003). Further, a biolistic approach was used to demonstrate that the amyloid form of the recombinant HET-s [218–295] peptide can induce the [Het-s] prion when introduced *in vivo* into *P. anserina* cells (Nazabal *et al.*, 2005a); this recombinant forms bears a six-histidine extension at its C-terminus. In this context, MALDI-HXMS has been used to probe the structure of soluble and amyloid forms of HET-s.

Materials and Methods

General

Deuterium oxide (99%) was from Eurisotop (Gif-sur-Yvette, France). Peptides used for external calibration of the mass spectrometer were from Sigma (St. Quentin Fallavier, France). Sequence grade trifluoroacetic acid (TFA) and pepsin immobilized on agarose were from Pierce (Rockford, IL). C18 ZipTips were from Millipore, and gradient high-performance liquid chromatography (HPLC) grade acetonitrile was from Mallinckrodt Baker (Deventer, The Netherlands).

Water was purified over a MilliQ apparatus (Millipore, Bedford, MA) and is hereafter referred to as MQ grade.

HET-s and HET-s[218-295] Purification and Amyloid Formation

Full-length HET-s was expressed from BL21 (DE3)pLyss cells and purified as previously described (Nazabal *et al.*, 2003a) The histidine-tagged HET-s[218-295] peptide was expressed and purified from inclusion bodies under denaturing conditions (Balguerie *et al.*, 2003). Protein concentration was adjusted to 125 μM. To initiate the spontaneous amyloid aggregation of HET-s and HET-s[218-295], pH was brought to 8 by addition of Tris base. Amyloid formation was monitored by thioflavine T binding and by electron microscopy as previously described (Balguerie *et al.*, 2003). After amyloid aggregation, aggregates were recovered by centrifugation (20 min at 12,000*g*) and suspended in H_2O.

H/D Exchange, Pepsin Digestion, and Sample Preparation for MALDI-HXMS

Aggregated HET-s (1 ml, 30 μM, H_2O MQ) was centrifuged for 20 min at 10,000*g*. The pellet was resuspended with 1 ml H_2O MQ. A 5-μl aliquot (30 μM) of the aggregated solution was diluted 1:20 in D_2O and vortexed before incubation at pH 7, 25° for variable times (between 5 and 120 min).

After incubation, 5 μl of the labeled solution was treated under strong agitation in 4 M urea with immobilized pepsin (10 μl, 45 units, washed three times with 0.1% aqueous TFA before use) at pH 2 and 0° (exchange quenching conditions) for 5 min. Digestion was followed by a short centrifugation (10,000g, 30 s), and a 5-μl aliquot of the supernatant containing the peptic digest was submitted to C18 ZipTip purification by means of a gradient in three steps. The three fractions were eluted with 1 μl of a solution of 10 mg/ml α-cyano-4-hydroxycinnamic acid in 2:8:1, 4:6:1, 7:3:1 acetonitrile/ethanol/0.1% aqueous TFA, respectively, and were directly spotted on the target under nitrogen flow. After drying, the target was transferred as quickly as possible (less than 1 min) to the mass spectrometer. The time from sample spotting to data collection was kept constant for each sample analysis.

Controls: In- and Back-Exchange

For back-exchange control, pepsin digests of HET-s were diluted 1:19 in D_2O to a volume of 100 μl and incubated for 12 h at 25°, pH 7 to achieve complete exchange of backbone amide protons for deuterium atoms. After incubation, the exchange was quenched at 0° by addition of 1 μl of a solution of 10% TFA (Pierce). After quenching, the peptide mixture was loaded on a C18 ZipTip (Millipore). The peptide fraction eluted from the ZipTip with a solution containing 6:4:1 acetonitrile/H_2O MQ with 0.1% TFA kept at 0° was submitted to mass analysis. The relative back-exchange (%) was determined for each peptide by comparing the maximal number of exchangeable amide hydrogen atoms and the corresponding experimental value.

For in-exchange control, the aggregated solution of HET-s protein (90 μM) was diluted 19:1 in D_2O, 0.1% TFA in-exchange quenching conditions (pH 2.2, 0°), and a 5-μl aliquot of this solution was treated with pepsin (10 μl, 45 units) for 5 min (pH 2.2, 0°). The peptide mixture was loaded on a C18 ZipTip and eluted with a solution containing 6:4:1 acetonitrile/D_2O with 0.1% TFA kept at 0°. The incorporation of deuterium for this in-exchange control was determined in the same way as for the back-exchange control.

Data Analysis

The spectra were baseline corrected, and the number of deuterium atoms incorporated in a given peptide at time t, D(t), was determined from the centroid value of its isotopic peak cluster using the formula given in Eq. (1) (Liu and Smith, 1994):

$$D(t) = \frac{m(t) - m(0)}{m(100) - m(0)} \times N \tag{1}$$

with m(t) being the observed centroid mass of the peptide at incubation time point t, m(0) being the observed mass at time point 0 (for unlabeled peptides, see in-exchange control procedure), m(100) being the observed mass for a fully exchanged peptide (see back-exchange control procedure), and N being the total number of exchangeable amide protons in the peptide.

The percentage of deuterium incorporation was calculated for each peptide of the mass fingerprint according to Eq. (2):

$$\% \text{ of deuterium incorporation} = D(t)/N \tag{2}$$

Mass Spectrometry

MALDI mass spectra were acquired on a Bruker Reflex III mass spectrometer equipped with a nitrogen laser with an emission wavelength of 337 nm. Spectra were obtained by accumulating an average of 100 shots in the positive ion mode while the laser spot was manually scanned over a surface area of about 0.2 mm^2. Deflection of the low mass ions was used to enhance the target protein signal. Mass spectra of HET-s digests were acquired in the reflectron mode, using an external calibration with a mixture of eight peptides covering a 900–3500-Da mass range. In this case, measured masses were monoisotopic $[M+H]^+$. For postsource decay analysis, the reflectron voltage was stepped down in 10–12 steps to record all fragment ions, including immonium ions.

Liquid chromatography (LC)-MS/MS analysis of HET-s peptides was performed by means of an ion trap mass spectrometer operated in the electrospray mode (LCQ DecaXP; Thermo Finnigan) and interfaced to a Dionex-LC Packings chromatographic system (C18 column, 75-μm internal diameter, 150 mm long). After analysis, the TurboSequest program was used to assign peptide sequences from their fragment ions with choosing "none" for the reference enzyme used in the algorithm.

Notes

After a long incubation time (e.g., overnight) of a full-length protein in a deuterated solvent, its analysis by MALDI-MS may reveal a number of deuterium atoms incorporated within the protein superior to the total number of amino acids minus prolines. This can only be explained by the detection of deuterium incorporated in side chains, although amide hydrogen atoms belonging to side chains exchange at a fast rate that is incompatible with mass spectrometric detection. The detection of such H/D exchange can be attributable to a protection effect related to the protein structure that is capable of strongly reducing the back-exchange rate of side chain amide hydrogen atoms. To avoid any confusion between side chain and backbone amides for the calculation of deuterium incorporation, before each mass analysis, we performed a fast microchromatographic purification step with protiated solvents that eliminates the H/D exchange on side chains (Bai *et al.*, 1993). This microchromatography (only a 5-μl sample volume is needed) on ZipTips allows a fast (about 45 s) purification step, eliminates urea when required, and can be conducted in a step-gradient mode in order to simplify the peptide mixtures.

Sample Preparation for MALDI-HXMS

The major constraint for this step is the need for a fast preparation to minimize H/D back-exchange. Mandell *et al.* (1998) proposed to keep the matrix solution as well as the MALDI target plate refrigerated. However, refrigerating the sample plate was shown to have adverse effects on both analysis time and H/D back-exchange, because of atmospheric water condensation. Thus, our recommendation is to keep the MALDI target plate at room temperature and to refrigerate only the matrix solution. High-speed crystallization is favored by the use of low sample volumes (less than 0.2 μl) and a volatile solvent, such as ethanol or ethanol-acetonitrile mixture, for the matrix solution. Further, solvent evaporation is accelerated by means of a flow of nitrogen. In this way, short target preparation times are achieved; that is, a target may be loaded in the mass spectrometer in less than 2 min.

In- and Back-Exchange

In- and back-exchange controls allow one to adjust the measurement of deuterium incorporation by taking into account the loss and the gain of deuterium attributable to sample preparation. After quenching the exchange reaction by lowering both the pH and temperature, 15–50% of the deuterium back-exchanges to the environment. In-exchange can occur even under quenched conditions and has also to be evaluated.

In the case of our ZipTip fractionation, samples were mixed with the elution solvent (containing the matrix) for 5 s before spotting on the MALDI target plate. Time between mixing and crystallization on the plate was about 15 s. Under these conditions, deuterium incorporation of peptides did not vary with the solvent composition used for peptide fractionation on ZipTips. Thus, in the case of our control experiments, a single solvent composition was used for the elution of peptide mixtures.

We have used in- and back-exchange controls on the peptides utilized for the study and not directly on HET-s amyloid aggregates. Indeed, in our experimental conditions, we did not observe full exchange of the aggregates even after an incubation time of 1 week in deuterium oxide. If the sample is not fully exchanged, the resulting correction for back-exchange will mask structural information from the investigated sample. Under our experimental conditions, the resulting in- and back-exchanges were in the range of 12–16%. Once loaded in the mass spectrometer, on-target H/D back-exchange can be neglected because of the short spectral acquisition time in the MALDI mode. Indeed, for a peptide having undergone full exchange of amide protons, several hours are needed to observe 50% H/D back-exchange in the mass spectrometer source.

Peptide Mapping

Making peptide mapping as exhaustive as possible is a major concern in the case of HXMS investigations. MALDI, in association with a time-of-flight (ToF) analyzer, provides highly accurate mass measurements. To be reliably used for MALDI-HXMS, a set of monoisotopic masses unambiguously assigned to unique sequences must be selected. Furthermore, the corresponding peaks must be sufficiently resolved from adjacent ones to avoid an overlap of isotopic clusters after H/D exchange. However, because of the lack of specificity of proteases that can be used under conditions of quenched H/D exchange, and more particularly pepsin, this accuracy is usually insufficient for an unambiguous assignment of a given mass of peptide to a sequence stretch. This fact becomes critical in the case of large proteins studied by HXMS. Indeed, selective desorption-ionization effects occurring in a MALDI ion source lead to a strong bias in favor of arginine-containing peptides, with possible spectral suppression effects of other peptides. To improve sequence coverage of proteins studied by HXMS, fungal proteases, such as protease XIII, that are also active at low pH represent an interesting alternative to pepsin and may help to increase sequence coverage of proteins studied by MALDI-HXMS (Cravello *et al.*, 2003; Englander *et al.*, 2003). Another useful way to improve sequence coverage consists of fractionation by reversed-phase LC of peptide mixtures obtained after proteolytic cleavage, using a step-gradient mode. With sample loads in the subpicomole range, this fractionation step was miniaturized by use of chromatographic supports packed into pipette tips (ZipTips with 0.6 μl C18 grafted silica; Belghazi *et al.*, 2001; Nazabal *et al.*, 2003a). In this way, back-exchange was minimal because of the short analysis time, small amount of stationary phase, and small volume of eluent, and a gradient elution in three steps led to a sequence coverage superior to 90% in the case of HET-s (Nazabal *et al.*, 2003b). However, in this latter case, it must be stated that safe sequence assignments were obtained from spectral data that combined accurate mass measurement of peptide ions with sequence information extracted from postsource decay analysis (MALDI) and MS/MS fragmentation (ESI-LC/MS coupling). Such a process may be complicated by selective ionization of peptic peptides in both ESI and MALDI modes. Thus, for MALDI-HXMS studies aimed at large proteins, it is recommended to make use of the capability of tandem instruments (MALDI-ToF-ToF or MALDI-Q-ToF) to deliver both exact mass measurements and sequence tags in order to simplify this step of the HXMS process.

It is worth mentioning that when HXMS is applied to amyloids, maps containing the same peptides for both soluble and aggregated forms of the protein of interest must be obtained. Because immobilized pepsin was used at a high enzyme/substrate ratio for proteolytic cleavage, we could add urea in the digestion buffer of the amyloid form of HET-s to a final concentration as high as 4 M, and were thus able to produce a very similar digestion pattern for both forms of the protein, as shown in Fig. 1.

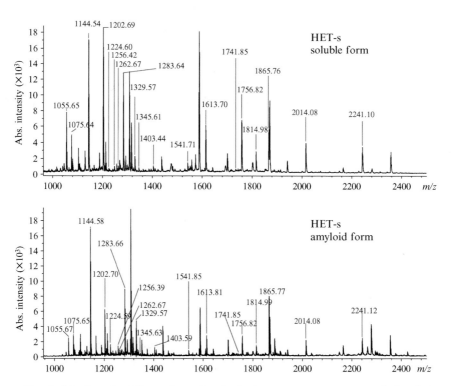

FIG. 1. Matrix-assisted laser desorption ionization-mass spectrometry analysis of the peptide mixture obtained after pepsin digestion of HET-s (no hydrogen/deuterium exchange) in soluble and amyloid forms, where the latter was digested in the presence of 4 M urea. In both cases, the protein digest was submitted to microchromatography over a C18 ZipTip; these spectra correspond to the fraction eluted with 2:8:1 acetonitrile/ethanol/0.1% aqueous trifluoroacetic acid containing 10 mg/ml α-cyano-4-hydroxycinnamic acid.

Results

Analysis by MALDI-HXMS of the full-length HET-s protein in its soluble form revealed that the C-terminal domain spanning residues 218–289 displayed a high solvent accessibility. This is illustrated by examples of the time course analysis of H/D exchange in Fig. 2. Peptides belonging to the N-terminal domain exhibit similar deuterium incorporation levels in both soluble and amyloid forms, whereas peptides belonging to the C-terminal domain show a strikingly different behavior. Indeed, solvent accessibility was shown to be drastically reduced in that region for the aggregated form, indicating that amyloid aggregation of HET-s involves a conformational transition of this region of the protein (Nazabal *et al.*, 2003b).

The [218–289] C-terminal domain tagged with a six-histidine tail was expressed and purified, and its ability to aggregate was tested. This HET-s [218–295] form was able to form amyloid fibers *in vitro*, and a biolistic assay

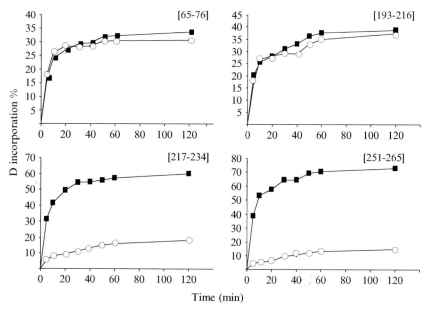

FIG. 2. Time course of hydrogen/deuterium exchange for HET-s peptic peptides from the soluble (squares) and aggregated form (circles). In both cases, HET-s was incubated in D_2O at 25° for various times up to 120 min and digested with immobilized pepsin (pH 2, 0°, 10 min), and the resulting peptide mixture was separated on a ZipTip and analyzed by matrix-assisted laser desorption ionization-mass spectrometry. The percentage (%) of deuterium incorporation was calculated by means of Eq. (2) (see section on materials and methods).

indicated the infectious character of this peptide (Nazabal *et al.*, 2005a). This peptide, which was first thought to be poorly structured, was studied by HXMS using sequence overlaps to achieve a good resolution. To achieve safe sequence assignments from the peptide mass fingerprint of the protein, we have taken benefit of an ion trap operating in the LC-MS/MS mode. MALDI-ToF-MS was then used to determine deuterium incorporation within each peptide along the sequence of HET-s[218–295]. Sharp differences in H/D exchange were revealed in several positions in the sequence (Fig. 3) (Nazabal *et al.*, 2005b). Such sharp differences were also noticed by authors who have monitored H/D exchange of HET-s[218–289] by means of nuclear magnetic resonance (NMR), and thus obtained a single residue resolution (Ritter *et al.*, 2005). Some differences in the designation of residues that appear to be protected from H/D exchange exist between these two experiments and are most probably attributable to the difference in spatial resolution obtained with these different approaches. Nevertheless, both experiments clearly designate the same four segments of the HET-s[218–295] prion as highly protected from H/D exchange. The high-resolution HXMS study of HET-s[218–295] in its amyloid form

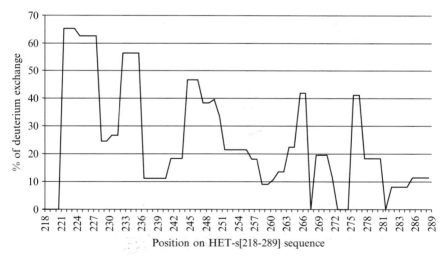

Fig. 3. Percentage (%) of deuterium incorporation along the sequence of HET-s[218–289]. Aggregated HET-s[218-289] was incubated for 7 h in deuterium oxide (pH 7, 25°). After quenching the reaction, the protein was digested using immobilized pepsin. Peptic peptides were then analyzed by matrix-assisted laser desorption ionization time-of-flight mass spectrometry. The % of deuterium incorporation is calculated comparing the number of deuterium atoms incorporated in a peptide with the number of hydrogen atoms exchangeable for this peptide.

might be useful in comparative studies of conformational variants, such as those underlying strain differences in yeast prions.

Future Perspective: Improving the Resolution of HXMS Experiments

The spatial resolution of deuterium incorporation detected by MS after hydrogen exchange experiments depends on the length of the peptides generated after pepsin proteolysis. Typically, proteolysis parameters are optimized in order to generate 6–20 amino acid-length fragments. Smaller fragments (i.e., less than 6 amino acids) cannot be easily utilized in MALDI-HXMS because of uncertainties in sequence assignments, frequent peak overlaps after exchange, and the predominance of cluster ions arising from the matrix in this spectral domain.

A way to improve the resolution of HXMS experiments has been proposed by Zhang and Smith (1993) and consists of considering the exchange of an overlapping region of 2 peptic peptides as an average of the deuterium incorporation in each peptide. This use of peptides with shared boundaries requires a large number of peptides but possibly allows reaching a single residue resolution. For instance, by taking into account the overlaps of a set of 61 peptic peptides generated after proteolysis of aggregated HET-s(218–295), a resolution of one to four amino acids has been obtained (Nazabal et al., 2005b).

The use of sets of peptides sharing boundaries considerably improves the resolution that can be obtained by standard MALDI-HXMS, but this procedure requires a careful, and thus time-consuming, assignment of peptide ions to their corresponding sequences. An interesting alternative to be considered in order to reach a single amino acid resolution of H/D exchange is the use of collision-induced dissociation (CID) of deuterium-labeled peptides. From a theoretical point of view, fragment ions could allow a direct calculation of deuterium incorporation into a precursor peptide, and the analysis of b and y ions series could yield a resolution of H/D exchange at the single amino acid level. Nevertheless, the applicability of this approach requires the absence of intramolecular migration (scrambling) of amide hydrogen atoms in the gas phase prior to dissociation. Controversial results have been reported for this matter: some authors have reported a low H/D scrambling level for b-fragment ions but not for y-fragment ions (Deng et al., 1999; Kim et al., 2001), whereas others found that neither b- nor y-fragment ions were affected by scrambling (Akashi et al., 1999; Hoerner et al., 2004). In sharp contrast, some authors have reported an extensive amide hydrogen scrambling effect for both b- and y-fragment ions (Demmers et al., 2002; Johnson et al., 1995). In a study conducted on a set of 10 peptic peptides from HET-s that generated

146 fragments (77 b and 69 y) under CID in an ion trap mass spectrometer, we have found that an extensive scrambling effect affected both y- and b-fragment ion types (Nazabal et al., 2005b). This result was clearly confirmed by an elegant study conducted with peptides forming a complex with a receptor, using a hybrid, quadrupole ToF mass spectrometer (Jørgensen et al., 2005a). Thus, with low-energy collisional activation, the spatial resolution of HXMS cannot be improved by means of CID spectra, but higher collisional energies used in a ToF-ToF instrument with MALDI ionization do not improve the situation (Jørgensen et al., 2005b). As proposed by Jørgensen et al. (2005a), ion cyclotron resonance with electron capture dissociation could perhaps better maintain the original H/D exchange pattern of a labeled peptide.

Conclusion

MALDI-HXMS provides direct insight into the structural rearrangements occurring upon amyloid aggregation of the HET-s prion protein. We have shown that modifications in solvent accessibility concern mainly the C-terminal region of the aggregated protein. A detailed study of deuterium incorporation in the C-terminal region of the aggregated protein allows distinguishing differences in solvent accessibility along the sequence with a resolution ranging from one to four amino acids. We have shown the advantage of combining both ion trap LC-MS/MS for assignment of a large number of peptides to obtain a complete mass fingerprint and MALDI-ToF-MS for efficient analysis of deuterated peptides. This method will be applied to the structural study of mammalian prion protein aggregates to evaluate the structural differences between prion strains. This study will open new technical challenges because of the infectivity and the heterogeneity of the mammalian prion protein aggregates.

Acknowledgments

The authors gratefully thank Sven Saupe (IBGC, CNRS-University Bordeaux 2) for several years of fruitful collaboration and Marc Bonneu (Pôle protéomique, Université Bordeaux 2) for his help with nano-ESI-LC/MS experiments. The contribution of the Conseil Régional d'Aquitaine (grant for A. Nazabal and financial support for the equipment) is gratefully acknowledged.

References

Akashi, S., Naito, Y., and Takio, K. (1999). Observation of hydrogen-deuterium exchange of ubiquitin by direct analysis of electrospray capillary-skimmer dissociation with Fourier transform ion cyclotron resonance mass spectrometry. Anal. Chem. 71, 4974–4980.

Bai, Y., Milne, J. S., Mayne, L., and Englander, S. W. (1993). Primary structure effects on peptide group hydrogen exchange. *Proteins* **17**, 75–86.

Balguerie, A., Dos Reis, S., Ritter, C., Chaignepain, S, Coulary-Salin, B., Forge, V., Bathany, K., Lascu, I., Schmitter, J. M., Riek, R., and Saupe, S. J. (2003). Domain organization and structure-function relationship of the HET-s prion protein of *Podospora anserina*. *EMBO J.* **22**, 2071–2081.

Belghazi, M., Bathany, K., Hountondji, C., Grandier-Vazeille, X., Manon, S., and Schmitter, J. M. (2001). Analysis of protein sequences and protein complexes by MALDI mass spectrometry. *Proteomics* **1**, 946–954; *Eur. J. Mass Spectrom.* **7**, 101–109.

Coustou, V., Deleu, C., Saupe, S., and Begueret, J. (1997). The protein product of the HET-s heterokaryon incompatibility gene of the fungus *Podospora anserina* behaves as a prion analog. *Proc. Natl. Acad. Sci. USA* **94**, 9773–9778.

Coustou-Linares, V., Maddelein, M. L., Begueret, J., and Saupe, S. J. (2001). *In vivo* aggregation of the HET-s prion protein of the fungus *Podospora anserina*. *Mol. Microbiol.* **42**, 1325–1335.

Cravello, L., Lascoux, D., and Forest, E. (2003). Use of different proteases working in acidic conditions to improve sequence coverage and resolution in hydrogen/deuterium exchange of large proteins. *Rapid Commun. Mass Spectrom.* **17**, 2387–2393.

Demmers, J. A. A., Rijkers, D. T. S., Haverkamp, J., Killian, J. A., and Heck, A. J. R. (2002). Factors affecting gas-phase deuterium scrambling in peptide ions and their implications for protein structure determination. *J. Am. Chem. Soc.* **124**, 11191–11198.

Deng, Y. Z., Pan, H., and Smith, D. L. (1999). Selective isotope labeling demonstrates that hydrogen exchange at individual peptide linkages can be determined by collision induced dissociation mass spectrometry. *J. Am. Chem. Soc.* **121**, 1966–1967.

Dos Reis, S., Coulary-Salin, B., Forge, V., Lascu, I., Begueret, J., and Saupe, S. J. (2002). The HET-s prion protein of the filamentous fungus *Podospora anserina* aggregates *in vitro* into amyloid-like fibrils. *J. Biol. Chem.* **277**, 5703–5706.

Englander, J. J., Del Mar, C., Li, W., Englander, S. W., Kim, J., Stranz, D. D., Hamuro, Y., and Woods, V. L., Jr. (2003). Protein structure change studied by hydrogen deuterium exchange, functional labeling and mass spectrometry. *Proc. Natl. Acad. Sci. USA* **100**, 7057–7062.

Hoerner, J. K., Xiao, H., Dobo, A., and Kaltashov, I. A. (2004). Is there hydrogen scrambling in the gas phase? Energetic and structural determinants of proton mobility within protein ions. *J. Am. Chem. Soc.* **126**, 7709–7717.

Hoshino, M., Katou, H., Hagihara, Y., Hasegawa, K., Naiki, H., and Goto, Y. (2002). Mapping the core of the β_2-microglobulin amyloid fibril by H/D exchange. *Nat. Struct. Biol.* **9**, 332–336.

Johnson, R. S., Krylov, D., and Walsh, K. A. (1995). Mobility within electrosprayed peptide ions. *J. Mass Spectrom.* **30**, 386–387.

Jørgensen, T. J. D., Gårdsvoll, H., Ploug, M., and Roepstorff, P. (2005a). Intramolecular migration of amide hydrogens in protonated peptides upon collisional activation. *J. Am. Chem. Soc.* **127**, 2785–2793.

Jørgensen, T. J., Bache, N., Roepstorff, P., Gardsvoll, H., and Ploug, M. (2005b). Collisional activation by MALDI tandem time-of-flight mass spectrometry induces intramolecular migration of amide hydrogens in protonated peptides. *Mol. Cell Proteomics* **4**, 1910–1919.

Kheterpal, I., Zhou, S., Cook, K. D., and Wetzel, R. (2000). Aβ amyloid fibrils possess a core structure highly resistant to hydrogen exchange. *Proc. Natl. Acad. Sci. USA* **97**, 13597–13601.

Kheterpal, I., Williams, A., Murphy, C., Bledsoe, B., and Wetzel, R. (2001). Structural features of the Aβ amyloid fibril elucidated by limited proteolysis. *Biochemistry* **40,** 11757–11767.

Kheterpal, I., Lashuel, H. A., Hartley, D. M., Waltz, T., Lansbury, P. T., and Wetzel, R. (2003). Aβ protofibrils possess a stable core structure resistant to hydrogen as chance. *Biochemistry* **92,** 14092–14098.

Kim, M. Y., Maier, C. S., Reed, D. J., and Deinzer, M. L. (2001). Site-specific amide hydrogen/deuterium exchange in *E. coli* thioredoxins measured by electrospray ionization mass spectrometry. *J. Am. Chem. Soc.* **123,** 9860–9866.

Kraus, M., Bienert, M., and Krause, E. (2003). Hydrogen exchange studies on Alzheimer's amyloid-β peptides by mass spectrometry using matrix-assisted laser desorption/ionization and electrospray ionization. *Rapid Commun. Mass Spectrom.* **17,** 222–228.

Liu, Y., and Smith, D. L. (1994). Probing high order structure of proteins by fast-atom bombardment mass spectrometry. *J. Am. Soc. Mass. Spectrom.* **5,** 19–28.

Maddelein, M. L., Dos Reis, S., Duvezin-Caubet, S., Coulary-Salin, B., and Saupe, S. J. (2002). Amyloid aggregates of the HET-s prion protein are infectious. *Proc. Natl. Acad. Sci. USA* **99,** 7402–7407.

Mandell, J. G., Falick, A. M., and Komives, E. A. (1998). Measurement of amide hydrogen exchange by MALDI-ToF mass spectrometry. *Anal. Chem.* **70,** 3987–3995.

Nazabal, A., Laguerre, M., Schmitter, J. M., Vaillier, J., Chaignepain, S., and Velours, J. (2003a). Hydrogen/deuterium exchange on yeast ATPase supramolecular protein complex analyzed at high sensitivity by MALDI mass spectrometry. *J. Am. Soc. Mass Spectrom.* **14,** 471–481.

Nazabal, A., Dos Reis, S., Bonneu, M., Saupe, S. J., and Schmitter, J. M. (2003b). Conformational transition occurring upon amyloid aggregation of the HET-s prion protein of Podospora anserina analyzed by hydrogen/deuterium exchange and mass spectrometry. *Biochemistry* **42,** 8852–8861.

Nazabal, A., Maddelein, M. L., Bonneu, M., Saupe, S. J., and Schmitter, J. M. (2005a). Probing the structure of the infectious amyloid form of the prion forming domain of HET-s using high-resolution hydrogen/deuterium exchange monitored by mass spectrometry. *J. Biol. Chem.* **280,** 13220–13228.

Nazabal, A., Bonneu, M., Saupe, S. J., and Schmitter, J. M. (2005b). High resolution H/D exchange studies on the HET-s[218–295] prion protein. *J. Mass Spectrom.* **40,** 580–590.

Resing, K. A., and Ahn, N. G. (1998). Deuterium exchange mass spectrometry as a probe of protein kinase activation. Analysis of wild-type and constitutively active mutants of MAP kinase kinase-1. *Biochemistry* **37,** 463–473.

Ritter, C., Maddelein, M-L., Siemer, A. B., Lühr, T., Ernst, M., Meier, B. H., Saupe, S. J., and Riek, R. (2005). Correlation of structural elements and infectivity of the HET-s prion. *Nature* **435,** 844–848.

Zhang, Z., and Smith, D. L. (1993). Determination of amide hydrogen exchange by mass spectrometry: A new tool for protein structure elucidation. *Protein Sci.* **2,** 522–531.

[10] Analysis of Amyloid Fibril Structure by Scanning Cysteine Mutagenesis

By Shankaramma Shivaprasad and Ronald Wetzel

Abstract

Introduction of Cys point mutations into amyloidogenic proteins provides several productive avenues to probing the structures of amyloid fibrils and other nonnative protein aggregates. We describe here the use of single and double Cys mutants to examine the structure of the amyloid $\beta(1–40)$ amyloid fibril. Single mutants provide information on local fibril environment around the side chain being investigated, including solvent accessibility within the fibril. Double Cys mutants provide information on whether the mutated side chains make intramolecular contacts within amyloid fibril structure.

Introduction

The challenge of amyloid fibril structure determination has been addressed by a variety of experimental approaches, none of which is capable of providing a complete structural picture. In fact, although great progress is being made (Makin *et al.*, 2005; Nelson *et al.*, 2005), high-resolution structures of complete amyloid fibrils continue to elude us. Of the various intermediate- and low-resolution methods that have been implemented, each has strengths and weaknesses. New methods thus continue to be welcome.

Scanning mutagenesis (Cunningham and Wells, 1989) has been widely used to study the molecular basis of folding, stability, and biological activity of native proteins. Mutagenesis is pointless, however, without the availability of methods for analyzing and comparing mutants. Thermal or chemical denaturation, for example, is commonly used for comparative analysis of globular proteins and their mutants (Alber, 1989; Pace *et al.*, 2005) and, in favorable cases, can also be applied to amyloid fibril structure (Narimoto *et al.*, 2004). Various amino acids have been used as scanning mutagenesis probes. Although the most commonly used is alanine (Couture *et al.*, 1979; Cunningham and Wells, 1989) (based on its being the amino acid in the normal repertoire with the smallest nonhydrogen side chain), proline (Williams *et al.*, 2004; Wood *et al.*, 1995) and cysteine (Kanaya *et al.*, 1991) have also been used to evaluate in a systematic manner the contributions of amino acid residues to protein structure and function. In another

METHODS IN ENZYMOLOGY, VOL. 413
0076-6879/06 $35.00
DOI: 10.1016/S0076-6879(06)13010-4

chapter in this volume, Margittai and Langen (see Chapter 7) describe the use of scanning cysteine mutagenesis to introduce a spin-labeled probe to access features of fibril structure. In this chapter, we describe several additional methods for exploiting cysteine mutations systematically introduced into an amyloid fibril precursor. We describe how the same series of mutants can be used for experiments to assess the solvent accessibility of each amino acid side chain and to investigate the local environment that the side chains are thrust into the fibril, even if it is inaccessible to solvent. Finally, we describe how double Cys mutants can be used to locate side chain interactions, for example, across the β-sheet interface in the amyloid fibril core.

Chemically Modified Cysteines to Explore Local Environment Within the Amyloid Fibril

Mutagenesis coupled with thermodynamic analysis of fibril stability (O'Nuallain *et al.*, 2005) can provide considerable information about the parts of a polypeptide sequence that are involved in the amyloid fibril core and how those residue side chains are aligned and interact with each other within the core. Proline replacements probe the compatibility of local φ,ψ angles with proline geometry, and hence the likelihood that a residue resides in a β-sheet or other restrictive structure (Williams *et al.*, 2004, 2006; Wood *et al.*, 1995). Alanine replacements probe the role of the wild-type side chain without altering in any major way the restrictions placed on φ,ψ angles by a nonhydrogen side chain (Williams *et al.*, 2006). Other amino acids, whether part of the standard repertoire or unnatural, might provide additional information about the local, packed environment that a particular residue side chain must be thrust into in the amyloid state. Exploring multiple amino acid replacements, however, can put great demands on resources. In order to explore amyloid fibril microenvironments with the maximum number of side chain probes, we developed a strategy of modifying single Cys mutants with sulfhydryl-selective reagents capable of creating different side chain chemistries. For example, modification of Cys with iodoacetic acid creates an amino acid isosteric with glutamic acid. Similarly, modification with iodoacetamide creates a glutamine isostere. Reaction with methyl iodide creates a methionine analogue. Many other modifications are possible using commercially available sulfhydryl reagents. In this application, the Cys-mutated protein is modified by the reagent of choice and the product is purified by high-performance liquid chromatography (HPLC). Before use, the purified peptide is rigorously disaggregated and then subjected to amyloid fibril growth conditions as described elsewhere in this volume (see Chapter 3 by O'Nuallain and colleagues). Seeding with wild-type fibrils is expected to direct the product fibrils into a conformation similar to that of wild-type fibrils, while accelerating the reaction at the same time to allow relatively clean detection of the endpoint plateau position, or critical concentration (C_r), from which the ΔG for fibril elongation can be calculated (O'Nuallain *et al.*, 2005). Fibril stabilities can then be related to the stability of the wild-type peptide fibrils, or fibrils formed from the unmodified Cys mutant, to gauge the compatibility of the added side chain within the fibril microenvironment.

Experimental Results

The Alzheimer's disease peptide amyloid β (Aβ)(1–40) has no cysteine residues, making it a good candidate for the above strategy. Data from the following single Cys mutants of Aβ(1–40) will be described here: F4C, L17C, L34C, M35C (Shivaprasad and Wetzel, 2006). Modified peptides were synthesized, purified, and characterized, and fibril formation reactions were initiated as described in the section on protocols. Figure 1 shows the results, arranged to compare fibrils of each mutant to wild-type Aβ(1–40) fibrils. Supporting many other studies that suggest that the N-terminal 10 residues of Aβ(1–40) are not involved in the amyloid core, it can be seen that all three states of residue 4 shown (free Cys, carboxymethyl-, and methyl-) exhibit wild-type-like stability. In contrast, residues 17, 34, and 35 exhibit great sensitivity to the alkylation state of a Cys residue at these positions, with each residue sweeping out the same basic pattern and with stabilities varying as follows: wild type > MeS- > HS- > $^-$OOC-CH$_2$-S-. The latter modification creates Aβ(1–40) peptides whose amyloid fibrils are destabilized by 2.5–3.0

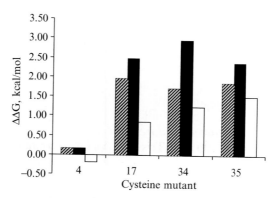

Fig. 1. $\Delta\Delta G$ values for Alzheimer's disease peptide amyloid β (Aβ)(1–40) amyloid fibril elongation for fibrils grown at 37° in phosphate-buffered saline without agitation, comparing mutant peptides with wild-type peptide. A positive $\Delta\Delta G$ indicates lower stability in the mutant. Free Cys peptides, cross-hatched bars; carboxymethyl Cys peptides, filled bars; methyl Cys peptides, open bars. Values are from Shivaprasad and Wetzel (2006b).

kcal/mol compared with wild-type, which is as high a destabilization as observed for Pro replacements at these β-sheet positions (Williams *et al.*, 2004). This suggests that these side chains are normally packed into hydrophobic environments in the fibril, consistent with Ala scanning data (Williams *et al.*, 2006). The ability of methylation of Cys to generate mutants that are more compatible in fibrils than are the unmodified Cys peptides further supports the view that these residues reside in hydrophobic environments. In later sections, it will be seen how this information largely agrees with other data also obtained from the same Cys mutations.

Protocols

Purification of Mutant Peptides. Wild-type Aβ(1–40) (H$_2$N-DAEFR HDSGY EVHHQKLVFF AEDVG SNKGA IIGLMVGGVV-COOH) used in this study is a chemically synthesized peptide purchased in pure form from Keck Biotechnology Center at Yale University. All the cysteine mutants of Aβ(1–40) are chemically synthesized peptides purchased from the same source as crude peptides and purified on Bio-Rad reverse phase (RP)-HPLC using a semipreparative Zorbax SB-C3 column (9.4 mm $\times$ 25 cm). For purification, the crude peptide was dissolved in 20% formic acid and eluted using a gradient of 1–51% aqueous acetonitrile containing 0.05% trifluoroacetic acid (TFA), a flow rate of 4 ml/min, and UV detection at 215 nm. The fractions collected were analyzed by mass spectrometry (MS; Agilent 1100 series LC/MSD), and the desired fractions were pooled and lyophilized. The pooled peptides were then checked for homogeneity by analytical RP-HPLC (SB-C3 column, 5 μm, 3.0 $\times$ 150 mm) at a flow rate of 1 ml/min using the same gradient and detection mentioned above. The pure peptides were stored at $-20°$ until further use.

The analytical HPLC conditions to determine the concentration of monomer in aggregation reactions, including the C_r, were identical to those described above. A similar gradient was used for the liquid chromatography mass spectrometry (LC/MS) analysis, except that 0.1% formic acid was used instead of TFA because of dampening of the mass spectra signals by the latter (Chapman, 2000).

Chemical Modification of Sulfhydryl Group of Cys Mutants. Purified peptides, in 8 M urea, 100 mM Tris HCl, pH 8.0, containing 10 mM Tris (2-carboxyethyl)phosphine (TCEP) and bubbled with argon, were incubated at 1 mg/ml with 10 mM iodoacetic acid or 10 mM iodomethane at 37°. Reaction progress was monitored by Ellman's reagent (Means and Feeney, 1971) and LC/MS as described above. After incubation for 20 min with iodoacetic acid, peptides were greater than 95% modified. Iodomethane required incubations of 60–90 min. Longer incubation times result in the formation of side products, mainly doubly alkylated peptide, because of the reaction of His or Lys side chains or the N^{α}-amino group. When reaction times are limited to the above times, however, secondary alkylation is less than 5%, and these doubly modified peptides are removed by HPLC purification. At the completion of the reaction, the solution was diluted with 1 ml 20% formic acid and immediately subjected to HPLC purification as described above. Typical recovered yields are in the range of 70–80%, based on starting purified Cys peptides. The final purity and identity of each modified peptide was determined by analytical HPLC/MS.

Amyloid Fibril Growth. Aβ(1–40) Cys mutants were subjected to the standard disaggregation protocol (see Chapter 3 by O'Nuallain and colleagues in this volume), with the exception that for the free Cys peptides, the PBSA buffer (phosphate-buffered saline [PBS] plus 0.05% sodium azide, pH 7.5) contained 10 mM TCEP and 1 mM ethylenediaminetetraacetic acid (EDTA) and was bubbled with argon prior to dissolution of the disaggregated peptide in order to keep the Cys sulfhydryls in the reduced state. After further removal of any aggregates by high-speed centrifugation (see Chapter 3 by O'Nuallain and colleagues in this volume), peptide concentration was determined by analytical HPLC and the concentration adjusted to 50–60 μM. For mutant peptides that exhibit no spontaneous aggregation at 50 μM, presumably because of an unusually high C_r, fibril assembly was accomplished by repeating the aggregation reaction at starting concentrations in the range of 100–200 μM. To encourage these mutant peptides to grow into amyloid fibrils resembling wild-type Aβ fibrils, all the reactions were seeded with 0.1% (wt/wt) sonicated wild-type Aβ(1–40) fibrils. Peptide solutions were incubated without agitation at 37° and monitored by the thioflavin T (ThT) reaction (LeVine, 1999). When the ThT value reached a plateau, 25 μl of the solution was removed at time intervals and centrifuged at 20,800g for 30 min; then, 10 μl of the supernatant was carefully removed, diluted with 110 μl 1% aqueous TFA, and injected onto analytical HPLC. The amount of monomer left in solution was determined using a wild-type Aβ(1-40) standard curve (see Chapter 3 by O'Nuallain and colleagues in this volume). The average molar concentration of monomer at equilibrium, the C_r, was used to calculate the free energy (ΔG) of elongation for the equilibrium constant for the elongation reaction ($K_{eq} = 1/C_r$) and the mutants using the relation $\Delta G = -RT\ln K_{eq}$. (O'Nuallain *et al.*, 2005).

Alkylation Accessibility of Free Cys Within Amyloid Fibril Structure

Most models of amyloid structure posit some kind of multilamellar arrangement of β-sheets. Unless the fibril is composed of a closed circular stack of sheets, however, there must inevitably be side chains in the β-sheet portion that are exposed to solvent. Residues not involved in sheet, and residues involved in turns between extended chain elements, may also be exposed to solvent. Identifying these residues can provide important information on amyloid structure. Solvent exposure of residue side chains can be assessed by challenging Cys residues with alkylating agents in the context of structure. This is a classic method for determining whether a Cys residue in a globular protein is exposed or buried (Means and Feeney, 1971). If one can independently establish whether a particular residue is in β-sheet, its exposure to solvent provides information not only on the target residue but on the other residues in the same extended chain element, given the alternating pattern of side chains in the extended chain.

The challenge in this type of experiment is in establishing control over thiol chemistry. To get meaningful quantitation, it is important to grow amyloid fibrils from Cys mutants in such a way that the Cys remains in the reduced state. It is likewise important to demonstrate that peptides not modified by alkylating agents in the fibril state remain in the reduced state throughout the reaction and analysis. Otherwise, the negative result of no modification might be attributable, for example, to oxidation of Cys during the modification reaction. A convenient approach to ensuring that results can be interpreted includes the use of a nonalkylatable reducing agent to prevent oxidation and the use of quantitative HPLC to determine a material balance at the end of the experiment so that most or all the peptide is accounted for.

Experimental Results

Figure 2 shows the results of challenging amyloid fibrils, grown under reducing conditions from Cys mutants of Aβ(1–40), with iodoacetamide. The data, which correspond to the same four residues analyzed in the section on chemically modified cysteines to explore local environment within the amyloid fibril, show a range of response. Position 4, as expected based on other data on Aβ(1–40) amyloid fibrils grown without agitation in PBS (Kheterpal *et al.*, 2001; Whittemore *et al.*, 2005; Williams *et al.*, 2004, 2006), appears to be completely exposed in the fibril, leading to 100% modification by alkylating agent. Fibrils containing Cys at positions 17 or 34, in contrast, are completely resistant to alkylation under these conditions. A Cys residue at position 35 is modified to about 50%. The data are consistent with other data suggesting that residues 17, 34, and 35 are involved in the two β-extended chain elements in the amyloid fibril. Complete protection from alkylation suggests that residues 17 and 34 are buried in packed sheet, and therefore are perhaps both directed into the same sheet-sheet interface. The ability of residue 35 to be alkylated in the fibril is consistent with its being on the solvent-exposed surface of a β-sheet, which is also expected because of its being adjacent within the same sheet to fully

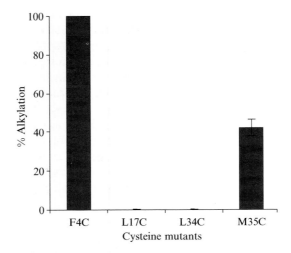

Fɪɢ. 2. Accessibility to alkylation of the cysteine sulfhydryl group in Alzheimer's disease peptide amyloid β (Aβ)(1–40) mutant peptides incorporated into amyloid fibrils. Cysteines at positions 17 and 34 are completely protected from alkylation under these conditions; no carboxymethyl cysteine derivatives are detected. Values are from Shivaprasad and Wetzel (2006).

buried residue 34. The explanation for incomplete alkylation is not clear. An obvious possibility is variable reaction rates at different sites, but extending reaction time normally does not improve yield, instead leading to secondary reactions. The most likely explanation is multiple environments, but there is no direct evidence for this.

Protocols

Solvent Accessibility Studies. Fibrils formed from unmodified single Cys mutants under reducing conditions (100 μl of 0.25 μg peptide/μl) were collected by centrifugation at 20,800g for 30 min at 4°, and the supernatant was removed. This removes any monomeric peptide, which, otherwise, would be readily alkylated and thus confound analysis. The pellet is then suspended in 50 μl alkylation buffer (argon-purged 50 mM Tris HCl, 150 mM NaCl, pH 8.0, 10 mM TCEP), adjusted to 10 mM iodoacetamide, and incubated at room temperature for 1 h. (Under these conditions, F4C mutant fibrils show essentially complete modification of the thiol group (Fig. 2).) The mixture was centrifuged at 20,800g for 30 min, the supernatant was discarded, and the pellet was washed twice with 100 μl PBS. The pellet was then dissolved in 20 μl 20% aqueous formic acid, diluted with 20 μl water, injected on a SB-C3 analytical HPLC column, and analyzed by

LC/MS as described above. The percentage of recovered peptide that was alkylated was reported. Under the conditions used, no oxidative cross-linking of sulfhydryl group or other side product was observed by LC/MS. It seemed possible that mutants exhibiting partial sensitivity to alkylation might be accessible but require longer incubation times or higher pH. However, when these measures were taken, little additional modification of the Cys was observed up to the point where side reactions (e.g., additional alkylation, dissociation of fibrils) limited the reactions (Shivaprasad and Wetzel, 2006). Because HPLC does not always cleanly separate alkylated from free Cys versions of the peptide, it is easier to obtain interpretable data by LC/MS, in which the MS software reports the relative signals of the different mass peaks obtained across a designated part of the HPLC gradient. The relative heights of the peaks corresponding to free Cys and alkylated peptide are then used to calculate the percent alkylation. The quality of the data obtained in this way was confirmed by integration of HPLC peaks for several peptides for which complete separation of the HPLC peaks for free Cys and alkylated peptides is obtained. The added advantage of using LC/MS data is the greater sensitivity of the MS compared with the ultraviolet (UV) detector, allowing analysis of smaller fibril samples.

We confirmed good to excellent recovery of $A\beta$ from the alkylated fibrils by doing the following experiment. Fibrils were grown from a known concentration of $A\beta$, and the approximate weight concentration of $A\beta$ in the fibril fraction of the reaction mixture was calculated from the HPLC-measured amount of monomer recovered at equilibrium. We then pelleted a 100-μl aliquot of the reaction mixture suspension at 20,800g for 30 min, decanted the supernatant, and added to the pellet 20 μl 20% formic acid followed by 20 μl water. After vortexing and injecting onto RP-HPLC, >80% of monomer was recovered. In contrast, using 40 μl 1% TFA instead of formic acid to dissolve the pellet, the recovery was only 40–60%.

Disulfide Cross-Linking of Amyloid Fibrils Containing Reduced Double Cys Mutants

Most models of the $A\beta(1–40)$ fibril posit the monomer undergoing a chain reversal when it engages the amyloid fold, such that β-extended chain elements in the N- and C-terminal extremes of the amyloid core segment form separate, parallel β-sheets that pack against each other in the fibril (Guo et al., 2004; Petkova et al., 2002). Proof of such hairpin models requires experimental support, however. Additionally, if a hairpin is involved, one wants to know which side chains in these two β-extended chain elements are facing in and toward each other and which of these

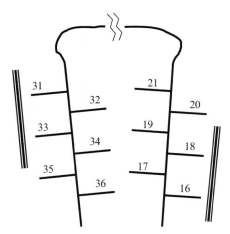

FIG. 3. A model of the cross section of the Alzheimer's disease peptide amyloid β (Aβ)(1–40) amyloid fibril shows the orientation of side chains pointing into and out of the packed β-sandwich protofilament structure. The double lines indicate possible packing surfaces for protofilament-protofilament contacts. Reproduced with permission from Shivaprasad and Wetzel (2006b).

internalized side chains are in direct packing contact with each other. These questions can be addressed using Cys mutagenesis. The Cys 35 (see the section on experimental results) and Cys 20 (not shown) mutants were already shown to be solvent-accessible and therefore likely pointing out into solution from a β-sandwich (Fig. 3). By extension, side chains of residues 33 (a Gly, however), 31, 18, and 16 should also be pointing out (Shivaprasad and Wetzel, 2006). The fact that these are protected from alkylation implies that the "outside" faces of the sheets at these positions must be docked against other filaments in extended sandwich-like structures (Fig. 3). Based on this analysis, it appeared possible that an appropriate double Cys mutant in two residue side chains both directed inward in the fibril core and in close contact with each other might, upon being exposed to oxidizing conditions, form a disulfide bond within the context of the fibril structure. With appropriate controls, this might be a way of exploring the interior structure of the amyloid fibril core.

As with single Cys mutant alkylation experiments, the key to double Cys oxidation experiments is an appreciation of thiol-disulfide chemistry and application of the appropriate controls. First, to obtain interpretable results, it is imperative that both residue positions are held rigidly in place in amyloid structure. As with any cross-linking experiment, flexible portions of chain, given sufficient time and sufficient lifetime of the reactive group, can become cross-linked, even though there is no special nearest neighbor relation between the probe residues. Second, thiol-disulfide exchange is notorious at giving artifacts in protein structure analysis, and appropriate controls and conservative interpretations are therefore critical. Third, as with many experiments on Aβ fibrils or other fibrils with significant critical concentrations, it is important at every stage of analysis to isolate the amyloid fraction from the monomer fraction; monomeric forms of double Cys mutants can oxidize to their cross-linked disulfide forms readily, as described in the section on thermodynamic stability of amyloid fibrils grown from cross-linked Cys mutants.

Experimental Results

Given the results of accessibility experiments, such as those reviewed in the section on alkylation accessibility of free cysteine within amyloid fibril structure, an experiment was designed to probe for nearest neighbor relations within the Aβ(1–40) amyloid fibril (Shivaprasad and Wetzel, 2004). Based on the above analysis (Fig. 3), we constructed three pairs of double Cys mutants, with residue 17 serving as the common residue from the N-terminal β-extended chain element of Aβ, whereas residues 34, 35, and 36 served as three possible nearest neighbor contacts on the C-terminal β-extended chain. Fibril formation reactions were performed for each of the 17/34, 17/35, and 17/36 double mutants under rigorous reducing conditions. Fibrils were isolated from unpolymerized monomer and confirmed, in all three cases, to contain only reduced peptide. These fibrils were then exposed to molecular oxygen and dimethyl sulfoxide (DMSO) for various periods, and aliquots were removed for analysis. For each time point, fibrils were isolated (from monomeric Aβ that might have dissociated from the fibrils during oxidation) by centrifugation, dissolved under alkylating conditions (to suppress thiol/disulfide interchange that could lead to artifactual results), and analyzed by HPLC to quantify cross-linked, oxidized peptide from reduced peptide. Disulfide cross-linked peptides generally elute faster than the reduced form in RP-HPLC (Perry and Wetzel, 1984) and sodium dodecyl sulfate (SDS) gels (Plunkett and Ryan, 1980) because of their lower amount of surface exposure and smaller hydrodynamic radii under denaturing conditions.

Figure 4 shows examples of HPLC chromatograms for oxidation of the 17/34, 17/35, and 17/36 fibrils after 24, 84, and 192 h. It can be seen that within this time frame, only the 17/34 peptide has undergone significant oxidative cross-linking. Exposure of the other two peptides, even for longer oxidation times, gave only a modest amount of intramolecular cross-linking for the 17/36 peptide and only some dimeric disulfide material for the 17/35 peptide. These results strongly suggest that (1) the method can discriminate favorable from unfavorable side chain interactions within protein structure and (2) Leu17 and Leu34 appear to be close in space and presumably

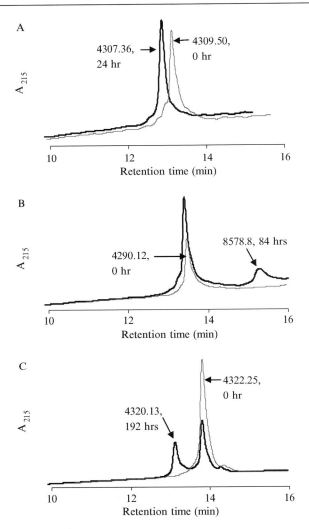

Fig. 4. High-performance liquid chromatography profiles show the analysis of oxidative cross-linking of double Cys mutants within the amyloid fibril. Peaks are labeled with average molecular mass derived from electrospray mass spectrometry. For closely eluting peaks, cross-linking is associated with the earlier eluting peak and a reduction in mass by 2 Da. A much later eluting peak with twice the mass is a disulfide-linked dimer. Traces of peptides obtained by disaggregating fibrils at oxidation time = 0 are shown in gray, and the corresponding traces after incubation of fibrils in 10% dimethyl sulfoxide are shown in black. L17C/L34C (A), L17C/M35C (B), and L17C/L36C (C). Reproduced with permission from Shivaprasad and Wetzel (2006).

packed against each other within the fibril core of the $A\beta(1–40)$ amyloid fibrils grown without agitation in PBS. Similar experiments also identified the side chains of Ile32 and Phe19 as interacting within the fibril (S. Shivaprasad and R. Wetzel, unpublished data).

Protocol

Oxidative Cross-Linking of Double Cys Mutant Peptide at Fibril Level. Each purified double Cys mutant was subjected to standard fibril growth assembly with added reducing agent as described in the section on chemically modified cysteines to explore local environment within the amyloid fibril. The three double Cys mutant fibrils grown under reducing conditions were collected by centrifugation for 30 min at 20,800g at 4°. The supernatant was removed, and the pellet was suspended in 10% DMSO in PBS, pH 8.0 (see the protocol section for the choice of this reagent for oxidation) to a final concentration of 0.45 mg/ml and incubated at room temperature. For monitoring oxidation, a 30-μl aliquot was removed and ultracentrifuged at 315,000g at 4° for 30 min, and the supernatant was carefully removed. (When working with small-volume aliquots, we used the table-top ultracentrifuge to ensure good recovery of the fibrils in a compact pellet; however, we have no data to suggest that a 20,800g spin would not be adequate.) The pellet was then dissolved in 20 μl 20% formic acid, diluted with 20 μl of water, and analyzed by analytical LC/MS as described above. The reaction products were conveniently monitored by the difference in molecular mass and elution time for the oxidized and reduced mutants (Fig. 4). Material recovery by HPLC analysis was consistently greater than 80%.

To confirm the reduced state of the sulfhydryl groups in the fibrils prior to oxidation, an aliquot of the suspension was centrifuged and the fibril pellet was treated with argon-purged 10 mM iodoacetic acid in 8 M urea, 100 mM Tris HCl, pH 8.0, to give a final concentration of 0.45 mg/ml, and this suspension was incubated for 1 h at room temperature. A 20-μl aliquot of the reaction mixture was removed and diluted with 20 μl 20% formic acid, mixed, and injected onto LC/MS. For each fibril, the product showed a single species with mass corresponding to the carboxamidomethyl-Cys form of $A\beta(1–40)$, with quantitative recovery of material.

To confirm further that the products from the cross-linking experiment were not influenced by thiol-disulfide exchange during the sample workup, a control experiment was run, in which the fibril pellet from a 30-μl aliquot, after oxidation and centrifugation, was resuspended in 20 mM iodoacetic acid in argon-purged 8 M urea, 100 mM Tris HCl, pH 8.0 (30 μl) and incubated for 1 h at room temperature (see the section on chemically modified cysteines to explore local environment with the amyloid fibril for chemical modification of sulfhydryl group of Cys mutants). The reaction was diluted with 20% formic acid and injected onto LC/MS. This treatment quantitatively yields alkylated Cys products when run on fibrils containing reduced Cys (see above). When run on fibrils that were exposed to the DMSO oxidation buffer, this treatment yielded distributions of cross-linked and non-cross-linked peptides identical to those obtained from analyses performed without a prior alkylation step.

Thermodynamic Stability of Amyloid Fibrils Grown from Cross-Linked Double Cys Mutants

The alternative approach to using double Cys mutants to gauge side chain nearest neighbor relations within fibril structure is to cross-link the Cys residues of double Cys mutants at the monomer level and then assess the ability of these cross-linked monomers to form fibrils, including the thermodynamic stabilities of the fibril products, compared with both wild-type fibrils and fibrils grown from the reduced double Cys mutants. The measurement of C_r values and their use to calculate K_{eq} and ΔG values for fibril formation are described elsewhere in this volume (see Chapter 3 by O'Nuallain and colleagues) and summarized briefly in the section on chemically modified cysteines to explore local environment within the amyloid fibril. Purified oxidized monomers were seeded with wild-type fibrils in an attempt to force fibril growth into a wild-type-like conformation, and the equilibrium point was quantified by determining the unpolymerized monomer concentration when the reaction reached plateau (O'Nuallain et al., 2005).

The contribution of disulfide bond formation to protein folding stability is complex (Wetzel, 1987). Classic polymer theory predicts a stabilizing effect on the folding equilibrium by an added cross-link because of the destabilization of the unfolded state attributable to a reduction in chain entropy. However, strain and steric clashes in the folded disulfide state, compared with the reduced state, might also destabilize the folded state, negating some of the chain entropy contribution. In fact, engineered disulfide bonds do not always contribute the amount of stability expected based on chain entropy considerations (Matsumura et al., 1989); hence, it was not clear that a disulfide bond introduced into nearest neighbor side chain positions within the fibril would necessarily provide stability with respect to the reduced state or to wild-type fibrils. This theoretical concern should be kept in mind if results are unclear or counterintuitive.

Experimental Results

The same three sets of double Cys mutants of $A\beta(1–40)$ as described in the section on disulfide cross-linking of amyloid fibrils containing reduced double Cys mutants were used for thermodynamic analysis. Peptides

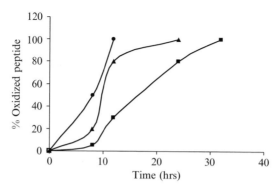

FIG. 5. Kinetics of oxidation of monomeric cysteine mutants in 10% dimethyl sulfoxide. The percentage of peptide converted to the oxidized, cross-linked state with respect to time of exposure to oxidizing conditions. L17C/L34C (●), L17C/M35C (■), and L17C/V36C (▲).

were cross-linked in the monomeric state by exposure to molecular oxygen in the presence and absence of DMSO. Interestingly, the kinetics of oxidative cross-linking differed significantly for the three peptides, in the order 17/34 > 17/36 > 17/35, suggesting that the Aβ(1–40) conformation in the monomer ensemble favors the oxidative cross-linking of different double mutants to different extents. We obtained the same relative reactivities whether or not DMSO was included in the oxidation reaction. Figure 5 shows the kinetics of oxidation of these three mutants.

These cross-linked peptides were disaggregated, subjected to fibril growth with wild-type fibril seeding, and analyzed for critical concentration as described elsewhere in this volume (see Chapter 3 by O'Nuallain and colleagues). The C_r values were converted to K_{eq}, and the resulting $\Delta\Delta G$ values comparing the relative stabilities of mutant and wild-type fibrils were calculated (O'Nuallain *et al.*, 2005; Shivaprasad and Wetzel, 2004). The results are shown in Fig. 6. Somewhat surprisingly, all three cross-linked peptides make reasonably stable amyloid fibrils. At the same time, the oxidized 17/34 double Cys mutant is clearly the most wild-type-like in its fibril formation by two separate criteria. First, the stability of the oxidized, cross-linked fibril is more like wild-type than any other double Cys mutant shown, oxidized or reduced. Second, the 17/34 double Cys mutant is the only double mutant showing an improvement of fibril stability upon oxidation. Time and experience with additional mutants and amyloid systems will determine whether one of these criteria is more discerning than the other as a gauge of wild-type-like fibril structure.

One aspect of these results is quite curious and surprising. When a disulfide bond is formed between residues 17 and 35 of Aβ(1–40), it necessarily alters the orientation of the extended chain containing residue 35, so that in

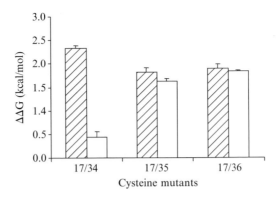

FIG. 6. $\Delta\Delta G$ values for Alzheimer's disease peptide amyloid β (Aβ)(1-40) amyloid fibril elongation for fibrils grown at 37° in phosphate-buffered saline without agitation, comparing oxidized and reduced double Cys mutant peptides with wild-type peptide. A positive $\Delta\Delta G$ indicates lower stability in the mutant. The shaded bars correspond to the reduced mutants, and the open bars correspond to the oxidized mutants. Reproduced with permission from Shivaprasad and Wetzel (2004).

the β-sandwich formed by the N-terminal and C-terminal sheets, the opposite face of the C-terminal extended chain must now be packed against the N-terminus. In other words, whereas the 17/34 result tells us that the I32/L34/V36 face must be directed into the packed core in the wild-type fibril (Fig. 3), the 17/35 cross-linked mutant must be forced to pack the I31/G33/M35 face into the core. Despite this seemingly radical alteration of fibril structure, the fibrils formed from the 17/35 cross-linked monomeric Aβ(1–40) are only destabilized by a small amount compared with wild-type fibrils or 17/34 fibrils. Perhaps this result is confirming what we already know—that many protein sequences can form amyloid. The practical lesson from these data would seem to be that mere formation of an amyloid fibril is not sufficient evidence for interpretation of a mutagenesis experiment. Other criteria, such as thermodynamic stability (O'Nuallain et al., 2005), antibody binding (O'Nuallain and Wetzel, 2002), cross-seeding efficiency (O'Nuallain et al., 2004), or hydrogen exchange protection (Kheterpal et al., 2000; Williams et al., 2004), will be important for confirming the resemblance of the mutant fibrils formed to those of the wild-type.

Protocols

Oxidative Cross-Linking of Double Cys Mutants at the Monomer Level. Two methods for oxidative cross-linking within the double Cys monomers were explored: PBS at pH 8.0 or 10% aqueous DMSO, both at room temperature and a peptide concentration of 100 μM. Both reactions gave intramolecularly cross-linked peptide as the major product; however, air oxidation in PBS required longer duration for completion compared with oxidation in 10% DMSO, which gave complete conversion in 12–48 h. Reactions were analyzed by both LC/MS and Ellman's reagent (Means and Feeney, 1971). DMSO is known to improve the rate of disulfide formation in proteins (Tam *et al.*, 1991). The L17C/L34C mutant goes to completion the fastest by both methods (within 12 h in aqueous DMSO, within 24 h by air oxidation in PBS).

Amyloid Fibril Assembly. Each double Cys mutant was used to make fibrils under two sets of conditions. Oxidized, cross-linked mutants were prepared as described previously (see section on chemically modified cysteines to explore local environment within the amyloid fibril) in 2 mM NaOH followed by adjustment to PBS buffer, without reducing agent. Reduced double Cys mutants were incubated as above but in buffer purged with argon and including 10 mM TCEP and 1 mM EDTA. The assembly reactions were carried out with a starting monomeric peptide concentration of about 50 μM. The progress of the reaction was monitored by ThT measurement of fibrils, and the C_r was determined by the HPLC sedimentation assay and used to calculate $\Delta\Delta G$ values (see Chapter 3 by O'Nuallain and colleagues).

Acknowledgments

This work was supported by grant R01 AG18416 (to R. Wetzel) from the National Institutes of Health. The authors thank B. O'Nuallain for helpful comments on the manuscript.

References

Alber, T. (1989). Mutational effects on protein stability. *Annu. Rev. Biochem.* **58,** 765–798.

Chapman, J. R. (2000). "Mass Spectrometry of Proteins and Peptides." Humana Press, Totowa, NJ.

Couture, R., Fournier, A., Magnan, J., St.-Pierre, S., and Regoli, D. (1979). Structure-activity studies on substance P. *Can. J. Physiol. Pharmacol.* **57,** 1427–1436.

Cunningham, B. C., and Wells, J. A. (1989). High-resolution epitope mapping of hGH-receptor interactions by alanine-scanning mutagenesis. *Science* **244,** 1081–1085.

Guo, J.-T., Wetzel, R., and Xu, Y. (2004). Molecular modeling of the core of Aβ amyloid fibrils. *Proteins* **57,** 357–364.

Kanaya, E., Kanaya, S., and Kikuchi, M. (1991). Introduction of a non-native disulfide bridge to human lysozyme by cysteine scanning mutagenesis. *Biochem. Biophys. Res. Commun.* **173,** 1194–1199.

Kheterpal, I., Williams, A., Murphy, C., Bledsoe, B., and Wetzel, R. (2001). Structural features of the Aβ amyloid fibril elucidated by limited proteolysis. *Biochemistry* **40,** 11757–11767.

Kheterpal, I., Zhou, S., Cook, K. D., and Wetzel, R. (2000). Abeta amyloid fibrils possess a core structure highly resistant to hydrogen exchange. *Proc. Natl. Acad. Sci. USA* **97,** 13597–13601.

LeVine, H. (1999). Quantification of β-sheet amyloid fibril structures with thioflavin T. *Methods Enzymol.* **309,** 274–284.

Makin, O. S., Atkins, E., Sikorski, P., Johansson, J., and Serpell, L. C. (2005). Molecular basis for amyloid fibril formation and stability. *Proc. Natl. Acad. Sci. USA* **102,** 315–320.

Matsumura, M., Becktel, W. J., Levitt, M., and Matthews, B. W. (1989). Stabilization of phage T4 lysozyme by engineered disulfide bonds. *Proc. Natl. Acad. Sci. USA* **86,** 6562–6566.

Means, G. E., and Feeney, R. E. (1971). "Chemical Modification of Proteins." Holden-Day, Inc, San Francisco, CA.

Narimoto, T., Sakurai, K., Okamoto, A., Chatani, E., Hoshino, M., Hasegawa, K., Naiki, H., and Goto, Y. (2004). Conformational stability of amyloid fibrils of beta2-microglobulin probed by guanidine-hydrochloride-induced unfolding. *FEBS Lett.* **576,** 313–319.

Nelson, R., Sawaya, M. R., Balbirnie, M., Madsen, A. O., Riekel, C., Grothe, R., and Eisenberg, D. (2005). Structure of the cross-beta spine of amyloid-like fibrils. *Nature* **435,** 773–778.

O'Nuallain, B., Shivaprasad, S., Kheterpal, I., and Wetzel, R. (2005). Thermodynamics of abeta(1-40) amyloid fibril elongation. *Biochemistry* **44,** 12709–12718.

O'Nuallain, B., and Wetzel, R. (2002). Conformational antibodies recognizing a generic amyloid fibril epitope. *Proc. Natl. Acad. Sci. USA* **99,** 1485–1490.

O'Nuallain, B., Williams, A. D., Westermark, P., and Wetzel, R. (2004). Seeding specificity in amyloid growth induced by heterologous fibrils. *J. Biol. Chem.* **279,** 17490–17499.

Pace, C. N., Grimsley, G. R., and Scholtz, J. M. (2005). Denaturation of proteins by urea and guanidine hydrochloride. *In* "Protein Folding Handbook, Part I" (J. M. Buchner and J. Kiefhaber, eds.), Vol. 1, pp. 45–69. Wiley-VCH, Weinheim.

Perry, L. J., and Wetzel, R. (1984). Disulfide bond engineered into T4 lysozyme: Stabilization of the protein toward thermal inactivation. *Science* **226,** 555–557.

Petkova, A. T., Ishii, Y., Balbach, J. J., Antzutkin, O. N., Leapman, R. D., Delaglio, F., and Tycko, R. (2002). A structural model for Alzheimer's beta-amyloid fibrils based on experimental constraints from solid state NMR. *Proc. Natl. Acad. Sci. USA* **99,** 16742–16747.

Plunkett, G., and Ryan, C. A. (1980). Reduction and carboxamidomethylation of the single disulfide bond of proteinase inhibitor I from potato tubers. Effects on stability, immunological properties, and inhibitory activities. *J. Biol. Chem.* **255,** 2752–2755.

Shivaprasad, S., and Wetzel, R. (2004). An intersheet packing interaction in Aβ fibrils mapped by disulfide crosslinking. *Biochemistry* **43,** 15310–15317.

Shivaprasad, S., and Wetzel, R. (2006). Scanning cysteine mutagenesis analysis of Aβ(1-40) amyloid fibrils. *J. Biol. Chem.* **281,** 993–1000.

Tam, J. P., Wu, C.-R., Liu, W., and Zhang, J.-W. (1991). Disulfide bond formation in peptides in dimethyl sulfoxide. Scope and applications. *J. Am. Chem. Soc.* **113,** 6657–6662.

Wetzel, R. (1987). Harnessing disulfide bonds using protein engineering. *Trends Biochem. Sci.* **12,** 478–482.

Whittemore, N. A., Mishra, R., Kheterpal, I., Williams, A. D., Wetzel, R., and Serpersu, E. H. (2005). Hydrogen-deuterium (H/D) exchange mapping of Aβ1-40 amyloid fibril secondary structure using NMR spectroscopy. *Biochemistry* **44,** 4434–4441.

Williams, A. D., Portelius, E., Kheterpal, I., Guo, J. T., Cook, K. D., Xu, Y., and Wetzel, R. (2004). Mapping abeta amyloid fibril secondary structure using scanning proline mutagenesis. *J. Mol. Biol.* **335,** 833–842.

Williams, A. D., Shivaprasad, S., and Wetzel, R. (2006). Alanine scanning mutagenesis of Aβ (1-40) amyloid fibril stability. *J. Mol. Biol.* **357,** 1283–1294.

Wood, S. J., Wetzel, R., Martin, J. D., and Hurle, M. R. (1995). Prolines and amyloidogenicity in fragments of the Alzheimer's peptide β/A4. *Biochemistry* **34,** 724–730.

[11]   Sedimentation Velocity Analysis of Amyloid Oligomers and Fibrils

By YEE-FOONG MOK and GEOFFREY J. HOWLETT

Abstract

The different aggregation states of amyloid oligomers and fibrils have been associated with distinct biological properties and disease pathologies. These various amyloid species are distinguished by their different molecular weights and sedimentation coefficients and can be consistently resolved, separated, and analyzed using sedimentation velocity techniques. We first describe the theoretical background and use of the preparative ultracentrifuge to separate amyloid fibrils and their oligomeric intermediates from monomeric subunits as well as the factors and limits involved in such methods. The approach can be used to monitor the kinetics of fibril formation as well as providing purified fractions for functional analysis. Secondly, we describe the use of analytical ultracentrifugation as a precise and robust system for monitoring the rate of sedimentation of amyloid fibrils under different solution conditions. Sedimentation velocity procedures to characterize the size, interactions, and tangling of amyloid fibrils as well as the binding of nonfibrillar components to form heterologous complexes are detailed.

Introduction

The self-association of proteins into amyloid fibrils involves the transient formation of a range of soluble oligomers and protofibrillar intermediates, some of which are implicated as pathogens in amyloid-linked diseases (Dahlgren *et al.*, 2002; El-Agnaf *et al.*, 2000; Kayed *et al.*, 2003; Reixach *et al.*, 2004; Roher *et al.*, 1996). Furthermore, the accumulation of mature amyloid fibrils as large insoluble deposits leads to organ dysfunction as manifest in the systemic amyloidoses (Hawkins and Pepys, 1995). This deposition process *in vivo* is accompanied by the accumulation of nonfibrillar components, such as serum amyloid P component (SAP), apolipoprotein (apo) E, proteoglycans, and lipids (Sipe and Cohen, 2000). Nonfibrillar components influence the interactions and tangling of amyloid fibrils (MacRaild *et al.*, 2004) and have the potential to exert regulatory effects on both proteolytic and innate immune surveillance mechanisms (Tennent *et al.*, 1995). Resolution and separation of the variously sized

METHODS IN ENZYMOLOGY, VOL. 413   0076-6879/06 $35.00
                    DOI: 10.1016/S0076-6879(06)13011-6

aggregates and complexes involved in amyloid diseases are therefore essential for understanding the mechanism of amyloidogenesis and for developing ways to control the process.

Options for the separation and structural characterization of amyloid fibrils and their solution behavior are limited by the extreme size and insoluble nature of amyloid fibrils. Because the oligomeric states of amyloid fibrils are distinguished by their different molecular weights and sedimentation coefficients, the preparative ultracentrifuge has been used extensively as a consistent way to separate mature fibrils from monomers and smaller oligomers. In addition, the analytical ultracentrifuge, which couples centrifugation to an optical system that determines the concentration distribution of a sample in solution as it sediments, provides a precise and robust system for monitoring the rate of sedimentation of amyloid fibrils under different solution conditions. Analysis of sedimentation velocity behavior offers a number of advantages over other classic procedures for analyzing interacting systems, such as dynamic light scattering, size-exclusion chromatography, or electrophoresis. For particles of constant friction coefficient, sedimentation rates depend on the {2/3}-power of the molar mass. This is in contrast to diffusion-based methods, such as chromatography and dynamic light scattering, which depend on the Stokes radius and where separation is based on the {1/3}-power of the mass. This explains the vastly increased resolution of sedimentation for size distributions compared with these other techniques. Furthermore, sedimentation analysis has a firm theoretical basis; involves no matrices, surfaces, or bulk flow; and is extremely versatile with respect to the size ranges of the interacting species under consideration. This chapter describes the use of the preparative ultracentrifuge to separate amyloid fibrils and their oligomeric intermediates from monomeric subunits. This approach can be used to monitor the kinetics of fibril formation as well as providing purified fractions for functional analysis. In addition, we describe the use of the analytical ultracentrifuge to characterize the size, interactions, and tangling of amyloid fibrils as well as the binding of nonfibrillar components to form heterologous complexes.

Preparative Centrifugation

Centrifugation has become an important component in the methodologies of many research workers for preparing amyloid samples, with the rationale being that the distinct size differences between soluble oligomers, protofibrils, and mature fibrils permit their separation by defined ultracentrifugation conditions. Some common centrifugation parameters used to prepare aggregates of different amyloid proteins are listed in Table I.

TABLE I
COMMON CENTRIFUGAL PARAMETERS USED TO PREPARE SAMPLES OF AMYLOID FIBRILS

Amyloid protein	Centrifugal parameters	Application	Reference
Aβ(1–40)	10,000g for 20 min	Pelleting of pH- or metal-induced aggregates	(Atwood et al., 1998)
	315,000g for 30 min	Pelleting of protofibrils	(Williams et al., 2005)
Aβ(1–42)	16,000g for 20 min through Centricon filters (Eppendorf 5415C microfuge)	Retention of low-molecular-weight intermediates	(Kirkitadze et al., 2001)
	14,000g for 10 min	Clearance of insoluble fibrils	(Wang et al., 2002)
	16,000g for 30 min, 4°, or 100,000g for 30 min, 4° (TLA 120.1 rotor)	Separation of oligomers from mature fibrils	(Stine et al., 2003)
α-Synuclein	50 μl at 16,000g for 5 min	Clearance of insoluble fibrils	(Conway et al., 2000)
	16,000g for 10 min	Clearance of insoluble fibrils	(Cole et al., 2005)
Transthyretin	14,000g for 30 min, 4°	Pelleting of mature fibrils	(Reixach et al., 2004)
Apolipoprotein C-II	350,000g for 20 min, 4° (TLA 100.1 rotor)	Pelleting of mature fibrils	(Hatters et al., 2002)

In general, it is thought that relatively short, low-speed spins around 14,000–16,000g sufficiently pellet insoluble fibrils of the major amyloid proteins from solution, whereas longer high-speed spins at 100,000g sediment protofibrils or even oligomers. This general tenet, however, needs to be used with caution, because studies with the β-amyloid peptide (Aβ) show that centrifugation for 30 min at 16,000g still leaves a significant proportion of mature fibrils in the supernatant (Stine et al., 2003). To prepare amyloid samples of a defined type consistently, an understanding of the parameters that affect the sedimentation of a protein is very important. This section provides a discussion of the factors that govern the sedimentation behavior of amyloid proteins in a preparative centrifuge and the limits of such preparative methods in separating different size classes of aggregates. The theoretical description of sedimentation that follows is aimed at general research workers in the field and is presented in terms of a simple, intuitive mechanical model of sedimentation. For more rigorous treatment of the derivations, the reader is directed to excellent texts elsewhere (Schachman, 1959; Tanford, 1961; VanHolde, 1985).

Theory

The movement of a macromolecule in a uniform solution within a centrifugal or gravitational field is determined by a gravitational force that is opposed by a frictional force. Because the frictional force depends directly on the velocity of movement, the particle initially accelerates but almost instantaneously reaches a constant velocity that is determined by the field strength. This constant velocity is expressed by the sedimentation coefficient (s), defined as the velocity of sedimentation per unit field strength:

$$s = \frac{\left(\frac{dr}{dt}\right)}{\omega^2 r} \tag{1}$$

where ω is the angular velocity of the rotor in rads per second, r is the radial position of the particle relative to the axis of rotation (Table II), and the sedimentation coefficient is expressed in Svedberg's units (S) of 10^{-3} s^{-1}. The gravitational force (F_g) is given by the buoyant mass of the particle multiplied by the gravitational field strength:

$$F_g = (m - vol.\rho) \cdot \omega^2 r \tag{2}$$

where the buoyant mass of the particle is defined as the mass of the particle (m) reduced by the mass of the displaced fluid, given as the product of the particle volume (vol) and the solution density (ρ). The opposing frictional force (F_f) is directly proportional to the velocity of the particle and related by a frictional coefficient f as follows:

$$F_f = -f\frac{dr}{dt} \tag{3}$$

Equating F_g and F_f and combining Eqs. (1–3) for a macromolecule of molecular mass (M) and partial specific volume $(\bar{v})$ in milliliters per gram yields:

$$s = \frac{\left(\frac{d\ln r}{dt}\right)}{\omega^2} = \frac{M(1 - \bar{v}\rho)}{Nf} \tag{4}$$

where N is Avogadro's number. Equation (4) relates the sedimentation coefficient of a sedimenting species to its molecular weight, partial specific volume, and the frictional coefficient. Thus, macromolecules with different molecular weights, partial specific volumes, and frictional coefficients will move at different velocities and yield different sedimentation coefficients in a given solution density and centrifugal field. Implicit in Eq. (4) is the dependence of the sedimentation coefficient on solution viscosity (η),

TABLE II
TYPICAL RATES OF SEDIMENTATION DURING PREPARATIVE CENTRIFUGATION

Angular velocity[a] (centrifugal field)	Sedimentation coefficient[b]	Time required to sediment	Radial values used
13,000 rpm (13,700g)	50 S	6.1 h	r_b = 8.0 cm
	500 S	36.8 min	r_m = 6.5 cm
	5000 S	3.7 min	45° · Rotor axis
100,000 rpm (364,000g)	50 S	14.3 min	r_b = 4.0 cm
	500 S	1.4 min	r_m = 2.5 cm
	5000 S	8.6 s	Rotor axis

[a] The times required to sediment species of the sedimentation coefficient indicated from the meniscus (r_m) to the base of the tube (r_b) were calculated using Eq. (6) assuming 13,700g for a standard microfuge and 364,000g for a high-speed preparative ultracentrifuge.
[b] A sedimentation coefficient of 50 S corresponds to a molecular mass of approximately 1.65×10^6 Da for a spherical particle of partial specific volume 0.73 ml/g in standard buffer conditions. The relation between sedimentation coefficients and molecular mass for typical amyloid fibrils is described in the section on interpreting sedimentation coefficient distributions.

which affects the frictional coefficient. Experimentally determined values for sedimentation coefficients are generally corrected to standard conditions of pure water at 20° as:

$$s_{20,w} = s \cdot \left(\frac{(1 - \bar{v}\rho)_{T,b}}{(1 - \bar{v}\rho)_{20,w}} \right) \left(\frac{\eta_{20,w}}{\eta_{T,b}} \right) \tag{5}$$

where the subscript 20,w denotes the value for water at 20° and the subscript T,b denotes the experimental conditions.

Equation (4) can be integrated to calculate the time required (t) for a species of known s to sediment from the sample meniscus, r_m, to the base of the tube, r_b, for any chosen angular velocity:

$$\ln r_b = \text{in } r_m = s\omega^2 t \tag{6}$$

Table II lists some typical values for the rate of sedimentation of species ranging in s from 50 to 5000 S using centrifugal fields representative of the standard bench-top microfuge and high-speed preparative ultracentrifuge. Data calculated using Eq. (6), such as those shown in Table II, can be used to choose experimental conditions for separating amyloid fibrils, oligomeric intermediates, and monomeric subunits. Conversely, in cases in which the sedimentation coefficients of specific species are not known, Eq. (6) can be used to provide a rough estimate of the relevant sedimentation coefficients based on the time required to deplete the meniscus and supernatant of the species of interest.

Several points regarding the use of Eq. (6) to calculate appropriate conditions for the sedimentation of amyloid fibrils should be noted. Firstly, the time needed for sedimentation depends on the difference between r_m and r_b. Thus, sample volume and tube geometry and orientation are factors that need to be taken into account. Secondly, as indicated in Eq. (5), sedimentation coefficients depend on solution viscosity and density. Effects attributable to common salts and buffer components are readily calculated from measured values of density and viscosity and are usually small. However, the use of concentrated solutes or low temperatures may require corrections for the effects of density and viscosity on sedimentation (Eq. 5). Thirdly, in using the preparative centrifuge for preparing fibril fractions, it should be kept in mind that the concentrations of sedimenting species increase significantly at the bottom of the tube, an effect that has the potential to alter the size and character of the fibrils. Fourthly, a major assumption in the use of a preparative centrifuge to separate oligomeric intermediates and fibrils is that the concentration gradients formed during centrifugation are stable and do not stir during deceleration of the rotor. This assumption depends on the design and setting of the braking system and the nature of the pellet formed during sedimentation. Previous studies on the use of a preparative ultracentrifuge to study sedimentation equilibrium behavior establish the importance of density stabilization during braking (Bothwell et al., 1978). For concentrated protein samples (e.g., >1 mg/ml), density stabilization provided by the protein concentration gradient itself is usually adequate. For less concentrated protein samples, however, the addition of an inert macromolecule, such as dextran T10 (2 mg/ml), may be required to stabilize the tube contents during braking of the rotor (Bothwell et al., 1978; Darawshe et al., 1993). A simple test of the need for density stabilization is to determine the fraction of sedimenting material in serially diluted samples.

Experimental Procedures

The results presented in Fig. 1 provide an example of the use of a bench-top preparative ultracentrifuge to monitor the kinetics of amyloid formation by human plasma apo C-II. Human apoC-II was expressed and purified from *Escherichia coli* and stored in 5 *M* GuHCl at 30 g/L. This solution was directly diluted to 1.0 g/L in 100 m*M* sodium phosphate, 0.1 % sodium azide, pH 7.4. Aliquots from this diluted solution (0.2 ml) were sedimented in polycarbonate tubes at 350,000g for 20 min using a Beckman TLA-100 ultracentrifuge (Beckman/Coulter, Fullerton, CA) at 20°. Our previous studies have shown that during amyloid formation, apoC-II exists as a bimodal population of monomers (s = 1 S) and large aggregates (s > 50 S) at any given time point, suggesting that nucleation is the rate-limiting step (Hatters *et al.*, 2000). Centrifugation under these conditions pellets aggregate but not monomer, thus providing a means to monitor aggregate formation. The concentration of apoC-II in the supernatant after centrifugation was monitored using fluorescamine reactivity with primary amine groups (Bohlen *et al.*, 1973). This involved rapidly mixing 0.15 ml of supernatant with 75 μl buffer (100 m*M* sodium phosphate, 0.1% sodium azide, pH 7.4) and 75 μl fluorescamine solution (0.5 g/L fluorescamine in acetonitrile). The fluorescence was measured using a fluorescence plate reader with a 350-nm/460-nm excitation/emission filter set. For comparison, fibril formation was also monitored by thioflavin T fluorescence measurements. Aliquots of 25 μl apoC-II (1 g/L) were added to final solution volumes

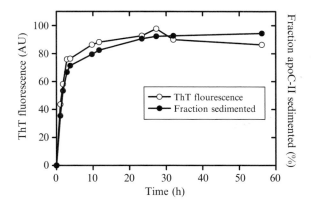

FIG. 1. Time course of amyloid formation by 1 mg/ml apoC-II as measured by thioflavin T reactivity (left axis, open circles) and fluorescamine assay of the pellet obtained after ultracentrifugation at 120,000g for 30 min (right axis, closed circles). Adapted and reproduced with permission from MacRaild *et al.* (2003).

of 0.2 ml containing 100 mM sodium phosphate, 0.1% (w/v) sodium azide, 5 μM thioflavin T, pH 7.4. The fluorescence was monitored using a fluorescence plate reader with a 444-nm/485-nm excitation/emission filter set. The results in Fig. 1 show that the fraction of apoC-II sedimenting as a function of incubation time increases with a kinetic profile that closely follows the development of thioflavin T reactivity, demonstrating good agreement between these two complementary methods for monitoring amyloid fibril formation.

Analytical Ultracentrifugation

Analytical ultracentrifugation involves the measurement and subsequent analysis of solute distributions formed during centrifugation. The two main techniques, sedimentation equilibrium and sedimentation velocity, use identical instrumentation but differing experimental protocols. In sedimentation equilibrium experiments, short solution columns and low speeds are used to ensure that equilibrium is reached. Analysis of the time-invariant concentration distribution yields information about molecular masses and interactions. A major advantage of this approach is that the analysis of equilibrium gradients makes no assumptions regarding hydrodynamics or molecular shape. Sedimentation equilibrium methods have been useful in defining the molecular nature of the subunits that form amyloid fibrils (Hammarstrom et al., 2001; Lashuel et al., 1998, 2002a). However, the large sizes of amyloid fibrils in general prevent their analysis by sedimentation equilibrium methods, because the fibrils fail to form a measurable equilibrium concentration gradient and quickly sediment to the bottom of the cell, even at very low angular velocities.

A significant advantage of sedimentation velocity, on the other hand, is the capacity to analyze large fibrillar structures. Sedimentation velocity methods are based on measurement of the rates of transport of solutes in a centrifugal field. Because the lowest angular velocity achievable in the analytical ultracentrifuge is approximately 1500 rpm or 164g, consideration of Eq. (6) for typical values of $r_m = 6.0$ cm and $r_b = 7.0$ cm indicates that sedimentation coefficients up to 10,000 S can be determined. Experimentally determined sedimentation coefficients have the potential to yield valuable information about the size, shape, and molecular interactions of amyloid fibrils, as indicated by several recent studies of amyloid fibrils composed of transthyretin, Aβ, α-synuclein, or apoC-II (Lashuel et al., 1998; Lashuel et al., 2002a; Lashuel et al., 2002b; MacRaild et al., 2003). Typical sedimentation velocity experiments using the analytical ultracentrifuge are completed in 1–2 h, allowing the growth and size distributions of fibrils to be monitored over periods such as those shown in Fig. 1. In considering the use of sedimentation velocity experiments for the analysis of amyloid fibrils, however, it should be borne in mind that some amyloid fibril systems form flocculent precipitates that settle in the absence of centrifugation. In such cases, the use of the analytical ultracentrifuge would be inappropriate.

Theory

The evolution of the concentration distribution, $\chi(r,t)$, of a single species of diffusing particles in a spinning rotor is given by the Lamm equation (Fujita, 1962):

$$\frac{d\chi}{dt} = \frac{1}{r}\frac{d}{dr}\left[rD\frac{d\chi}{dr} - s\omega^2 r^2 \chi\right] \tag{7}$$

where D is the translational diffusion coefficient. Analysis of experimental data in terms of both s and D has until recently been dependent on approximate solutions of Eq. (7) (Fujita, 1962; Holloday, 1979; Philo, 1997) or on graphic transformations of the raw data (VanHolde and Weischet, 1978). Although these approaches have proved valuable, they have imposed significant limitations on experimental design and on the levels of sample complexity that can be addressed. More recently, it has become practical to employ numeric solutions to the Lamm equation as fitting functions, enabling the direct fitting of the experimental data to models comprising sedimentation coefficient distribution (c(s)) or several noninteracting components (Demeler and Saber, 1998; Schuck, 1998, 2000). This approach is applied in a recent analysis of the polymerization of α1-antitrypsin to form dimers, trimers, and higher order oligomers (Devlin et al., 2002).

For large amyloid fibrils, the degree of diffusion during sedimentation is negligible and simpler analytical procedures pertain. The evolution of the concentration distribution of a population of identical nondiffusing particles with a uniform initial distribution is given by step functions:

$$U(r,t) = e^{-2\omega^2 st} \times \begin{cases} 0 \text{ for } r < r_{\mathrm{m}} e^{\omega^2 st} \\ 1 \text{ else} \end{cases} \tag{8}$$

Most notable among the methods used to analyze the sedimentation velocity behavior of nondiffusing particles is the recently developed ls-g*(s) approach, in which experimental concentration distributions are modeled in a least-squares fashion by a summation of step functions weighted according to a variable distribution of sedimentation coefficients (Schuck and Rossmanith, 2000). Analogous distributions are also available from time-derivative analysis methods in which apparent sedimentation coefficient distributions are obtained as a transform of the time-derivative of the experimental concentration distributions (Philo, 2000; Stafford, 1992). These approaches are especially valuable for the analysis of amyloid fibrils for which the assumption of no diffusion is a good approximation. The ls-g*(s) procedure is encapsulated in a program, SEDFIT, for analyzing sedimentation data (Schuck, 1998, 2000; Schuck and Demeler, 1999; Schuck and Rossmanith, 2000; Schuck et al., 2002). This program is freely available (http://www.analyticalultracentrifugation.com). The web site is, in addition, an excellent resource for information on the different approaches used in the sedimentation data analysis of both diffusing and nondiffusing particles using the c(s), ls-g*(s) and the VanHolde-Weischet method.

Experimental Procedures

Figure 2 shows sedimentation velocity profiles of apoC-II fibrils obtained using a Beckman XL-A analytical ultracentrifuge equipped with an An 60 Ti rotor at 20°. The data were obtained as follows: apoC-II fibrils were prepared by dilution of a concentrated stock of apoC-II in 5 M GuHCl into 100 mM sodium phosphate buffer, 0.1% sodium azide, pH 7.4, to final apoC-II concentrations of 0.3, 0.7, and 1.0 mg/ml, followed by incubation at room temperature for 1 week. Approximately 350 μl of each sample was loaded into one sector of a double-sector centerpiece, whereas reference buffer solution was loaded into the other sector. A small excess volume of reference is loaded to ensure that the reference meniscus does not obscure the sample meniscus. Data were collected at an angular velocity of 4000 rpm, with radial increments of 0.002 cm in continuous scanning mode, with scans taken at 10-min intervals at 292 nm. This wavelength was chosen to give initial starting optical densities in the range 0.2–1.2. Greater sensitivity can be achieved at a wavelength of 230 nm, where advantage is taken of the high absorbance of peptide bonds and high intensity of the light source at this wavelength. Under these conditions, it is important to avoid high concentrations of components like sodium azide, which absorb strongly at low wavelengths. The data in Fig. 2 indicate a significant increase in the rate of movement of the sedimenting boundaries as the apoC-II starting concentration for fibril formation is increased.

Measurements can also be made using the interference optics available on the model XL-I. In this case, it is important to match the meniscus closely to avoid artifacts caused by the sedimentation of buffer salts. Interference optics offer the advantage that scans can be collected quickly (less than 10 s) with high accuracy and sensitivity and are useful for solution conditions in which there is interference from strongly absorbing ligands or buffer components. In performing sedimentation velocity experiments, it is important to ensure that the solution contents are properly equilibrated at the chosen temperature, because temperature gradients cause convection and stirring of the sample. In addition, the viscosity of aqueous samples varies by

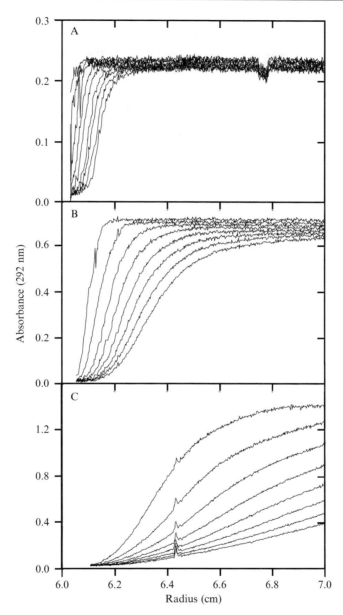

FIG. 2. Sedimentation velocity analysis of amyloid fibrils formed by 1-week incubation of apoC-II at 0.3 (A), 0.7 (B) and 1.0 (C) mg/ml. Ultracentrifugation was performed at 4000 rpm, and radial scans were taken every 10 min. For clarity, only every second scan is shown. Adapted and reproduced with permission from MacRaild *et al.* (2003).

approximately 3% per degree Centigrade, leading to a direct effect on sedimentation coefficients (Eq. 5) and data analysis.

Data Analysis

The sedimentation velocity data in Fig. 2 were analyzed using the program SEDFIT assuming nondiffusing particles to obtain ls-g*(s) sedimentation coefficient distributions (Fig. 3). Fitting the data requires an appropriate range of sedimentation coefficients to be chosen. In practice, this is achieved by initially choosing a wide range and then narrowing the selection. The data in Fig. 2 were fitted using 200 sediment coefficient increments within the ranges of 1–500 S (for Panel A), 20–2000 S (for Panel B), and 20–3000 S (for Panel C). An option is also available in SEDFIT to fit the data using logarithmic rather than linear spacing. This approach gives better coverage of sedimentation coefficients in the lower range. SEDFIT also contains an option to calculate the weight-average sedimentation coefficient of the population and the standard deviation about the mean. The distributions from the ls-g*(s) analysis (Fig. 3) indicate an increase in the weight-average sedimentation coefficient of the fibrils and a greater degree of heterogeneity for apoC-II fibrils formed at higher starting concentrations. The areas under the distributions in Fig. 3 are proportional to the amount of sedimenting solute and can be used to assess the fibrillar content of different samples.

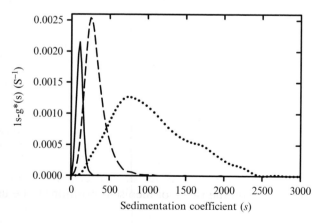

FIG. 3. Continuous sedimentation coefficient distributions of amyloid fibrils as determined by ls-g*(s) analysis of the data shown in Fig. 2. Distributions are for fibrils formed by 1-week incubation of apoC-II at 0.3 (solid line), 0.7 (dashed line), and 1.0 (dotted line) mg/ml. The analysis used 200 sedimentation coefficient (s)-values between 1 and 3000 S and a confidence level of 0.68. Adapted and reproduced with permission from MacRaild et al. (2003).

There are a number of other considerations for the analysis of sedimentation velocity distributions using the ls-g*(s) approach and SEDFIT:

1. The ls-g*(s) approach assumes that the species in solution sediments independently without significant reversible interactions. A test of this assumption is to establish that the distributions obtained are not affected by diluting the sample or by collecting the data at different angular velocities. For the apoC-II sample formed at 0.3 mg/ml, identical sedimentation coefficient distributions were obtained from separate experiments performed at 4000, 8000, and 10,000 rpm, whereas the data for the sample formed at 1 mg/ml gave similar distributions when diluted two- and fourfold (MacRaild et al., 2003).

2. Algorithms are included in SEDFIT to minimize time-independent and radial-independent noise. Time-independent noise may arise from optical imperfections in the cell windows, whereas radial-independent noise arises as a result of the entire scan moving up or down at different time intervals. This latter effect is particularly relevant to data collected using interference optics and is attributed to "fringe jitter." The use of time- and radial-independent noise-fitting options can lead to baseline corrections that are sloped or curved and not physically reasonable. This difficulty could arise from the presence of low molecular-weight impurities and can be addressed by thorough dialysis of the samples before centrifugation. This is particularly important for data collected using interference optics, where imbalances in the concentrations of buffer salts add to the measured signal. Other options in the fitting procedure are to fit the data assuming a single value for the baseline nonsedimenting absorbance or to fix the baseline correction to an experimentally determined value obtained at higher angular velocities.

3. The radial position of the meniscus is required for the determination of sedimentation coefficient distributions. This value cannot be determined with absolute precision from the experimental data using either absorbance or interference data and can be included in the fitting procedure to obtain optimal fits to the data. The use of the radial position of the meniscus as a fitting parameter may compensate for other uncertainties in fitting the data, such as the time taken to accelerate to the final rotor speed and minor variations in the initial temperature that affect the initial rates of sedimentation. Limits on the range of meniscus positions to be used in the fitting procedure can be applied.

4. In some cases, it is necessary to fit data that are composed of a range of amyloid fibrils in the presence of significant amounts of starting material of added nonfibrillar component. SEDFIT contains an option to use ls-g*(s) analysis with an additional diffusing component. The parameters for the

single diffusing solute can be fixed from values determined in separate experiments.

5. The SEDFIT program contains an option to fit data using the VanHolde and Weischet approach (VanHolde and Weischet, 1978). The approach has been used to characterize the distribution of amyloidogenic states of transthyretin, Aβ, and α-synuclein (Lashuel *et al.*, 1998, 2002a,b). In this approach, each scan of the sedimenting boundary is divided into equal concentration fractions and the position of each fraction is converted to apparent sedimentation coefficients. These are then calculated for sequential scans and combined with an extrapolation to infinite time to give the sedimentation coefficients of the fractions corrected for the effects of diffusion. For samples of mixed species, only boundary divisions that originate from positions in the solution in which the sample is homogeneous yield true apparent *s*-values (Schuck *et al.*, 2002), and as such, the sedimentation coefficient distribution should only be qualitatively interpreted in terms of the upper and lower limits of the distribution. The sedimentation data for apoC-II amyloid fibrils formed at 0.3 mg/ml (Fig. 2, Panel A) were analyzed using the VanHolde-Weischet model with a fraction resolution of 10 (Fig. 4). The relatively constant apparent *s*-values of each boundary fraction (Fig. 4, upper panel) are consistent with a model in which there is very little diffusion of the fibrils. The resultant integral sedimentation coefficient distribution (Fig. 4, lower panel) is in good agreement with the sedimentation coefficient distribution for 0.3 mg/ml apoC-II fibrils as shown in Fig. 3. The advantages of the ls-g*(s) approach presented in Fig. 3 are that it provides a more intuitive presentation of the distributions, fits the entire collection of data points, and permits other fitting parameters, including radial- and time-independent noise and the radial position of the meniscus.

Interpreting Sedimentation Coefficient Distributions

Sedimentation coefficient distributions, such as those shown in Figs. 3 and 4, provide a convenient way of comparing samples of amyloid fibrils formed under different conditions. The final distributions depend on a number of factors, including the molecular weight and shape of the fibrils as well as the degree of interactions or tangling of the fibrils. The shape dependence is encompassed in the frictional factor for the sedimenting species, which, for spherical particles, is given by:

$$f = 6\pi\eta a \tag{9}$$

where *a* is the radius of a sphere. Combination of Eqs. (4) and (9) relates the sedimentation coefficient and molecular weight of a spherical particle. No equivalent relation exists for amyloid fibrils. However, approximate

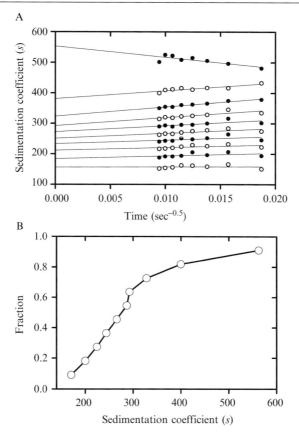

FIG. 4. Sedimentation coefficient distributions of amyloid fibrils as determined by VanHolde-Weischet analysis of the data shown in Fig. 2. (A) Sedimentation boundaries were divided into 10 fractions for each scan and converted to apparent sedimentation coefficient (s)-values (circles) and then extrapolated to infinite time (lines). (B) Integral distribution plot showing the sedimentation coefficient for each of the 10 boundary fractions in (A).

relations have been developed using simple bead models and worm-like chains (MacRaild *et al.*, 2003). The results in Fig. 5 show the relations between amyloid fibril length and sedimentation coefficient for apoC-II based on an assigned value for the mass-per-unit length and persistence length. The values obtained for simulated worm-like chains agree well with approximate analytical solutions (Hearst and Stockmayer, 1962). The capacity to relate the sedimentation coefficient distribution for amyloid fibrils to a molecular weight distribution is important in comparing different kinetic models for fibril formation.

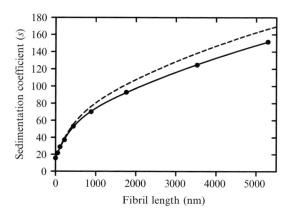

FIG. 5. Theoretical sedimentation coefficients of apoC-II amyloid fibrils using bead models of simulated worm-like chains. The approximate analytical solution of Hearst and Stockmayer (1962) is shown as a dashed line. Adapted and reproduced with permission from MacRaild *et al.* (2003).

Several observations suggest that the distributions for apoC-II fibrils presented in Figs. 3 and 4 are determined by both the size of the fibrils and fibril-fibril interactions (MacRaild *et al.*, 2004). First, the magnitude of the sedimentation coefficients obtained for fibrils formed at 1 mg/ml (Fig. 2, Panel C) would require single fibrils that are several centimeters long to account for the observed sedimentation coefficients. Second, the measured sedimentation coefficient distribution continues to increase, even after the depletion of monomeric apoC-II. Finally, subsequent rheometry experiments using a cone-and-plate viscometer show that apoC-II fibrils prepared at 1 mg/ml behave as a weak gel consistent with weak fibril–fibril interactions. Analysis of sedimentation coefficient distributions for apoC-II fibrils in the presence of added nonfibrillar components of amyloid deposits, such as SAP and apoE, indicate that these agents increase fibril tangling and network condensation (MacRaild *et al.*, 2004). These results illustrate the general utility of sedimentation velocity analysis in studies of amyloid fibrils, where fibril tangling and interactions and the binding of nonfibrillar components could significantly affect the metabolism and fate of *in vivo* amyloid deposits.

References

Atwood, C. S., Moir, R. D., Huang, X., Scarpa, R. C., Bacarra, N. M., Romano, D. M., Hartshorn, M. A., Tanzi, R. E., and Bush, A. I. (1998). Dramatic aggregation of Alzheimer abeta by Cu(II) is induced by conditions representing physiological acidosis. *J. Biol. Chem.* **273**, 12817–12826.

Bohlen, P., Stein, S., Dairman, W., and Udenfriend, S. (1973). Fluorometric assay of proteins in the nanogram range. *Arch. Biochem. Biophys.* **155,** 213–220.

Bothwell, M. A., Howlett, G. J., and Schachman, H. K. (1978). A sedimentation equilibrium method for determining molecular weights of proteins with a tabletop high speed air turbine centrifuge. *J. Biol. Chem.* **253,** 2073–2077.

Cole, N. B., Murphy, D. D., Lebowitz, J., Di Noto, L., Levine, R. L., and Nussbaum, R. L. (2005). Metal-catalyzed oxidation of alpha-synuclein: Helping to define the relationship between oligomers, protofibrils, and filaments. *J. Biol. Chem.* **280,** 9678–9690.

Conway, K. A., Lee, S. J., Rochet, J. C., Ding, T. T., Williamson, R. E., and Lansbury, P. T., Jr. (2000). Acceleration of oligomerization, not fibrillization, is a shared property of both alpha-synuclein mutations linked to early-onset Parkinson's disease: Implications for pathogenesis and therapy. *Proc. Natl. Acad. Sci. USA* **97,** 571–576.

Dahlgren, K. N., Manelli, A. M., Stine, W. B., Jr., Baker, L. K., Krafft, G. A., and LaDu, M. J. (2002). Oligomeric and fibrillar species of amyloid-beta peptides differentially affect neuronal viability. *J. Biol. Chem.* **277,** 32046–32053.

Darawshe, S., Rivas, G., and Minton, A. P. (1993). Rapid and accurate microfractionation of the contents of small centrifuge tubes: Application in the measurement of molecular weight of proteins via sedimentation equilibrium. *Anal. Biochem.* **209,** 130–135.

Demeler, B., and Saber, H. (1998). Determination of molecular parameters by fitting sedimentation data to finite-element solutions of the Lamm equation. *Biophys. J.* **74,** 444–454.

Devlin, G. L., Chow, M. K., Howlett, G. J., and Bottomley, S. P. (2002). Acid denaturation of alpha1-antitrypsin: Characterization of a novel mechanism of serpin polymerization. *J. Mol. Biol.* **324,** 859–870.

El-Agnaf, O. M., Mahil, D. S., Patel, B. P., and Austen, B. M. (2000). Oligomerization and toxicity of beta-amyloid-42 implicated in Alzheimer's disease. *Biochem. Biophys. Res. Commun.* **273,** 1003–1007.

Fujita, H. (1962). "Mathematical Theory of Sedimentation Analysis." Academic Press, New York.

Hammarstrom, P., Jiang, X., Deechongkit, S., and Kelly, J. W. (2001). Anion shielding of electrostatic repulsions in transthyretin modulates stability and amyloidosis: Insight into the chaotrope unfolding dichotomy. *Biochemistry* **40,** 11453–11459.

Hatters, D. M., MacPhee, C. E., Sawyer, W. H., and Howlett, G. J. (2000). Human apolipoprotein C II forms twisted amyloid ribbons and closed loops. *Biochemistry* **39,** 8276–8283.

Hatters, D. M., Minton, A. P., and Howlett, G. J. (2002). Macromolecular crowding accelerates amyloid formation by human apolipoprotein C-II. *J. Biol. Chem.* **277,** 7824–7830.

Hawkins, P. N., and Pepys, M. B. (1995). Imaging amyloidosis with radiolabelled SAP. *Eur. J. Nucl. Med.* **22,** 595–599.

Hearst, J. E., and Stockmayer, W. H. (1962). Sedimentation constants of broken and wormlike coils. *J. Chem. Phys.* 1425–1433.

Holloday, L. (1979). An approximate solution of the Lamm equation. *Biophys. Chem.* **10,** 187–190.

Kayed, R., Head, E., Thompson, J. L., McIntire, T. M., Milton, S. C., Cotman, C. W., and Glabe, C. G. (2003). Common structure of soluble amyloid oligomers implies common mechanism of pathogenesis. *Science* **300,** 486–489.

Kirkitadze, M. D., Condron, M. M., and Teplow, D. B. (2001). Identification and characterization of key kinetic intermediates in amyloid beta-protein fibrillogenesis. *J. Mol. Biol.* **312,** 1103–1119.

Lashuel, H. A., Lai, Z., and Kelly, J. W. (1998). Characterization of the transthyretin acid denaturation pathways by analytical ultracentrifugation: Implications for wild-type, V30M, and L55P amyloid fibril formation. *Biochemistry* **37,** 17851–17864.

Lashuel, H. A., Hartley, D. M., Balakhaneh, D., Aggarwal, A., Teichberg, S., and Callaway, D. J. (2002a). New class of inhibitors of amyloid-beta fibril formation. Implications for the mechanism of pathogenesis in Alzheimer's disease. *J. Biol. Chem.* **277,** 42881–42890.

Lashuel, H. A., Petre, B. M., Wall, J., Simon, M., Nowak, R. J., Walz, T., and Lansbury, P. T., Jr. (2002b). Alpha-synuclein, especially the Parkinson's disease-associated mutants, forms pore-like annular and tubular protofibrils. *J. Mol. Biol.* **322,** 1089–1102.

MacRaild, C. A., Hatters, D. M., Lawrence, L. J., and Howlett, G. J. (2003). Sedimentation velocity analysis of flexible macromolecules: Self-association and tangling of amyloid fibrils. *Biophys. J.* **84,** 2562–2569.

MacRaild, C. A., Stewart, C. R., Mok, Y. F., Gunzburg, M. J., Perugini, M. A., Lawrence, L. J., Tirtaatmadja, V., Cooper-White, J. J., and Howlett, G. J. (2004). Non-fibrillar components of amyloid deposits mediate the self-association and tangling of amyloid fibrils. *J. Biol. Chem.* **279,** 21038–21045.

Philo, J. S. (1997). An improved function for fitting sedimentation velocity data for low-molecular-weight solutes. *Biophys. J.* **72,** 435–444.

Philo, J. S. (2000). A method for directly fitting the time derivative of sedimentation velocity data and an alternative algorithm for calculating sedimentation coefficient distribution functions. *Anal. Biochem.* **279,** 151–163.

Reixach, N., Deechongkit, S., Jiang, X., Kelly, J. W., and Buxbaum, J. N. (2004). Tissue damage in the amyloidoses: Transthyretin monomers and nonnative oligomers are the major cytotoxic species in tissue culture. *Proc. Natl. Acad. Sci. USA* **101,** 2817–2822.

Roher, A. E., Chaney, M. O., Kuo, Y. M., Webster, S. D., Stine, W. B., Haverkamp, L. J., Woods, A. S., Cotter, R. J., Tuohy, J. M., Krafft, G. A., Bonnell, B. S., and Emmerling, M. R. (1996). Morphology and toxicity of Abeta-(1-42) dimer derived from neuritic and vascular amyloid deposits of Alzheimer's disease. *J. Biol. Chem.* **271,** 20631–20635.

Schachman, H. K. (1959). "Ultracentrifugation in Biochemistry." Academic Press, New York.

Schuck, P. (1998). Sedimentation analysis of noninteracting and self-associating solutes using numerical solutions to the Lamm equation. *Biophys. J.* **75,** 1503–1512.

Schuck, P. (2000). Size-distribution analysis of macromolecules by sedimentation velocity ultracentrifugation and Lamm equation modeling. *Biophys. J.* **78,** 1606–1619.

Schuck, P., and Demeler, B. (1999). Direct sedimentation analysis of interference optical data in analytical ultracentrifugation. *Biophys. J.* **76,** 2288–2296.

Schuck, P., and Rossmanith, P. (2000). Determination of the sedimentation coefficient distribution by least-squares boundary modeling. *Biopolymers* **54,** 328–341.

Schuck, P., Perugini, M. A., Gonzales, N. R., Howlett, G. J., and Schubert, D. (2002). Size-distribution analysis of proteins by analytical ultracentrifugation: Strategies and application to model systems. *Biophys. J.* **82,** 1096–1111.

Sipe, J. D., and Cohen, A. S. (2000). Review: History of the amyloid fibril. *J. Struct. Biol.* **130,** 88–98.

Stafford, W. F., 3rd (1992). Boundary analysis in sedimentation transport experiments: A procedure for obtaining sedimentation coefficient distributions using the time derivative of the concentration profile. *Anal. Biochem.* **203,** 295–301.

Stine, W. B., Jr., Dahlgren, K. N., Krafft, G. A., and LaDu, M. J. (2003). *In vitro* characterization of conditions for amyloid-beta peptide oligomerization and fibrillogenesis. *J. Biol. Chem.* **278,** 11612–11622.

Tanford, C. (1961). "Physical Chemistry of Macromolecules." Wiley, New York.

Tennent, G. A., Lovat, L. B., and Pepys, M. B. (1995). Serum amyloid P component prevents proteolysis of the amyloid fibrils of Alzheimer disease and systemic amyloidosis. *Proc. Natl. Acad. Sci. USA* **92,** 4299–4303.

VanHolde, K., and Weischet, W. (1978). Boundary analysis of sedimentation velocity experiments with monodisperse and paucidisperse solutes. *Biopolymers* **17,** 1387–1403.

VanHolde, K. (1985). Sedimentation. *In* "Physical Biochemistry" (K. VanHolde, ed.), pp. 110–136. Prentice Hall, Englewood Cliffs, NJ.

Wang, H. W., Pasternak, J. F., Kuo, H., Ristic, H., Lambert, M. P., Chromy, B., Viola, K. L., Klein, W. L., Stine, W. B., Krafft, G. A., and Trommer, B. L. (2002). Soluble oligomers of beta amyloid (1-42) inhibit long-term potentiation but not long-term depression in rat dentate gyrus. *Brain Res.* **924,** 133–140.

Williams, A. D., Sega, M., Chen, M., Kheterpal, I., Geva, M., Berthelier, V., Kaleta, D. T., Cook, K. D., and Wetzel, R. (2005). Structural properties of Abeta protofibrils stabilized by a small molecule. *Proc. Natl. Acad. Sci. USA* **102,** 7115–7120.

[12] Structural Study of Metastable Amyloidogenic Protein Oligomers by Photo-Induced Cross-Linking of Unmodified Proteins

By GAL BITAN

Abstract

Oligomers of amyloidogenic proteins are believed to be key effectors of cytotoxicity and cause a variety of amyloid-related diseases. Dissociation or inhibition of formation of the toxic oligomers is thus an attractive strategy for the prevention and treatment of these diseases. In order to develop reagents capable of inhibiting protein oligomerization, the structures and mechanisms of oligomer formation must be understood. However, structural studies of oligomers are difficult because of the metastable nature of the oligomers and their existence in mixtures with monomers and other assemblies. A useful method for characterization of oligomer size distributions *in vitro* is photo-induced cross-linking of unmodified proteins (PICUP) (Fancy and Kodadek, 1999). By providing "snapshots" of dynamic oligomer mixtures, PICUP enables quantitative analysis of the relations between primary and quaternary structures, offering insights into the molecular organization of the oligomers. This chapter discusses the photochemical mechanism; reviews the scope, usefulness, and limitations of PICUP for characterizing metastable protein assemblies; and provides detailed experimental instructions for performing PICUP experiments.

METHODS IN ENZYMOLOGY, VOL. 413   0076-6879/06 $35.00
                    DOI: 10.1016/S0076-6879(06)13012-8

Introduction

The Role of Protein Oligomers in Amyloidosis

Amyloidogenic proteins are characterized by their tendency to aggregate into β-sheet-rich amyloid fibrils, leading to a variety of pathologic conditions. Diseases characterized by accumulation of amyloid fibrils are termed *amyloidoses* (Buxbaum, 1996). These diseases can be systemic (Buxbaum, 2004) (e.g., light-chain amyloidosis), or affect particular tissues, such as the pancreas in type II diabetes mellitus (Marzban *et al.*, 2003). Some of the most devastating amyloidoses affect the central nervous system, including Alzheimer's disease (AD), Parkinson's disease (PD), Huntington's disease (HD), prion diseases (e.g., "mad cow" disease), and amyotrophic lateral sclerosis (ALS, Lou Gehrig disease) (Trojanowski and Mattson, 2003). The amyloidogenic proteins that cause these diseases have diverse sequences, origins, and structures. Nevertheless, they all share the tendency to aggregate into amyloid fibrils. Fibrils isolated from diseased tissues or prepared from recombinant or synthetic amyloidogenic proteins (e.g., amyloid-β protein [Aβ], α-synuclein, transthyretin, islet amyloid polypeptide [IAPP]), are cytotoxic *in vitro* and *in vivo* (Gambetti and Russo, 1998). In view of these data, for many years, the prevailing paradigm, known as the "amyloid cascade hypothesis" (Hardy and Higgins, 1992), mandated that aggregation of amyloidogenic proteins into fibrils caused the respective amyloidoses. However, accumulating evidence from studies in humans, normal rodents, transgenic mice, cultured cells, and *in vitro* systems now suggests that soluble, oligomeric assembly intermediates of amyloidogenic proteins are the primary pathogenetic effectors in amyloidoses (Kirkitadze *et al.*, 2002; Thirumalai *et al.*, 2003; Walsh and Selkoe, 2004b). The majority of the data regarding oligomer assembly and toxicity have been obtained in studies of Aβ, the primary cause of AD (Mattson, 2004; Walsh and Selkoe, 2004a), which is often considered an archetype of amyloidogenic proteins (Lazo *et al.*, 2005). The evidence is not limited to Aβ or AD, however. Abundant data obtained for other proteins demonstrate that oligomer formation may be a common mechanism by which amyloidogenic proteins cause disease (Conway *et al.*, 2000; Demuro *et al.*, 2005; El-Agnaf *et al.*, 2001; Malisauskas *et al.*, 2005; Reixach *et al.*, 2004). In addition, protein-folding studies have shown that under suitable conditions, globular proteins that do not normally aggregate and are not associated with amyloidosis also form oligomers and fibrils similar to those formed by amyloidogenic proteins (Chiti *et al.*, 2002). Interestingly, oligomers formed by such proteins were found to be cytotoxic, whereas the counterpart fibrils were benign (Bucciantini *et al.*, 2002). Taken together, these data have supported a paradigm shift (Kirkitadze *et al.*, 2002) and a revision of the amyloid cascade hypothesis (Hardy, 2002; Hardy and Selkoe, 2002) that de-emphasize the role of fibrils and ascribe pathogenetic primacy to oligomeric assemblies. Thus, protein oligomers are new key targets of strategies developed to treat diseases associated with protein misfolding and aggregation.

Challenges in Biophysical Characterization of Amyloidogenic Protein Oligomers

In order for efforts toward disrupting protein oligomers to be successful, the oligomer structures and assembly processes must be understood. However, structural and biophysical characterization of oligomers of amyloidogenic proteins is difficult, because the oligomers are metastable and often exist in dynamically changing mixtures comprising monomers, oligomers of different sizes, and polymers. Classic, high-resolution structural biology methods, such as X-ray crystallography and solution-phase nuclear magnetic resonance (NMR), are not suitable for study of metastable oligomers. Therefore, a variety of lower resolution biochemical, biophysical, immunologic, and computational techniques have been employed for oligomer characterization (Bitan *et al.*, 2005; also see Chapter 11 by Mok and Howlett, and Chapter 17 by Kayed and Glabe in this volume). Each of these methods generates a limited set of data. Therefore, current views of oligomer structure and assembly are synergistic syntheses of multiple data sets obtained using a variety of strategies and techniques.

An important aspect of the structural characterization of protein oligomers is determination of oligomer order. Attempts to characterize the oligomer order of amyloidogenic proteins in general, and $A\beta$ in particular, using various biophysical and biochemical methods have not yielded a consensus (Bitan *et al.*, 2001). Reasons for lack of consensus have included using methods with limited resolution (e.g., dynamic light scattering, electron microscopy, size-exclusion chromatography, ultracentrifugation) or prone to artifacts (e.g., sodium dodecyl sulfate polyacrylamide gel electrophoresis [SDS-PAGE]) (Bitan *et al.*, 2005). An ideal method for determining oligomer size in a situation in which metastable oligomers exist in dynamically changing mixtures would provide accurate, quantitative "snapshots" of the distributions. Because oligomers dissociate back into monomers and associate into larger assemblies over time, the method should be applicable within intervals significantly shorter than the lifetime of the assemblies under study. In addition, in order to reveal accurately the native oligomerization state of the protein under investigation, the method should require no *pre facto* protein modifications and be applicable under physiological conditions. Photo-induced cross-linking of unmodified proteins (PICUP), a method originally developed by Fancy and Kodadek for study of stable protein complexes (Fancy and Kodadek, 1999), has most of the characteristics of an ideal method for this task. PICUP enables cross-linking of proteins within time intervals of 1 s without *pre facto* modification of the native sequence and is applicable within wide pH and temperature ranges, including physiological values. Other cross-linking methods, such as chemical cross-linking using bifunctional linkers (Das and Fox, 1979; Kluger and Alagic, 2004) or benzophenone/arylazide-based photoaffinity labeling (Knorre and Godovikova, 1998; Kotzyba-Hibert *et al.*, 1995) require substantially longer reaction times. In addition, some chemical cross-linking reactions necessitate nonphysiological pH, and photoaffinity labeling relies on incorporation of nonnative functional groups into the protein. Therefore, PICUP is superior to these methods for studying native, metastable protein oligomers.

PICUP

PICUP Photochemistry

The photochemistry of PICUP is based on photo-oxidation of Ru^{2+} in a tris(bipyridyl)Ru(II) complex (Ru(Bpy)) to Ru^{3+} by irradiation with visible light in the presence of an electron acceptor. Ru(Bpy) is a common, commercial chemical used in a variety of photochemical reactions (Bjerrum *et al.*, 1995). In Ru(Bpy), Ru^{2+} can become excited upon absorption of photons with $\lambda_{max} = 452$ nm ($\varepsilon = 14{,}600$ M^{-1} [Kalyanasundaram, 1982]) (Reaction 1):(1)

$$Ru^{2+} \xrightarrow[\lambda_{max}=452 \text{ nm}]{h\nu} Ru^{2+*} \tag{1}$$

If a suitable electron acceptor, A, is available, the Ru^{2+*} ion will donate the excited electron to the acceptor and become oxidized to Ru^{3+} (Reaction 2). A common electron acceptor in PICUP chemistry is ammonium persulfate (APS). An alternative acceptor is $Co(III)(NH_3)_5Cl^{2+}$ (Fancy *et al.*, 2000):(2)

$$Ru^{2+*} + A \longrightarrow Ru^{3+} + A^{\bullet -} \tag{2}$$

(Note that $A^{\bullet -}$ represents the oxidation state of a generic electron acceptor after Reaction 2. The actual ionization state of the reduced acceptor following reaction with Ru^{2+*} depends on its initial oxidation state. For example, following reduction, the persulfate anion ($S_2O_8^{2-}$) decomposes into $SO_4^{2-} + SO_4^{\bullet -}$, whereas $Co(III)(NH_3)_5Cl^{2+}$ is reduced to $Co(II)(NH_3)_5Cl^{+}$.)

Ru^{3+} is a strong (+1.24 V) one-electron oxidizer capable of abstracting an electron from a neighboring protein molecule, generating a protein radical (Reaction 3). As long as irradiation continues and sufficient electron acceptor is available, Ru^{2+} can be recycled into Reaction 1, get oxidized again to Ru^{3+}, and generate more protein radicals:(3)

$$\text{(structure)} + R^3 \xrightarrow{\;-H^-\;} \text{(structure)} + Ru^{2}1 \qquad (3)$$

Radicals are unstable, highly reactive species and therefore disappear rapidly through a variety of intra- and intermolecular reactions. One route a radical may utilize to relieve the high energy caused by an unpaired electron is to react with another protein monomer to form a dimeric radical, which may subsequently lose a hydrogen atom and form a stable, covalently cross-linked dimer (Reaction 4). The dimer then may react further through a similar mechanism with monomers or other dimers, leading to the formation of higher order oligomers:

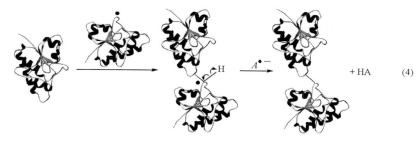

The potential for a particular functional group in a protein to react with Ru^{3+}, donate an electron, and form a radical or to react with another protein radical depends on a number of parameters, including the capability of the group to stabilize an unpaired electron, the proximity of the group to the Ru^{3+} ion or to a radical on a neighboring protein, and the structure of the protein. Stabilization of a radical can occur via mechanisms like resonance, hyperconjugation, neighboring group effect, or a combination of the three. Based on electronic considerations, the probability for the side chains of the amino acids Trp and Tyr to be sites of radical formation and/ or reaction is highest, whereas that of the side chain of Ala is lowest among the 20 natural amino acids. A radical also can form, in principle, on the protein backbone. However, this is unlikely, because steric interference hinders backbone atoms from being in close proximity to the Ru^{3+} ion or to a neighboring protein radical. For the same reason, the probability of radical formation/reaction on the α-carbon (C_α) of Gly is low. The surrounding environment of each functional group strongly influences the potential for radical formation on, or reaction with, this group. Tyr is highly prone to form a radical upon reaction with Ru^{3+} (Reaction 3) (Fancy, 2000; Fancy et al., 2000). The human amyloidogenic peptides, $A\beta(1–40)$ (40 residues), calcitonin (CT, 32 residues), and IAPP (37 residues) each contain a single Tyr residue (residue 10 in $A\beta$, residue 12 in CT, and residue 37 in IAPP). Because Trp is not present in these peptides, Tyr likely is the most reactive residue in each of these peptides in PICUP chemistry. When subjected to PICUP, ≈80% of Aβ and ≈75% of CT monomers react to form cross-linked oligomers (Bitan *et al.*, 2001), whereas only ≈30% of IAPP monomers form oligomers (G. Bitan, unpublished results), demonstrating the strong influence of the environment of the Tyr residue in each peptide on its reactivity in PICUP chemistry. The difference in reactivity is not merely an effect of the C-terminal position of Tyr^{37} in IAPP, because when Tyr^{10} in Aβ is repositioned at the C-terminus, as in the analogue $[Phe^{10},Tyr^{40}]Aβ(1–40)$, ≈80% of the monomer reacts to form oligomers, similar to wild-type (WT) Aβ(1–40) (S. K. Maji and D. B. Teplow, personal communication). When neither Trp nor Tyr is present in a peptide, the overall cross-linking efficiency is substantially lower than even in the presence of a single Tyr. For example, when Tyr^{10} in Aβ is substituted by Phe as in $[Phe^{10}]Aβ(1–40)$ or $[Phe^{10}]Aβ(1–42)$, the cross-linking yield (monomer conversion into oligomers) decreases from ≈80% for both WT Aβ(1–40) and Aβ(1–42), to 51% and 33%, respectively (S. K. Maji and D.B. Teplow, personal communication). Similarly, when residues 1–10 of Aβ are deleted, as in Aβ(11–40) and Aβ(11–42), the cross-linking yields of the N-terminally truncated peptides are 43% and 38%, respectively (Bitan *et al.*, 2003c). Notably, Aβ alloforms lacking Tyr always form abundant dimers and, in some cases, trimers and tetramers as well, indicating that amino acid residues other than Tyr and Trp are reactive in PICUP chemistry. It will be important and interesting to determine the reactivity of each of the 20 natural amino acids in PICUP chemistry, both in forming a radical (Reaction 3) and in reacting with one (Reaction 4) in different protein conformations. Such data will enable making predictions about the feasibility and usability of PICUP for particular protein systems. Until such data become available, reaction conditions must be optimized empirically for each experimental system.

Optimizing the Experimental System

For optimization of an experimental system, it is important to consider the factors that determine the result of a PICUP experiment, which include the reactivity of the protein under study, the steady state concentration of Ru^{3+} ions, $[Ru^{3+}]^{\ddagger}$, and the protein/Ru(Bpy) ratio (the Ru(Bpy)/APS ratio should be kept at 1:20). $[Ru^{3+}]^{\ddagger}$ is a function of the initial concentration of Ru^{2+}, the characteristics of the irradiation system, and the time of irradiation. Practically, for optimization of cross-linking yield, it is convenient to maintain constant protein and Ru(Bpy) concentrations and modify the irradiation time systematically. Using this protocol, we found that for 60 μM Ru(Bpy) and a Ru(Bpy)/Aβ(1–40) concentration ratio of 2:1, efficient cross-linking occurred with 0.5–8 s of illumination using a 150-W incandescent lamp positioned 10 cm from the reaction vessel (Bitan *et al.*, 2001). Within this time range, irradiation time had only a moderate effect on the observed oligomer size distribution of $A\beta(1-40)$. At shorter irradiation times, formation of trimer and tetramer decreased substantially. At higher irradiation times, extensive radical reactions caused protein degradation, and "fading away" of entire lanes (Bitan et al., 2001). A similar effect was observed using excess (fivefold) Ru(Bpy) and 1-s irradiation (G. Bitan, unpublished results).

For studies of the relation between protein concentration and oligomerization state, once an optimal irradiation period has been determined, it is important to maintain a constant protein/Ru(Bpy) ratio. For example, under the experimental conditions described above, we determined that the oligomer size distributions of $A\beta(1-40)$ and $A\beta(1-42)$ at 30 and 300 μM were essentially unchanged, whereas a shift in abundance toward smaller oligomers was observed when $A\beta(1-40)$ was diluted below 10 μM or when $A\beta(1-42)$ was diluted below 3 μM (G. Bitan and D.B. Teplow, unpublished results). The distributions observed for dilute (<3 μM) $A\beta(1-42)$ were similar to those observed by other investigators who used nanomolar concentrations of $A\beta(1-42)$ (Crouch et al., 2005; LeVine, 2004).

The choice of detection method for protein oligomers following PICUP depends on the starting protein preparation and the protein concentration. In the examples mentioned in the previous paragraph, SDS-PAGE and silver staining were used in our laboratory, whereas LeVine (2004) and Crouch et al. (2005) used Western blot analysis for visualization of $A\beta$ (1–42) oligomers. The results were qualitatively similar. Immunodetection must be used for biological samples in which the protein of interest exists in a mixture with other proteins (e.g., in cell extracts or conditioned cell culture media). Caution must be exercised when Western blot analysis is used for detection of PICUP products, because antigenic epitopes may be modified by radical reactions and such modifications may affect certain products more than others. This would complicate data interpretation, because it would be difficult to distinguish between a situation in which certain oligomers form with a low yield because of inherent instability and low detection of stable oligomers because of modification of antigenic epitopes. This potential problem may be overcome by using several antibodies recognizing different epitopes of the same protein.

PICUP products may be analyzed without fractionation using a variety of morphological and spectroscopic methods (e.g., Bitan et al., 2003a). Fractionation using size exclusion chromatography (SEC) provides lower resolution than SDS-PAGE but enables further analysis of isolated oligomers individually, without the need to remove SDS from the isolated fractions (Bitan et al., 2003a). Analyzing cross-linking products using mass spectrometry (MS) would offer advantages relative to SDS-PAGE, because oligomers can be assigned unambiguously based on their mass rather than their electrophoretic mobility, which does not always correlate directly to mass (Bitan *et al.*, 2005). However, detection of oligomers by MS following PICUP has been difficult. We have attempted to analyze $A\beta40$ that had been subjected to PICUP using both matrix-assisted laser desorption ionization (MALDI) and electrospray ionization (ESI) techniques. Crude PICUP reaction mixtures yielded no signal in either technique. Fractionation of the mixtures by high-performance liquid chromatography (HPLC) or SEC interfaced with an ESI source produced predominantly monomer signals. Detection of $A\beta$ oligomers (dimer through hexamer) by MALDI time of flight (TOF) was enabled eventually following purification of the oligomers by SEC, using ammonium acetate as the mobile phase, and lyophilization of this volatile buffer (G. Bitan, D. Teplow, R. Loo, and J. Loo, unpublished results).

The type of protein preparation dictates not only the choice of method for analysis of the PICUP products but also the reaction conditions and the way the data are interpreted. When pure proteins are studied, the Ru (Bpy)/protein stoichiometry should be maintained at $\sim2:1$. As mentioned above, lower ratios will decrease the cross-linking yield and may lead to misrepresentation of higher order oligomers, whereas higher stoichiometric ratios increase formation of artifactual, diffusion-controlled cross-linking products and may promote protein degradation. When the protein preparation is more complex, (e.g., cell culture medium or cell extract), other reactive molecules, including proteins and carbohydrates, compete for reaction with Ru^{3+}. Therefore, substantially larger (10–100-fold) amounts of cross-linking reagents are required. In these preparations, in addition to cross-linking of oligomers, if they exist, cross-linking of the protein of interest to other proteins (or nonproteinaceous molecules) also may be observed. This provides an opportunity to study interactions of a protein of interest with its binding partners but may complicate interpretation of the data (Lin and Kodadek, 2005). For example, it may be difficult to distinguish between a cross-linked dimer and a cross-linked complex of two different proteins of similar size.

For experiments using biological samples, it should be noted that APS and $Co(III)(NH_3)_5Cl^{2+}$ are not cell-permeable. Therefore, cross-linking of intracellular proteins using these reagents is not feasible unless the cells are permeabilized artificially.

Scope and Limitations of PICUP

PICUP was originally developed for studies of stable protein assemblies. Proof of concept was given using UvsY, a native protein hexamer involved in phage T4 recombination (Beernink and Morrical, 1998). When UvsY was cross-linked using PICUP, the main product was a hexamer (Fancy and Kodadek, 1999). Similar results were obtained for the enzymes glutathione *S*-transferase (Fancy *et al.*, 2000), glyoxylate aminotransferase (Lumb and Danpure, 2000), muscle acylphosphatase (Paoli *et al.*, 2001), hormone-sensitive lipase (Shen *et al.*, 2000), the prokaryotic RNA-editing enzyme tadA (Wolf *et al.*, 2002), and the yeast transcription factor Pho4 (Fancy *et al.*, 2000), all of which form stable dimers. Other studies found the predicted oligomerization patterns for the yeast mating-type proteins SMTA-1 and SMTa-1, which form homo- and heterodimers (Jacobsen *et al.*, 2002), and for Cowpea mosaic virus subunit, which is a stable pentamer (Meunier *et al.*, 2004). PICUP also has been applied successfully to characterization of protein-ligand interactions, including mapping the interaction of signal recognition particle (SRP) with various signal sequences (Cleverley and Gierasch, 2002), binding of the transcription factor ETS-1 to stromelysin-1 promoter (Baillat *et al.*, 2002), and affinity labeling of G-protein-coupled receptors for bioactive peptide hormones, including bradykinin, angiotensin, vasopressin, and oxytocin, using agonists and antagonists derived from the native hormones (Duroux-Richard *et al.*, 2005). The latter study demonstrated the usefulness of PICUP not only for cross-linking of proteins in buffers or cell-extracts but for studies of membrane-bound proteins. Additional uses of PICUP included "fishing out" specific interactions in mixtures of peptides and proteins (Lin and Kodadek, 2005) and modulation of cell adhesion to glass (Luebke *et al.*, 2004).

The studies listed above demonstrate the usefulness of PICUP in stabilizing protein oligomers for analysis using denaturing methods (e.g., SDS-PAGE). In addition, important features of the method itself were gleaned. In all cases, in addition to the predicted stable oligomer(s), monomers and, where appropriate, lower order oligomers, were observed following PICUP and SDS-PAGE analysis. These products reflect the fact that the cross-linking efficiency is <100% and non-cross-linked oligomers can dissociate in the presence of SDS. An opposite effect also was observed in certain cases—diffusion-controlled cross-linking of pre-existing oligomers with monomer yielded artifactual, higher order oligomers.

An important question for studies of oligomer size distributions of metastable protein oligomers is whether artifactual oligomers formed by diffusion-controlled cross-linking can be distinguished from bona fide pre-existing oligomers. To answer this question, we applied PICUP to two amyloidogenic peptides, $A\beta(1\text{--}40)$ and CT, and two peptides of similar size, growth hormone-releasing factor (GRF) and pituitary adenylate cyclase-activating polypeptide (PACAP), which have not been reported to oligomerize or form amyloid under physiological conditions. In all cases, oligomers were observed following cross-linking (Bitan *et al.*, 2001). To distinguish pre-existing oligomers from those formed by diffusion-controlled cross-linking of monomers, the observed distributions were compared with theoretical distributions produced using a mathematical model, which assumes no association among molecules except for random, diffusion-controlled elastic collision (see Bitan *et al.*, 2001 for details).

Figure 1 shows SDS-PAGE analysis of the four cross-linked peptides (Fig. 1A), a modeled distribution obtained under high-efficiency conditions (Fig. 1B), and densitometric analysis of each lane (Fig. 1C–F). In the

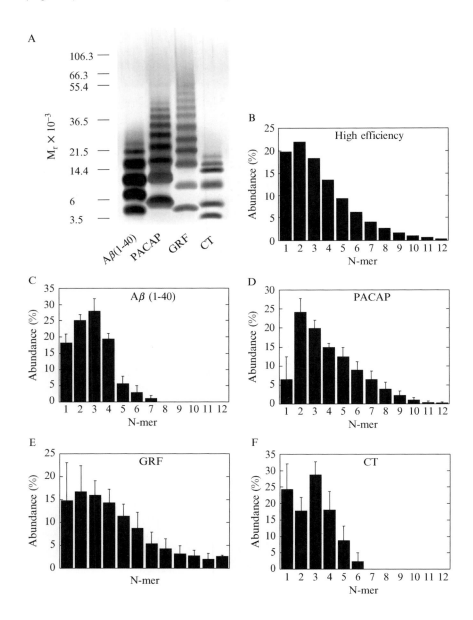

absence of pre-existing oligomers and under high-efficiency cross-linking conditions, $\approx$80% of monomer is converted to oligomers ranging from dimer through dodecamer (Fig. 1B). The experimental oligomer size distributions obtained for PACAP (Fig. 1D) and GRF (Fig. 1E) were similar to this theoretical distribution, with the exception of a higher cross-linking efficiency observed for PACAP, leading to consumption of $\approx$95% of the monomer. Both distributions were characterized by formation of a "ladder" of oligomers extending up to a dodecamer and by an exponential decline in oligomer abundance. These distributions were distinct from those observed for Aβ(1–40) (Fig. 1C) or CT (Fig. 1F). Importantly, for both Aβ(1–40) and CT, the cross-linking efficiency was similar to that of the model and of GRF, consuming $\approx$75–80% of the monomer. As discussed above, cross-linking efficiency depends on the local environment of each reactive group. Therefore, when oligomer size distributions of different peptides are compared, it is important that the cross-linking yield is similar for all peptides. Aβ(1–40) and CT yielded oligomer size distributions that did not extend beyond hexamer (CT) or heptamer (Aβ (1–40)). In addition, the abundance of monomer through tetramer for Aβ (1–40) and of monomer through trimer for CT diverged from an exponential pattern. These differences indicated that the solutions of Aβ(1–40) and CT contained species other than peptide monomers and suggested that these species were pre-existing oligomers. In both cases, the abundance of higher oligomers (pentamer through heptamer for Aβ(1–40) and tetramer through hexamer for CT) declined exponentially, demonstrating that diffusion-controlled cross-linking of pre-existing oligomers to monomers is an inevitable side reaction. A conservative interpretation of these data is that oligomers whose abundance diverges from an exponential decline pattern are bona fide pre-existing oligomers, whereas oligomers whose abundance declines exponentially likely are generated by diffusion-controlled cross-linking.

FIG. 1. Sodium dodecyl sulfate polyacrylamide gel electrophoresis (SDS-PAGE) and densitometric analysis of photo-induced cross-linking of unmodified proteins (PICUP) products of amyloidogenic and nonamyloidogenic peptides. (A) Low-molecular-weight (LMW) preparations of amyloidogenic (amyloid β protein (Aβ)(1–40) and calcitonin [CT]) and nonamyloidogenic (pituitary adenylate cyclase-activating polypeptide [PACAP] and growth hormone releasing factor [GRF]) peptides were prepared by filtration through a 10-kDa molecular-weight cutoff filter (Bitan and Teplow, 2005) and cross-linked immediately. A silver-stained gel is shown. Positions of molecular weight standards are shown on the left. (B) Theoretical distribution of monomers in the absence of preassociation under high-efficiency cross-linking conditions. (C–F) densitometric analysis of the gel bands in panel A. Reproduced with permission from Bitan *et al.* (2001).

PICUP as a Tool for Structural Studies

PICUP enables quantitative study of metastable, quaternary protein structures. Thus, by studying the effect of amino acid sequence modifications on the quaternary structure, the relation between primary and quaternary structures of metastable protein oligomers can be delineated. These relations can have a great impact on protein bioactivity. For example, certain amyloidoses are caused by mutations, resulting in single amino acid substitutions in the respective amyloid protein (Buxbaum and Tagoe, 2000). Studying the effect of such substitutions on protein oligomerization may be crucial to understanding disease mechanism.

We have applied PICUP to the study of primary-quaternary structure relations of $A\beta$ (Bitan and Teplow, 2004). The predominant $A\beta$ alloforms in the brain are $A\beta(1–40)$ and $A\beta(1–42)$. $A\beta(1–40)$ is ≈ 10 times more abundant than $A\beta(1–42)$. Nevertheless, genetic, pathologic, and biochemical evidence demonstrates that $A\beta(1–42)$ is linked most strongly to the etiology of AD (Selkoe, 2001). Oligomers of $A\beta(1–42)$ have been shown to be more neurotoxic than those of $A\beta(1–40)$ (Dahlgren et al., 2002; Hoshi et al., 2003), but the mechanistic basis for these toxicity differences is not known. Using PICUP, we found that $A\beta(1–40)$ and $A\beta(1–42)$ form distinct oligomer size distributions. $A\beta(1–40)$ forms a roughly equimolar, quasi-equilibrium mixture of monomer, dimer, trimer, and tetramer, whereas $A\beta(1–42)$ preferentially forms pentamer/hexamer units (Fig. 2), which self-associate into larger assemblies, including dodecamers and octadecamers, and hence were termed *paranuclei* (Bitan et al., 2003a). Consistent with the PICUP data, distinct particle size distributions of $A\beta(1–40)$ and $A\beta(1–42)$ were observed by dynamic light scattering (Bitan et al., 2003a). Morphological studies showed that $A\beta(1–40)$ oligomers were amorphous, whereas $A\beta(1–42)$ paranuclei appeared as spheroids ≈ 5 nm in diameter (Bitan et al., 2003a). These differences in oligomer size distribution and morphology between $A\beta(1–40)$ and $A\beta(1–42)$ offer a plausible explanation for the differences in neurotoxicity observed for the two alloforms.

Insight into the mechanism(s) controlling the distinct oligomerization behavior of $A\beta(1–40)$ and $A\beta(1–42)$ was obtained by examination of PICUP-derived oligomer size distributions of $A\beta$ analogues ending in positions 39–43. With the exception of $A\beta(1–41)$, these alloforms are found in $A\beta$ samples from cultured cells (Wang et al., 1996) and AD patients (Mori et al., 1992; Wiltfang et al., 2002). The oligomer size distribution of $A\beta(1-39)$ was essentially identical to that of $A\beta(1–40)$, but the distributions obtained for $A\beta(1–41)$, $A\beta(1–42)$, and $A\beta(1–43)$ were distinct and demonstrated that paranucleus formation did not occur in the absence of Ile-41 (Fig. 2) (Bitan et al., 2003a). Subsequent studies demonstrated that the side

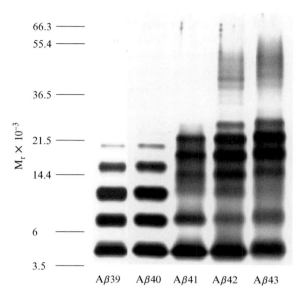

FIG. 2. C-terminal length-dependence of amyloid β protein (Aβ) oligomer size distribution. Low-molecular-weight Aβ(1–39), Aβ(1–40), Aβ(1–41), Aβ(1–42), and Aβ(1–43) were cross-linked individually and analyzed by sodium dodecyl sulfate polyacrylamide gel electrophoresis. Positions of molecular weight standards are shown on the left. Reproduced with permission from Bitan *et al.* (2003a).

chain in position 41 and the C-terminal carboxylate group of Aβ(1–42) are critical modulators of paranucleus assembly (Bitan *et al.*, 2003c). Study of clinically relevant alloforms containing substitutions in the midregion of Aβ and of N-terminally truncated Aβ analogues, which are found in plaques from AD patients, demonstrated that Aβ(1–40) oligomerization is largely affected by charge alterations at the N-terminus and in positions 22 and 23, whereas oligomer formation by Aβ(1–42) is controlled primarily by hydrophobic interactions and is highly sensitive to conformational changes at the central hydrophobic region (Bitan *et al.*, 2003c). Further study showed that oxidation of Met-35, a modification often found in Aβ extracted from AD brain (Nordstedt *et al.*, 1994), abolishes formation of Aβ(1–42) paranuclei but has no effect on early oligomerization of Aβ (1–40) (Bitan *et al.*, 2003b). Thus, structural data obtained using PICUP demonstrated that modification of as little as one atom can induce dramatic effects on Aβ assembly and provided important insights into the mechanism by which Aβ assembles into neurotoxic oligomers relevant to AD pathogenesis.

Experimental Protocol

Materials

1. Light source. Both 150-W Xe lamps and 150-W incandescent lamps have been used successfully (Bitan *et al.*, 2001; Fancy *et al.*, 2000) (Notes 1 and 2).
2. Reaction apparatus allowing controlled exposure and positioning of samples a fixed distance from the light source. We and others have used a 35-mm single lens reflex (SLR) camera body to control exposure time (Bitan *et al.*, 2001; Fancy and Kodadek, 1999). In our setting, a bellows attached to the camera in place of a lens provides a convenient means to place the sample and control its distance from the light source (Fig. 3). The data described above were obtained with the light source at a distance of 10 cm from the sample.
3. Clear, thin-walled plastic tubes (Note 3).
4. Tris(2,2′-bipyridyl)dichlororuthenium(II) hexahydrate (Ru(Bpy), Sigma), 1 mM, in 10 mM sodium phosphate, pH 7.4 (Notes 4–6).

Fig. 3. Schematic cross-linking system. The reaction mixture is prepared in a polymerase chain reaction (PCR) tube (see Note 3) immediately before irradiation. A glass vial is used to hold the PCR tube within the dark chamber (bellows). The sample is illuminated through the open back of a camera body using the camera shutter mechanism to control the illumination time.

5. APS (Sigma), 20 mM, in 10 mM sodium phosphate, pH 7.4 (Notes 5 and 6).
6. Low-molecular-weight Aβ (Note 7).
7. Quenching reagent: 5% (v/v) β-mercaptoethanol (β-ME; Sigma) in 2× Sample Buffer (Invitrogen), or 1 M dithiothreitol (DTT; Fisher) in water (Note 8).

Method

General Instructions. The method described here is applicable to samples volume of 20–120 μl. For volumes larger than 120 μl, the cross-linking efficiency declines with increasing sample volume. Using a ratio of 2:40:1 for Ru(Bpy), APS, and LMW Aβ, respectively, the cross-linking yield is relatively insensitive to changes in protein concentration between 10–50 μM. Lower or higher concentrations may require empirical adjustment of the Ru(Bpy)/protein ratio. The Ru(Bpy)/APS ratio should be kept at 1:20. Longer irradiation may be necessary for highly diluted samples for the same cross-linking yield to be obtained. If larger amounts of cross-linked protein are desired, several samples can be pooled together following cross-linking and quenching of each sample.

Specific Steps

1. Prepare the peptide or protein sample as appropriate. Here, LMW Aβ was isolated according to published protocols (Note 7).
2. Transfer an 18-μl aliquot to a polymerase chain reaction (PCR) tube (Note 3).
3. Add 1 μl Ru(Bpy) and 1 μl APS, and mix by drawing up and expelling solution several times from a pipette tip (Notes 9–11).
4. Place in the illumination chamber (bellows), and irradiate for 1 s (Note 12).
5. Quench immediately by mixing with either 10 μl β-ME in Sample Buffer or 1 μl DTT (Note 8).
6. Cross-linked samples may be stored in a $-20°$ freezer for 7–10 days prior to analysis. Longer storage of samples may result in decreased resolution on a gel.

Notes

1. Other lower intensity sources of light can be used (Fancy and Kodadek, 1999). Irradiation time must be adjusted empirically to maximize cross-linking efficiency. Care should be taken, because long irradiation may induce protein degradation.

2. Filtering (IR) radiation by using distilled water to prevent sample overheating has been used by some researchers (Fancy *et al.*, 2000). We have found this to be unnecessary for short ($\leq$8 s) irradiation times.

3. We have used clear, 0.2-ml PCR tubes (Eppendorf) for sample preparation and a flat-bottom, 1.8-ml glass vial (Kimble Chromatography) as a sample holder (Fig. 3). Others have used larger (1–2 ml) sample tubes. We find that the tube size is not a critical parameter for successful cross-linking as long as the sample can be placed reproducibly at a fixed distance and angle relative to the light source. In our setting, this distance is 10 cm directly in front of the light source. Reproducibility is of critical importance. The absolute distance and angle of the sample from the light source are of lesser importance, with the understanding that cross-linking yield decreases as a function of the distance.

4. Palladium (II) porphyrins also have been used as photoactivators in PICUP chemistry (Fancy *et al.*, 2000; Kim *et al.*, 1999).

5. Buffers other than sodium phosphate can be used, but the efficiency of the cross-linking reaction in different buffers must be determined empirically. The cross-linking yield of Aβ(1–40) in different solvents is $H_2O \approx NaCl > Na_2HPO_4 > NaHCO_3 > NaBO_4$ (10 mM of each buffer or salt was used, buffer pH was 7.4, H_2O and NaCl solution pH was 3.3) (G. Bitan, unpublished results).

6. Dissolution of Ru(Bpy) requires vortexing for $\sim$1 min until the solution is transparent to the eye. The Ru(Bpy) solution is light sensitive and must be protected from ambient light. A simple method is to use aluminum foil to wrap the tube containing the Ru(Bpy) solution. The APS and Ru(Bpy) reagent solutions can be used for up to 48 h following preparation.

7. The method described here uses low-molecular-weight (LMW) Aβ, an aggregate-free preparation described elsewhere (Bitan and Teplow, 2005; Fezoui *et al.*, 2000; Walsh *et al.*, 1997). However, the method is readily applied to the analysis of other peptides and proteins, with appropriate optimization of reaction conditions. The most important factors to consider are the reagent stoichiometry, irradiation time, and sample preparation procedure (see the section on optimizing the experimental system). The former two issues require empirical optimization. The latter issue largely determines how the experimental data are to be interpreted. For amyloidogenic proteins in particular, determination of native oligomerization states requires using aggregate-free starting preparations.

8. The choice of a quenching reagent depends upon the purpose of the cross-linking experiment. Samples analyzed using PAGE are quenched with the appropriate sample buffer containing 5% β-ME. Samples analyzed by chromatography or other methods may be quenched with

1 M DTT. Lower concentrations of DTT (as low as 200 mM) also can be used if preferred.

9. Do not vortex samples containing amyloidogenic proteins, because vortexing may promote their aggregation.

10. In order to prevent cross-linking induced by ambient light, the procedure may be performed in a dark room. However, the efficiency of ambient light-induced cross-linking is low. In our experience, a nonirradiated mixture of Aβ(1–40) and cross-linking reagents yields a very faint dimer band following exposure to ambient light for the same time that normally is required to cross-link such a sample.

11. Proteins also can be cross-linked in biological fluids, such as cultured cell media or cell extracts. Because Ru^{3+} is a nonselective oxidizer, it will react with susceptible components of biological solutions. Therefore, cross-linking of these types of samples requires higher concentrations of reagents, up to 100 mM Ru(Bpy) and 2 M APS. Upon addition of reagents at these high concentrations to the sample, some precipitate may form. This precipitate does not appear to interfere with cross-linking and can be removed by centrifugation or dissolved upon addition of sample buffer after the cross-linking process is complete.

12. Irradiation time should be kept to a minimum and should be optimized empirically (see the section on optimizing the experimental system).

Acknowledgments

The author thanks Drs. David Teplow, Noel Lazo, Erica Fradinger, and Samir Maji for critical reading of the manuscript. This work was supported by American Foundation for Aging grant A04084 and by Larry L. Hillblom Foundation grant 20052E.

References

Baillat, D., Begue, A., Stehelin, D., and Aumercier, M. (2002). ETS-1 transcription factor binds cooperatively to the palindromic head to head ETS-binding sites of the stromelysin-1 promoter by counteracting autoinhibition. *J. Biol. Chem.* **277,** 29386–29398.

Beernink, H. T., and Morrical, S. W. (1998). The uvsY recombination protein of bacteriophage T4 forms hexamers in the presence and absence of single-stranded DNA. *Biochemistry* **37,** 5673–5681.

Bitan, G., Fradinger, E. A., Spring, S. M., and Teplow, D. B. (2005). Neurotoxic protein oligomers—what you see is not always what you get. *Amyloid* **12,** 88–95.

Bitan, G., Kirkitadze, M. D., Lomakin, A., Vollers, S. S., Benedek, G. B., and Teplow, D. B. (2003a). Amyloid β-protein (Aβ) assembly: Aβ40 and Aβ42 oligomerize through distinct pathways. *Proc. Natl. Acad. Sci. USA* **100,** 330–335.

Bitan, G., Lomakin, A., and Teplow, D. B. (2001). Amyloid β-protein oligomerization: Prenucleation interactions revealed by photo-induced cross-linking of unmodified proteins. *J. Biol. Chem.* **276,** 35176–35184.

Bitan, G., Tarus, B., Vollers, S. S., Lashuel, H. A., Condron, M. M., Straub, J. E., and Teplow, D. B. (2003b). A molecular switch in amyloid assembly: Met^{35} and amyloid β-protein oligomerization. *J. Am. Chem. Soc.* **125,** 15359–15365.

Bitan, G., and Teplow, D. B. (2004). Rapid photochemical cross-linking—a new tool for studies of metastable, amyloidogenic protein assemblies. *Acc. Chem. Res.* **37,** 357–364.

Bitan, G., and Teplow, D. B. (2005). Preparation of aggregate-free, low molecular weight Aβ for assembly and toxicity assays. *In* "Amyloid Proteins—Methods and Protocols" (E. M. Sigurdsson, ed.), Vol. 299, pp. 3–10. Humana Press, Totawa, NJ.

Bitan, G., Vollers, S. S., and Teplow, D. B. (2003c). Elucidation of primary structure elements controlling early amyloid β-protein oligomerization. *J. Biol. Chem.* **278,** 34882–34889.

Bjerrum, M. J., Casimiro, D. R., Chang, I. J., Di Bilio, A. J., Gray, H. B., Hill, M. G., Langen, R., Mines, G. A., Skov, L. K., Winkler, J. R., and Wuttke, D. S. (1995). Electron transfer in ruthenium-modified proteins. *J. Bioenerg. Biomembr.* **27,** 295–302.

Bucciantini, M., Giannoni, E., Chiti, F., Baroni, F., Formigli, L., Zurdo, J. S., Taddei, N., Ramponi, G., Dobson, C. M., and Stefani, M. (2002). Inherent toxicity of aggregates implies a common mechanism for protein misfolding diseases. *Nature* **416,** 507–511.

Buxbaum, J. (1996). The amyloidoses. *Mt. Sinai J. Med.* **63,** 16–23.

Buxbaum, J. N. (2004). The systemic amyloidoses. *Curr. Opin. Rheumatol.* **16,** 67–75.

Buxbaum, J. N., and Tagoe, C. E. (2000). The genetics of the amyloidoses. *Annu. Rev. Med.* **51,** 543–569.

Chiti, F., Calamai, M., Taddei, N., Stefani, M., Ramponi, G., and Dobson, C. M. (2002). Studies of the aggregation of mutant proteins *in vitro* provide insights into the genetics of amyloid diseases. *Proc. Natl. Acad. Sci. USA* **99,** 16419–16426.

Cleverley, R. M., and Gierasch, L. M. (2002). Mapping the signal sequence–binding site on SRP reveals a significant role for the NG domain. *J. Biol. Chem.* **277,** 46763–46768.

Conway, K. A., Lee, S. J., Rochet, J. C., Ding, T. T., Williamson, R. E., and Lansbury, P. T. (2000). Acceleration of oligomerization, not fibrillization, is a shared property of both α-synuclein mutations linked to early-onset Parkinson's disease: Implications for pathogenesis and therapy. *Proc. Natl. Acad. Sci. USA* **97,** 571–576.

Crouch, P. J., Blake, R., Duce, J. A., Ciccotosto, G. D., Li, Q. X., Barnham, K. J., Curtain, C. C., Cherny, R. A., Cappai, R., Dyrks, T., Masters, C. L., and Trounce, I. A. (2005). Copper-dependent inhibition of human cytochrome c oxidase by a dimeric conformer of amyloid-β1-42. *J. Neurosci.* **25,** 672–679.

Dahlgren, K. N., Manelli, A. M., Stine, W. B., Jr., Baker, L. K., Krafft, G. A., and LaDu, M. J. (2002). Oligomeric and fibrillar species of amyloid-b peptides differentially affect neuronal viability. *J. Biol. Chem.* **277,** 32046–32053.

Das, M., and Fox, C. F. (1979). Chemical cross-linking in biology. *Annu. Rev. Biophys. Bioeng.* **8,** 165–193.

Demuro, A., Mina, E., Kayed, R., Milton, S. C., Parker, I., and Glabe, C. G. (2005). Calcium dysregulation and membrane disruption as a ubiquitous neurotoxic mechanism of soluble amyloid oligomers. *J. Biol. Chem.* **280,** 17294–17300.

Duroux-Richard, I., Vassault, P., Subra, G., Guichou, J. F., Richard, E., Mouillac, B., Barberis, C., Marie, J., and Bonnafous, J. C. (2005). Crosslinking photosensitized by a ruthenium chelate as a tool for labeling and topographical studies of G-protein-coupled receptors. *Chem. Biol.* **12,** 15–24.

El-Agnaf, O. M., Nagala, S., Patel, B. P., and Austen, B. M. (2001). Non-fibrillar oligomeric species of the amyloid ABri peptide, implicated in familial British dementia, are more potent at inducing apoptotic cell death than protofibrils or mature fibrils. *J. Mol. Biol.* **310,** 157–168.

Fancy, D. A. (2000). Elucidation of protein-protein interactions using chemical cross-linking or label transfer techniques. *Curr. Opin. Chem. Biol.* **4**, 28–33.

Fancy, D. A., Denison, C., Kim, K., Xie, Y. Q., Holdeman, T., Amini, F., and Kodadek, T. (2000). Scope, limitations and mechanistic aspects of the photo-induced cross-linking of proteins by water-soluble metal complexes. *Chem. Biol.* **7**, 697–708.

Fancy, D. A., and Kodadek, T. (1999). Chemistry for the analysis of protein-protein interactions: Rapid and efficient cross-linking triggered by long wavelength light. *Proc. Natl. Acad. Sci. USA* **96**, 6020–6024.

Fezoui, Y., Hartley, D. M., Harper, J. D., Khurana, R., Walsh, D. M., Condron, M. M., Selkoe, D. J., Lansbury, P. T., Fink, A. L., and Teplow, D. B. (2000). An improved method of preparing the amyloid β-protein for fibrillogenesis and neurotoxicity experiments. *Amyloid* **7**, 166–178.

Gambetti, P., and Russo, C. (1998). Human brain amyloidoses. *Nephrol. Dial. Transplant* **13**, 33–40.

Hardy, J. (2002). Testing times for the "amyloid cascade hypothesis." *Neurobiol. Aging* **23**, 1073–1074.

Hardy, J., and Selkoe, D. J. (2002). The amyloid hypothesis of Alzheimer's disease: Progress and problems on the road to therapeutics. *Science* **297**, 353–356.

Hardy, J. A., and Higgins, G. A. (1992). Alzheimer's disease: The amyloid cascade hypothesis. *Science* **256**, 184–185.

Hoshi, M., Sato, M., Matsumoto, S., Noguchi, A., Yasutake, K., Yoshida, N., and Sato, K. (2003). Spherical aggregates of β-amyloid (amylospheroid) show high neurotoxicity and activate tau protein kinase I/glycogen synthase kinase-3β. *Proc. Natl. Acad. Sci. USA* **100**, 6370–6375.

Jacobsen, S., Wittig, M., and Poggeler, S. (2002). Interaction between mating-type proteins from the homothallic fungus Sordaria macrospora. *Curr. Genet.* **41**, 150–158.

Kalyanasundaram, K. (1982). Photophysics, photochemistry and solar energy conversion with tris(bipyridyl)ruthenium(II) and its analogues. *Coord. Chem. Rev.* **46**, 159–244.

Kim, K., Fancy, D. A., Carney, D., and Kodadek, T. (1999). Photoinduced protein cross-linking mediated by palladium porphyrins. *J. Am. Chem. Soc.* **121**, 11896–11897.

Kirkitadze, M. D., Bitan, G., and Teplow, D. B. (2002). Paradigm shifts in Alzheimer's disease and other neurodegenerative disorders: The emerging role of oligomeric assemblies. *J. Neurosci. Res.* **69**, 567–577.

Kluger, R., and Alagic, A. (2004). Chemical cross-linking and protein-protein interactions—a review with illustrative protocols. *Bioorg. Chem.* **32**, 451–472.

Knorre, D. G., and Godovikova, T. S. (1998). Photoaffinity labeling as an approach to study supramolecular nucleoprotein complexes. *FEBS Lett.* **433**, 9–14.

Kotzyba-Hibert, F., Kapfer, I., and Goeldner, M. (1995). Recent trends in photoaffinity labeling. *Angew. Chem. Int. Ed. Engl.* **34**, 1296–1312.

Lazo, N. D., Maji, S. K., Fradinger, E. A., Bitan, G., and Teplow, D. B. (2005). The Amyloid β-Protein. *In* "Amyloid Proteins—The β-Sheet Conformation and Disease" (J. D. Sipe, ed.), Vol. 1, pp. 385–448. Wiley-VCH, Weinheim.

LeVine, H., 3rd (2004). Alzheimer's β-peptide oligomer formation at physiologic concentrations. *Anal. Biochem.* **335**, 81–90.

Lin, H. J., and Kodadek, T. (2005). Photo-induced oxidative cross-linking as a method to evaluate the specificity of protein-ligand interactions. *J. Pept. Res.* **65**, 221–228.

Luebke, K. J., Carter, D. E., Garner, H. R., and Brown, K. C. (2004). Patterning adhesion of mammalian cells with visible light, tris(bipyridyl)ruthenium(II) chloride, and a digital micromirror array. *J. Biomed. Mater. Res. A* **68**, 696–703.

Lumb, M. J., and Danpure, C. J. (2000). Functional synergism between the most common polymorphism in human alanine:glyoxylate aminotransferase and four of the most common disease-causing mutations. *J. Biol. Chem.* **275**, 36415–36422.

Malisauskas, M., Ostman, J., Darinskas, A., Zamotin, V., Liutkevicius, E., Lundgren, E., and Morozova-Roche, L. A. (2005). Does the cytotoxic effect of transient amyloid oligomers from common equine lysozyme *in vitro* imply innate amyloid toxicity? *J. Biol. Chem.* **280**, 6269–6275.

Marzban, L., Park, K., and Verchere, C. B. (2003). Islet amyloid polypeptide and type 2 diabetes. *Exp. Gerontol.* **38**, 347–351.

Mattson, M. P. (2004). Pathways towards and away from Alzheimer's disease. *Nature* **430**, 631–639.

Meunier, S., Strable, E., and Finn, M. G. (2004). Crosslinking of and coupling to viral capsid proteins by tyrosine oxidation. *Chem. Biol.* **11**, 319–326.

Mori, H., Takio, K., Ogawara, M., and Selkoe, D. J. (1992). Mass spectrometry of purified amyloid β protein in Alzheimer's disease. *J. Biol. Chem.* **267**, 17082–17086.

Nordstedt, C., Näslund, J., Tjernberg, L. O., Karlstrom, A. R., Thyberg, J., and Terenius, L. (1994). The Alzheimer Aβ peptide develops protease resistance in association with its polymerization into fibrils. *J. Biol. Chem.* **269**, 30773–30776.

Paoli, P., Giannoni, E., Pescitelli, R., Camici, G., Manao, G., and Ramponi, G. (2001). Hydrogen peroxide triggers the formation of a disulfide dimer of muscle acylphosphatase and modifies some functional properties of the enzyme. *J. Biol. Chem.* **276**, 41862–41869.

Reixach, N., Deechongkit, S., Jiang, X., Kelly, J. W., and Buxbaum, J. N. (2004). Tissue damage in the amyloidoses: Transthyretin monomers and nonnative oligomers are the major cytotoxic species in tissue culture. *Proc. Natl. Acad. Sci. USA* **101**, 2817–2822.

Selkoe, D. J. (2001). Alzheimer's disease: Genes, proteins, and therapy. *Physiol. Rev.* **81**, 741–766.

Shen, W. J., Patel, S., Hong, R., and Kraemer, F. B. (2000). Hormone-sensitive lipase functions as an oligomer. *Biochemistry* **39**, 2392–2398.

Thirumalai, D., Klimov, D. K., and Dima, R. I. (2003). Emerging ideas on the molecular basis of protein and peptide aggregation. *Curr. Opin. Struct. Biol.* **13**, 146–159.

Trojanowski, J. Q., and Mattson, M. P. (2003). Overview of protein aggregation in single, double, and triple neurodegenerative brain amyloidoses. *Neuromolecular Med.* **4**, 1–6.

Walsh, D. M., Lomakin, A., Benedek, G. B., Condron, M. M., and Teplow, D. B. (1997). Amyloid β-protein fibrillogenesis—detection of a protofibrillar intermediate. *J. Biol. Chem.* **272**, 22364–22372.

Walsh, D. M., and Selkoe, D. J. (2004a). Deciphering the molecular basis of memory failure in Alzheimer's disease. *Neuron* **44**, 181–193.

Walsh, D. M., and Selkoe, D. J. (2004b). Oligomers on the brain: The emerging role of soluble protein aggregates in neurodegeneration. *Protein Pept. Lett.* **11**, 213–228.

Wang, R., Sweeney, D., Gandy, S. E., and Sisodia, S. S. (1996). The profile of soluble amyloid β protein in cultured cell media. Detection and quantification of amyloid β protein and variants by immunoprecipitation-mass spectrometry. *J. Biol. Chem.* **271**, 31894–31902.

Wiltfang, J., Esselmann, H., Bibl, M., Smirnov, A., Otto, M., Paul, S., Schmidt, B., Klafki, H. W., Maler, M., Dyrks, T., Bienert, M., Beyermann, M., Ruther, E., and Kornhuber, J. (2002). Highly conserved and disease-specific patterns of carboxyterminally truncated Aβ peptides 1–37/38/39 in addition to 1-40/42 in Alzheimer's disease and in patients with chronic neuroinflammation. *J. Neurochem.* **81**, 481–496.

Wolf, J., Gerber, A. P., and Keller, W. (2002). tadA, an essential tRNA-specific adenosine deaminase from Escherichia coli. *EMBO J.* **21**, 3841–3851.

[13]   High-Pressure Studies on Protein Aggregates and Amyloid Fibrils

By Yong-Sung Kim, Theodore W. Randolph, Matthew B. Seefeldt, and John F. Carpenter

Abstract

High hydrostatic pressure (HHP) modulates protein–protein and protein–solvent interactions through volume changes and thereby affects the equilibrium of protein conformational species between native and denatured forms as well as monomeric, oligomeric, and aggregated forms without the addition of chemicals or use of high temperature. Because of this unique property, HHP has provided deep insights into the thermodynamics and kinetics of protein folding and aggregation, including amyloid fibril formation. In particular, HHP is a useful tool to stabilize and populate specific folding intermediates, the characterization of which provides thorough understanding of protein folding and aggregation pathways. Furthermore, recent application of HHP for dissociation of protein aggregates, such as inclusion bodies (IBs), into native proteins in a single step facilitates protein preparation for structural and functional studies. This chapter overviews recent HHP studies on the population and characterization of folding intermediates associated with protein aggregation and protein refolding from protein aggregates of amyloid fibrils and IBs. Finally, we describe overall experimental procedures of HHP-mediated protein refolding and provide a detailed discussion of each operating parameter to optimize the refolding.

Introduction

Recently, high hydrostatic pressure (HHP) has been used to study proteins associated with amyloid diseases, such as lysozyme (Niraula *et al.*, 2004; Sasahara *et al.*, 1999), transthyretin (TTR) (Ferrao-Gonzales *et al.*, 2000, 2003), prions (Martins *et al.*, 2003; Torrent *et al.*, 2003, 2004), α-synuclein (Foguel *et al.*, 2003), insulin (Dzwolak *et al.*, 2003; Jansen *et al.*, 2004), and immunoglobulin light chain variable domain (Kim *et al.*, 2002, 2003). These investigations have provided new insights into protein unfolding (Foguel and Silva, 2004; Kim *et al.*, 2002; Sasahara *et al.*, 1999; Silva and Weber, 1993; Silva *et al.*, 2001), formation of aggregation-prone folding intermediates (Ferrao-Gonzales *et al.*, 2000, 2003; Foguel and Silva, 2004; Silva *et al.*, 2001), and assembly processes of proteins into aggregates

METHODS IN ENZYMOLOGY, VOL. 413
0076-6879/06 $35.00
DOI: 10.1016/S0076-6879(06)13013-X

(Foguel and Silva, 2004; Kim *et al.*, 2002). HHP has also been used as a powerful tool for protein dissociation and refolding from nonnative aggregates, including amyloid fibrils and inclusion bodies (IBs) (Foguel and Silva, 2004; Randolph *et al.*, 2002; Silva *et al.*, 2001; St. John *et al.*, 1999).

Pressure is a physical parameter that modulates protein–solvent interactions through volume changes, because the pressure derivative of the difference in Gibbs' free energy (ΔG) between two states is equal to the difference in partial molar volume between the respective states (Hawley, 1971; Silva and Weber, 1993; Zipp and Kauzmann, 1973). Thus, by Le Chatelier's principle, applied HHP shifts the equilibrium toward states with lower volume. Similar analysis based on transition state theory predicts that reaction kinetics slow with increasing pressure for reactions with positive activation volumes (the difference in partial molar volume between the transition state and ground state for the reaction) and accelerate for those with negative activation volumes (Gross and Jaenicke, 1994; Mozhaev *et al.*, 1996). In aqueous solutions of proteins, loss of intra- and intermolecular cavities of proteins, hydration of hydrophobic residues, and electrostriction attributable to charged residues all reduce system volume. Thus, protein states containing minimal cavity space, states with increased exposure of hydrophobic groups to solvent, and more highly ionized states (e.g., by dissociation of ion-pairs) are favored by higher pressures (Gross and Jaenicke, 1994; Mozhaev *et al.*, 1996; Randolph *et al.*, 2002; Seefeldt *et al.*, 2004; Silva *et al.*, 2001; Van Eldik *et al.*, 1989). In contrast, the volume change associated with formation of hydrogen bonds is approximately 0, making hydrogen bonding insensitive to pressure (Silva and Weber, 1993; Silva *et al.*, 2001; Van Eldik *et al.*, 1989). Accordingly, elimination of cavities, hydration of newly exposed hydrophobic residues, and disruption of salt bridges are the most likely results of application of HHP to proteins and protein aggregates (Gross and Jaenicke, 1994; Mozhaev *et al.*, 1996; Silva and Weber, 1993; Silva *et al.*, 2001; Van Eldik *et al.*, 1989).

Moderate pressures (100–300 MPa) (1 MPa = 10 bar = 145 psi = 9.869 atm) often are effective for dissociating protein oligomers and aggregates, whereas relatively higher hydrostatic pressures (>300 MPa) are typically required for the denaturation of proteins (Hawley, 1971; Randolph *et al.*, 2002; Silva and Weber, 1993; Silva *et al.*, 2001; Zipp and Kauzmann, 1973). The partial specific volumes of globular proteins at temperatures near room temperature typically range between 0.69 and 0.76 cm^3 g^{-1}, and the partial specific adiabatic compressibilities are within the range of -1×10^{-6} to 10×10^{-6} cm^3 g^{-1} bar^{-1} (Chalikian *et al.*, 1996; Gekko and Hasegawa, 1986; Richards, 1977). Of 25 globular proteins that have been studied, 23 have shown positive adiabatic compressibility, which was not related to protein size but increased with increasing partial specific volume and hydrophobicity of proteins (Chalikian *et al.*, 1996; Gekko and Hasegawa, 1986;

Richards, 1977). These results suggest that the volume decreases that drive pressure-induced protein unfolding mainly come from elimination of water-excluded interior cavities and hydration of hydrophobic surfaces (Gekko and Hasegawa, 1986). The magnitude of the volume decrease upon unfolding usually is only about 10–250 ml mol^{-1}, which is less than 2% of the total specific volume of the protein (Gross and Jaenicke, 1994; Mozhaev et al., 1996; Silva and Weber, 1993).

Pressure-Induced Population of Folding Intermediates of Amyloidogenic Proteins

Many studies have shown that protein aggregation occurs through aggregation-prone partially folded intermediates rather than through fully denatured random-coil states (Randolph et al., 2002; Silva et al., 2001). Thus, characterization of the specific intermediates is crucial to the understanding of protein aggregation. HHP has provided a unique tool to stabilize and populate folding intermediates, which have more perturbed and solvated structures, and thus a smaller volume, than the native state (Randolph et al., 2002; Silva et al., 2001). Application of HHP to the characterization of folding intermediates using high-pressure spectroscopic techniques has been reported for amyloidogenic proteins, such as lysozyme (Nash and Jonas, 1997) and prion protein (Kuwata et al., 2002; Martins et al., 2003). Direct nuclear magnetic resonance (NMR) measurements under HHP (200 MPa) on hamster prion protein detected a metastable intermediate conformer of cellular prion protein (PrP^C) under roughly physiological conditions (pH 5.2 and 30°) (Kuwata et al., 2002). For murine prion protein, HHP (350 MPa) combined with low temperature (−9°) populated folding intermediates exhibiting strong 4,4′ dianilino-1-1′-binaphthyl-5,5disufonic acid (bis-ANS) fluorescence (Martins et al., 2003). These folding intermediates trapped by HHP might be intermediates associated with conformational conversion of the monomeric PrP^C into a neurotoxic oligomeric species (PrP^{SC}) (Kuwata et al., 2002; Martins et al., 2003; Prusiner, 1998).

For p53 protein (Ishimaru et al., 2003), metmyoglobin (Smeller et al., 1999), and TTR (Ferrao-Gonzales et al., 2000), a pressurization-depressurization technique also has been used to populate aggregation-competent conformations from native states at HHP and then to study aggregation and fibrillogenesis upon reduction to atmospheric pressure (Ferrao-Gonzales et al., 2000). With TTR, exposure of the native tetramer to HHP caused population of nonnative tetrameric states without aggregation. Upon depressurization at atmospheric pressure and 37°, these species rapidly formed amyloid fibrils (Ferrao-Gonzales et al., 2000). However, amyloid fibril formation was prevented when the temperature was maintained at 1°, so that the amyloidogenic intermediate could be further characterized by other methods (Ferrao-Gonzales et al., 2000).

Pressure-Induced Aggregation and Amyloid Fibril Formation

Under solution conditions favoring protein aggregation, HHP-induced partially unfolded conformations have been shown to aggregate for a few proteins. For example, interferon-γ (Webb *et al.*, 2001), interleukin-1 receptor antagonist (Seefeldt *et al.*, 2005), and β-lactoglobulin (Panick *et al.*, 1999) directly formed aggregates under HHP. Recently, Torrent and co-workers (Torrent *et al.*, 2004) showed that incubation of hamster prion protein at 600 MPa converted PrP^C into a misfolded conformer with β-sheet–rich structure, which subsequently formed amyloid fibril at HHP. Interestingly, for the same protein, moderate pressures (200–400 MPa) reversed temperature-induced aggregates, recovering the original native protein (Torrent *et al.*, 2003). The pressure-dependent determination of preferred conformational species was also observed for bovine insulin at pH 1.9. Pressures as low as 30 MPa completely inhibited bovine insulin aggregation (Dzwolak *et al.*, 2003), but higher pressure (150 MPa) induced amyloid fibrils with a unique circular morphology (Jansen *et al.*, 2004).

With the dimeric immunoglobulin light chain variable domain (Kim *et al.*, 2002), direct monitoring of protein conformation and aggregation with HHP ultraviolet (UV)-visible (Vis) spectroscopy provided quantitative information about the transition states for nucleation and growth of amyloid fibrils. HHP-induced aggregation monitored by optical density at 320 nm showed kinetics typical of nucleation-dependent growth, characterized by an initial lag phase followed by an exponential growth phase (Kim *et al.*, 2002). The HHP-induced aggregation and amyloid fibril formation showed first-order kinetics in protein concentration and could be seeded, which are behaviors typical of amyloid fibrillogenesis at atmospheric pressure (Rochet and Lansbury, 2000). The activation volume changes required for the formation of the nucleation transition state and the fibril growth transition state were about 11 and 26%, respectively, of the volume change required for equilibrium dissociation of the homodimer. Thus, the transition states for fibril nucleation and growth are structurally similar to the native state, and their formation requires only small structural perturbations (Kim *et al.*, 2002). These observations are consistent with results for other proteins that have been shown to form amyloid fibrils, even under conditions greatly favoring the native state (Dobson, 2003; Kim *et al.*, 2000). Thus, in these cases, the precursors for aggregation are species within the native state ensemble (Dobson, 2003; Kim *et al.*, 2000; Randolph *et al.*, 2002).

Pressure-Induced Dissolution of Amyloid Fibrils

Amyloid fibrils show well-defined ordered structures characterized by cross-β-sheet structures (Ohnishi and Takano, 2004; Rochet and Lansbury, 2000). However, amyloid fibrils might exhibit a larger partial specific volume than that of the composing proteins, because the assembly process, which involves various conformational transitions of native structures to nonnative β-sheet structures, could create new water-excluded cavities and induce burial of hydrophobic surface areas (Foguel and Silva, 2004; Ohnishi and Takano, 2004; Randolph *et al.*, 2002; Silva and Weber, 1993; Silva *et al.*, 2001; St. John *et al.*, 1999). In fact, a recent study of sound velocity and protein-specific density revealed that the transition of α-helix to β-sheet of poly-L-lysine resulted in increases in volume and compressibility of 0.016 cm^3 g^{-1} and 3.5×10^{-6} cm^3 g^{-1} bar^{-1} (Noudeh *et al.*, 2003), respectively. Thus, HHP has a potential to dissociate amyloid fibrils by solvating hydrophobic surfaces and eliminating solvent-free cavities within the aggregates.

Recent studies have shown that moderate pressures (100–300 MPa) can dissociate amyloid fibrils of amyloid A (Dubois *et al.*, 1999), α-synuclein (Foguel *et al.*, 2003), TTR (Foguel *et al.*, 2003), and prion protein (Cordeiro *et al.*, 2004; Torrent *et al.*, 2003), as well as soluble oligomeric precursor to amyloid fibrils of lysozyme (Niraula *et al.*, 2004). For α-synuclein, fibrils formed from the Parkinson's disease–linked mutants (A53T and A30P) were more sensitive to high pressure than the wild-type fibrils, suggesting that those mutations affect the hydrophobic interactions and packing of amyloid fibrils (Foguel *et al.*, 2003). The temperature-induced β-sheet aggregates of murine prion protein (β-PrP) were dissociated by much lower pressures (200–400 MPa) than those (>500 MPa) that caused denaturation of the native α-helical prion protein (α-PrP) (Cordeiro *et al.*, 2004). The much greater pressure sensitivity of β-PrP was interpreted to be indicative of the presence of more water-excluded cavities than those exhibited by α-PrP (Cordeiro *et al.*, 2004).

Sometimes, the conformation of the protein molecules obtained from amyloid fibrils by HHP treatment is not the same as that of the initial native proteins. For example, the dissociated soluble species from murine β-PrP aggregates were structurally perturbed compared with native structures but also different from the denatured conformation of α-PrP (Cordeiro *et al.*, 2004). Monomers of α-synuclein and TTR obtained by HHP treatment of fibrils reformed fibrils upon returning to atmospheric pressure, with kinetics that were much faster than those for untreated native proteins (Ferrao-Gonzales *et al.*, 2000; Foguel *et al.*, 2003) .

Pressure-Induced Refolding of Active, Native Proteins from Aggregates and Inclusion Bodies

Traditional techniques of refolding proteins from aggregates and IBs use chaotropic agents, such as GdnHCl and urea. Aggregated proteins are first solubilized in a high concentration of the chaotropes (e.g., 6–8 M GdnHCl or 6–9 M urea), which usually results in complete unfolding of the protein. Then, the unfolded proteins are refolded by reducing the concentration of the chemical denaturants using dialysis, dilution, or solid phase methods (Clark, 2001). However, these conventional refolding methods often generate new protein aggregates during the process of chaotrope removal, forcing them to be operated at very low protein concentrations ($\approx$10–50 μg ml^{-1}) (Clark, 2001).

HHP is an attractive alternative for protein refolding from aggregates compared with the traditional processes, in terms of refolding yield, operating protein concentrations, and process simplicity and cost (Randolph et al., 2002). Moderate pressures (100–300 MPa) have been shown to be effective for disaggregation and refolding of proteins from insoluble aggregates prepared in vitro (Foguel et al., 1999; Randolph et al., 2002; St. John et al., 1999, 2001, 2002) and IBs formed in bacteria (St. John et al., 1999). For example, HHP (100–200 MPa) fostered high recovery (>90%) of native protein from aggregates of recombinant human growth hormone (rhGH), IBs of β-lactamase, and covalently cross-linked aggregates of lysozyme, even at high protein concentrations of up to several milligrams per milliliter^{-1} (St. John et al., 1999, 2001, 2002).

The volume change involved in the pressure-induced dissociation and refolding of soluble aggregates to the native conformation was about -28 ml mol^{-1} for bikunin (Seefeldt et al., 2004) and about -53 ml mol^{-1} for lysozyme (Niraula et al., 2004). Those volume changes are smaller than those (about -100 to -250 ml mol^{-1}) typical for the denaturation of proteins. Thus, with the pressure refolding method, pressure conditions can be chosen so as to disfavor nonnative aggregates while still favoring the native protein conformation. Thus, unlike the case with chemical denaturants, with high pressure, it is not necessary to induce complete unfolding before beginning to refold the protein.

Optimization Parameters for Protein Refolding from Aggregates and Amyloid Fibrils

Overview of Approach

HHP-induced refolding of active, native proteins from aggregates, including IBs and amyloid fibrils, can be affected by many parameters, such as pressure, temperature, protein concentration, pH, redox shuffling agents, ionic strength, co-solutes, and depressurization rates (Clark, 2001; Randolph et al., 2002; Seefeldt et al., 2004; St. John et al., 1999). This section briefly describes overall experimental procedures of HHP-mediated protein refolding, followed by a detailed discussion for each operating parameter to optimize the refolding.

Sample Preparation

The concentrations of insoluble aggregates can be measured by first solubilizing them in a high concentration of chaotropes (e.g., 6–9 M urea) and then subjecting the resulting solutions to UV spectroscopy, Bradford, or bicinchoninic acid (BCA) within compatible chaotrope concentrations by serial dilutions. With Bradford and BCA assays, the contribution of chaotropes to the assays should be corrected by adding the same amount of chaotropes to the standard samples. Prior to pressurization, aggregates should be centrifuged (12,000g for 15 min), decanted to remove soluble species, and resuspended in refolding buffers. The refolding buffers should be formulated considering some properties of aggregates and native proteins as described in detail below. Samples can be prepared in sterile, disposable syringes (one end heat-sealed and the other sealed with the rubber plunger) and placed in a high-pressure bomb (St. John *et al.*, 2001) or placed in quartz tubing (Kim *et al.*, 2002; Seefeldt *et al.*, 2004; St. John *et al.*, 2001) with both ends sealed by custom-fabricated rubber seals and inserted into an on-line spectroscopy cell. To monitor structural transitions of aggregates by on-line spectroscopy, aggregates can be dispersed in solution using a 20% solution of hydroxyethyl starch (pentastarch with an average molecular weight of 294 kg mol^{-1}; Fresenius, Austria) to inhibit sedimentation of aggregates (St. John *et al.*, 2001).

Pressurization and Depressurization

Sealed samples are placed in a high-pressure vessel equipped with a heating controller. The pressure vessel can be custom-built, assembled from parts available from High Pressure Equipment Co. (HiP Co., Erie, PA), or purchased as a unit from other manufacturers, such as BaroFold, Inc. (BaroFold, Inc., Boulder, CO) (see section on high-pressure equipment). Pressure can be generated by a high-pressure crank generator from HiP Co. with water as a pressure-transmitting fluid (Seefeldt *et al.*, 2004; St. John *et al.*, 2001). Samples are slowly pressurized (e.g., 10 MPa-increase every 10 min) to the desired pressure to minimize pressurization-induced heating of the sample (St. John *et al.*, 1999, 2001). Pressurization at 200 MPa and room temperature (20°–25°) for 24 h is a practical starting condition, with subsequent optimization experiments varying pressure in the range of 100–400 MPa and temperature in the range of 0°–50°. In parallel studies, pressure-induced structural transitions of protein molecules in aggregates can be monitored by collecting spectra, such as UV-Vis, fluorescence, and infrared, at regular intervals. After pressurization, samples are slowly depressurized (e.g., 10-MPa decrease every 10 min) to atmospheric pressure.

Analysis of the Recovered Proteins

The recovered samples are centrifuged (e.g., 12,000g for 15 min) to remove remaining insoluble aggregates. The soluble species can be characterized by various methods, such as total protein assays, functional assays, size-exclusion and reverse-phase chromatography, various spectroscopies (e.g., fluorescence, circular dichroism, infrared), and native and non-native polyacrylamide gel electrophoresis (PAGE), to investigate whether the recovered, soluble species is monomeric or oligomeric and functional or nonfunctional.

High-Pressure Equipment

To explore structural, conformational, and functional changes of proteins under HHP, a HHP cell has been combined with various optical detection methods, such as UV-Vis, light scattering, fluorescence, circular dichroism, infrared, and NMR spectroscopies (Gross and Jaenicke, 1994; Lange and Balny, 2002; Mozhaev *et al.*, 1996; St. John *et al.*, 2001; Webb *et al.*, 2001). Each spectroscopic technique has provided insight into the effects of HHP on proteins. Schematic designs of HHP spectroscopic cells have been shown for circular dichroism (Harris *et al.*, 1976), fluorescence (Gross and Jaenicke, 1994), and NMR (Kamatari *et al.*, 2004) spectroscopies.

UV-Vis spectroscopy can detect tertiary structural changes of proteins by the second- and fourth-derivative spectra in the near-UV region (260–310 nm) and aggregation and dissociation by the light scattering in the range of 320–500 nm (Kim *et al.*, 2002; Lange and Balny, 2002; Seefeldt *et al.*, 2004; St. John *et al.*, 2001; Webb *et al.*, 2001). Figure 1 shows the schematic diagram of an HHP cell adapted for on-line UV-Vis spectroscopy (Webb *et al.*, 2001). The cell is constructed from 316 SS stainless steel and rated to 650 MPa. The dimension of the cell is such that it fits into the sample compartment of a PerkinElmer Lambda 35 UV-Vis spectrometer. The sealing system is composed of a metal back-up ring and rubber O-ring made out of 70-D polyurethane (Custom Seal & Rubber Products, Polo, IL), which works for a wide range of temperatures. The high-pressure windows are UV grade sapphire (Spectral System, Hopewell Junction, NY), which are attached to the metal by an epoxy glue (DP420; 3*M*). The sample container is made from UV grade quartz tubing of 12.5 mm internal diameter × 1 mm wall × 25 mm length (Wilmad Labglass, Buena, NJ), both ends of which can be tightly sealed after sample loading. The temperature of the cell can be controlled to within ±0.5° with a circulating water bath connected to a heat exchanger mounted on the bottom of the cell. The pressure generator system connected to the cell is a manually

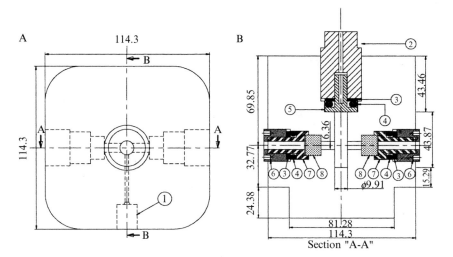

FIG. 1. Schematic diagram of a high hydrostatic pressure cell designed for adopting on-line ultraviolet (UV)-Visible (Vis) spectroscopy. Top view (A) and side view (B) of the cells are shown. The dimension of the cell was shown in millimetric units and designed to be fitted into the sample compartment of a PerkinElmer Lambda 35 UV-Vis spectrometer. 1, pressure inlet; 2, top plug for sample handling; 3, back-up metal ring; 4, rubber O-ring; 5, inner top plug; 6, side plug for light transmission; 7, inner side plug; 8, sapphire window.

operated piston screw pump rated to ≈700 MPa (Model 25-5.75-100; High Pressure Equipment Co.) and connected to the cell by 0.25-inch outer diameter tubing coned and threaded with XF4 type threads.

Pressure and Temperature

When solution conditions are held constant, the stability of proteins is a function of temperature and pressure, as characterized by an elliptical phase diagram (Gross and Jaenicke, 1994; Hawley, 1971; Randolph et al., 2002). The shape and position of the phase diagram are determined by the temperature and pressure dependence of the enthalpy and volume changes between the native and denatured states (Gross and Jaenicke, 1994; Hawley, 1971; Randolph et al., 2002). Refolding efficiency usually will be optimized at the conditions of pressure and temperature thermodynamically favoring native state proteins, that is, inside ($\Delta G > 0$) the elliptical pressure-temperature phase diagram (so-called "refolding window") (Hawley, 1971; Randolph et al., 2002). Native state proteins are usually stable up to 400 MPa (Mozhaev et al., 1996; Randolph et al., 2002; Silva and Weber, 1993). Thus, HHP ranges of 100–300 MPa have been effective in the disaggregating and refolding of proteins (Randolph *et al.*, 2002; Seefeldt *et al.*, 2004; Smeller *et al.*, 1999). In this pressure range, aggregates are readily dissociated but the native state is favored over the denatured state.

An increase in temperature can facilitate disaggregation under pressure, because at higher temperatures, hydrogen bonds that play a role in stabilizing intermolecular contacts in the aggregate will be weakened (St. John *et al.*, 2001). In practice, the sample pressure is first increased to the desired value (e.g., 200 MPa) while the temperature is maintained, for example, at 25°. The sample temperature is then increased to facilitate disaggregation and subsequently decreased to allow refolding. The sample is then depressurized. This approach was shown to greatly increase the yield of native rhGH obtained from insoluble aggregates (St. John *et al.*, 2001).

Conversely, a reduction in temperature can augment the pressure-induced disruption of intermolecular hydrophobic interactions in aggregates (Hui Bon Hoa *et al.*, 1982; Silva and Weber, 1993). Because it is not possible to ascertain directly the relative contributions of hydrophobic interactions and hydrogen bonds to stabilizing aggregates, it is important to explore a range of temperatures during optimization of HHP-mediated disaggregation and refolding.

Protein Concentration

The concentration of protein can also be an important parameter, because the correct refolding pathway competes with misfolding and aggregation pathways (Kiefhaber *et al.*, 1991; Randolph *et al.*, 2002). Protein refolding involves intramolecular interactions, and thus follows first-order kinetics, as shown, for example, in the refolding of rhGH from aggregates (St. John *et al.*, 2001). However, protein aggregation involves intermolecular interactions, and thus shows second- or higher order kinetics (Kiefhaber *et al.*, 1991; Randolph *et al.*, 2002). Therefore, aggregation increases and refolding yields usually decrease with increasing initial protein concentrations. For example, the refolding yield of bikunin from soluble aggregates decreased from 96% at 0.06 mg/ml to 70% at 0.5 mg/ml incubated under 200 MPa at 24° for 24 h (Seefeldt *et al.*, 2004). In contrast, the high-pressure refolding yield was independent of initial protein concentrations for rhGH up to 8.7 mg ml^{-1} and lysozyme up to 2 mg ml^{-1} (St. John *et al.*, 1999).

It should be noted that the protein concentrations for refolding by HHP were several orders of magnitude higher than those typically used for refolding by traditional chemical chaotropes, where initial protein concentrations usually ranged about 10–50 μg ml^{-1} (Kiefhaber *et al.*, 1991; Randolph *et al.*, 2002). However, even with HHP processing, it is important to test a range of protein concentrations during optimization of the process. There may be a need to compromise between the yield of native protein and the volume of solution that must be processed.

Pressurization and Depressurization Rates

Pressurization and depressurization rates, like heating and cooling rates, are also important parameters in the refolding from aggregates to achieve equilibrium among protein species at each pressure during compression and decompression (Kim *et al.*, 2002; Randolph *et al.*, 2002; Silva and Weber, 1993; St. John *et al.*, 2001). Hysteresis phenomena have been observed in several oligomeric proteins (Foguel and Silva, 2004; Ruan and Weber, 1989; Ruan *et al.*, 2003; Silva and Weber, 1993; Silva *et al.*, 1986), where the dissociation and association curves obtained on compression and decompression were different, indicative of incomplete equilibrium among various conformational species during the cycle of pressurization and depressurization. These phenomena were attributed to slow interconversions between the species, so-called "conformational drift" (Silva and Weber, 1993). Thus, it is recommended to determine if conformational equilibrium has been achieved after pressure shifts by monitoring the conformational changes of proteins at each pressure using on-line spectroscopy. For a dimeric immunoglobulin variable domain, it took about 10 min every 10-MPa interval from 0–330 MPa to achieve complete conformational equilibrium during HHP-induced folding-unfolding (Kim *et al.*, 2002).

No differences in final protein refolding were observed for P22 tailspike protein between "instantaneous" release and a depressurization rate of 17 MPa min^{-1} (Lefebvre and Robinson, 2003) and for bikunin between "instantaneous" release and a depressurization rate of 0.33 MPa min^{-1} (Seefeldt *et al.*, 2004). However, in many cases, it has been observed that aggregates form as a result of rapid depressurization rates (e.g., >4.5 MPa s^{-1}) (Randolph *et al.*, 2002). Thus, the depressurization rate should be optimized for the maximal refolding yield, because rapid depressurization rates can trap nonequilibrium aggregation-prone species.

How long should samples be maintained at the target HHP? Refolding time constants for the first-order dissolution were ≈5 h for rhGH (St. John *et al.*, 2001), ≈8 h for lysozyme (St. John *et al.*, 2002), and ≈4 h for bikunin (Seefeldt *et al.*, 2004). Postpressure refolding yield of P22 tailspike proteins did not show any significant differences in HHP treatments between 5 and 360 min (Lefebvre *et al.*, 2004). Thus, about 24 h seems to be enough to reach maximum disaggregation at HHP. However, the optimal incubation times for maximal refolding yield at HHP have to be determined empirically for each protein.

Co-Solutes

Addition of chemical compounds, such as low concentrations of chaotropes (GdnHCl and urea), amino acids (arginine), surfactants (Triton X-100, Tween 20, and 3-[3-cholamidopropyl dimethyl ammonie]-1-propane sulfonate [CHAPS]), and compatible osmolytes (sucrose, glycerol, and sorbitol), is a very useful tool to modulate thermodynamics and kinetics of reactions of proteins and protein aggregates under HHP (Clark, 2001; Randolph *et al.*, 2002; Seefeldt *et al.*, 2004; St. John *et al.*, 1999, 2001). Low, nondenaturing concentrations of GdnHCl (0.5–1 M) and urea (0.5–2 M) have shown synergistic effects on the refolding from aggregates for many proteins (Kim *et al.*, 2002; Seefeldt *et al.*, 2004; St. John *et al.*, 1999, 2001, 2002) and facilitated protein dissociation and unfolding (Kim *et al.*, 2002; Silva *et al.*, 2001). Those significant effects are mainly attributed to breakage of intra- and intermolecular hydrogen bonding in proteins by chaotropes, complementing the disruption of hydrophobic and electrostatic interactions by HHP (Seefeldt *et al.*, 2004; St. John *et al.*, 2001).

In particular, L-arginine with a guanidinium group has been known as a suppressor of protein aggregation and has been most widely used as an additive in the range of 0.4–0.8 M in the refolding studies (Clark, 2001; Tsumoto *et al.*, 2004). We have also observed that L-arginine showed significant synergistic effects on the improvement of HHP-mediated refolding yield of various proteins from IBs and aggregates (unpublished data). Even though the exact mechanism is unknown, the interactions between the guanidinium group of arginine and side chains of proteins might be responsible for the suppression of aggregations (Tsumoto *et al.*, 2004).

Compatible osmolytes, such as sugars and polyols, stabilize proteins by the preferential exclusion from the protein's surface, which shifts the equilibrium toward compact states, such as the native state (Kendrick *et al.*, 1997; Timasheff, 1998). Compatible osmolytes have been shown to inhibit protein aggregation at atmospheric and high pressures (Kim *et al.*, 2002; Randolph *et al.*, 2002; Webb *et al.*, 2001). Glycerol (4 M) blocked the formation of unproductive aggregates of rhodanese (Panda *et al.*, 2000), improving the refolding yield, and sucrose (0.5–1 M) efficiently inhibited aggregation under HHP by inhibiting the formation of aggregation-competent transition states of proteins (Kim *et al.*, 2002; Webb *et al.*, 2001). It should be noted, however, that by favoring compact states, excluded solutes may also inhibit the dissociation and dissolution of aggregates.

Redox Shuffling Agents

Addition of redox shuffling agents facilitates protein refolding from covalently cross-linked aggregates of disulfide-containing proteins to break nonnative intermolecular disulfide bonds under conditions that allow native disulfide bonds to form (Seefeldt *et al.*, 2004; St. John *et al.*, 2002). Typically used oxido-shuffling agents are reduced/oxidized glutathione (GSH/GSSG), dithiothreitol (DTT)/GSSG, cysteine/cystine, cysteamine/ cystamine, and dihydroxyethyl disulfide/2-mercaptoethanol (Clark, 2001). Molar ratios of reduced to oxidized thiol of 1:1 to 1:4 and total thiol concentrations between 5 and 15 mM have been found to be optimal (Clark, 2001; St. John *et al.*, 2002). The pH of buffer is also critical for the efficient shuffling of reduction and oxidation (Clark, 2001). An alkaline pH of 7.5–8.7 is required to allow formation and reshuffling of disulfide bonds, because pKa = 8.5 for the formation of thiolate anion, the nucleophilic reacting species (Fernandes and Ramos, 2004). Chelating agents, such as ethylenediaminetetraacetic acid (EDTA), are added to buffer solution to prevent metal-catalyzed air oxidation of free cysteine and methionine.

Buffer Conditions

Buffers are often chosen to minimize pH shifts under HHP, which occur because HHP favors ionization as a result of the electrostrictive effect of the separated charges (Disteche, 1972; Zipp and Kauzmann, 1973). For example, phosphate buffer has a high ionization volume (i.e., $\Delta V = +25$ ml mol^{-1}), causing large pressure-induced pH shifts. Thus, its use adds complexity to studies of the effects of pressure on thermodynamic equilibria (Disteche, 1972; Zipp and Kauzmann, 1973). In contrast, protonation of the bases of N-Tris(hydroxymethyl)methyl-3-aminopropanesulfonic acid (TAPS) is associated with small ionization volumes, resulting in pH values that are largely insensitive to pressure (Disteche, 1972; Zipp and Kauzmann, 1973). We recommend acetate buffer for the range of pH 4–5.5; cacodylate, 2-morpholinoethanesulfonic acid (MES), or histidine for pH 5.5–7.0; TAPS for pH 7–9; N-cyclohexyl-2-aminoethanesulfonic acid (CHES) for pH 8.4– 10.4; N-cyclohexyl-3-aminopropanesulfonic acid (CAPS) for pH 9.6–11.6; and NaOH for pH 11–12 (Zipp and Kauzmann, 1973).

The pH is also a critical factor for optimal solubilization and refolding of proteins. A pH value near the protein's isoelectric point (pI) can favor protein aggregation, which can be minimized at a pH away from the pI (Mozhaev *et al.*, 1996; Randolph *et al.*, 2002; Seefeldt *et al.*, 2004; Smeller *et al.*, 1999). Ionic strength should also be considered in buffer formulation because it perturbs both attractive and repulsive electrostatic interactions by shielding the charged ions. For example, the refolding yield of bikunin at 200 MPa increased from 66–79% by the addition of 150 mM NaCl, possibly because of disruption of salt bridges within the aggregates (Seefeldt *et al.*, 2004).

Perspectives

HHP is a robust tool to study various processes of unfolding of proteins, formation of folding intermediates, assembly of aggregates, and refolding from aggregates. The capacity of HHP to populate folding intermediates directly involved in various human diseases provides a potentially useful new tool for drug discovery. HHP can also accelerate protein aggregation within hours (Kim *et al.*, 2002; Seefeldt *et al.*, 2005; Webb *et al.*, 2001), providing another high-throughput screening tool in the development of protein formulations. Finally, HHP has shown efficient refolding from various protein aggregates, which is very useful in the preparation of native proteins for therapeutics and structural genomics.

Acknowledgment

This work was supported by Korea Research Foundation Grant (KRF-2004-003-D00102) (to Y-S. Kim).

References

Chalikian, T. V., Totrov, M., Abagyan, R., and Breslauer, K. J. (1996). The hydration of globular proteins as derived from volume and compressibility measurements: Cross correlating thermodynamic and structural data. *J. Mol. Biol.* **260**, 588–603.

Clark, E. D. (2001). Protein refolding for industrial processes. *Curr. Opin. Biotechnol.* **12**, 202–207.

Cordeiro, Y., Kraineva, J., Ravindra, R., Lima, L. M., Gomes, M. P., Foguel, D., Winter, R., and Silva, J. L. (2004). Hydration and packing effects on prion folding and beta-sheet conversion. High pressure spectroscopy and pressure perturbation calorimetry studies. *J. Biol. Chem.* **279**, 32354–32359.

Disteche, A. (1972). Effects of pressure on the dissociation of weak acids. *Symp. Soc. Exp. Biol.* **26**, 27–60.

Dobson, C. M. (2003). Protein folding and misfolding. *Nature* **426**, 884–890.

Dubois, J., Ismail, A. A., Chan, S. L., and Ali-Khan, Z. (1999). Fourier transform infrared spectroscopic investigation of temperature- and pressure-induced disaggregation of amyloid A. *Scand. J. Immunol.* **49**, 376–380.

Dzwolak, W., Ravindra, R., Lendermann, J., and Winter, R. (2003). Aggregation of bovine insulin probed by DSC/PPC calorimetry and FTIR spectroscopy. *Biochemistry* **42**, 11347–11355.

Fernandes, P. A., and Ramos, M. J. (2004). Theoretical insights into the mechanism for thiol/disulfide exchange. *Chemistry* **10**, 257–266.

Ferrao-Gonzales, A. D., Palmieri, L., Valory, M., Silva, J. L., Lashuel, H., Kelly, J. W., and Foguel, D. (2003). Hydration and packing are crucial to amyloidogenesis as revealed by pressure studies on transthyretin variants that either protect or worsen amyloid disease. *J. Mol. Biol.* **328**, 963–974.

Ferrao-Gonzales, A. D., Souto, S. O., Silva, J. L., and Foguel, D. (2000). The preaggregated state of an amyloidogenic protein: Hydrostatic pressure converts native transthyretin into the amyloidogenic state. *Proc. Natl. Acad. Sci. USA* **97**, 6445–6450.

Foguel, D., Robinson, C. R., de Sousa, P. C., Jr., Silva, J. L., and Robinson, A. S. (1999). Hydrostatic pressure rescues native protein from aggregates. *Biotechnol. Bioeng.* **63,** 552–558.

Foguel, D., and Silva, J. L. (2004). New insights into the mechanisms of protein misfolding and aggregation in amyloidogenic diseases derived from pressure studies. *Biochemistry* **43,** 11361–11370.

Foguel, D., Suarez, M. C., Ferrao-Gonzales, A. D., Porto, T. C., Palmieri, L., Einsiedler, C. M., Andrade, L. R., Lashuel, H. A., Lansbury, P. T., Kelly, J. W., and Silva, J. L. (2003). Dissociation of amyloid fibrils of alpha-synuclein and transthyretin by pressure reveals their reversible nature and the formation of water-excluded cavities. *Proc. Natl. Acad. Sci. USA* **100,** 9831–9836.

Gekko, K., and Hasegawa, Y. (1986). Compressibility-structure relationship of globular proteins. *Biochemistry* **25,** 6563–6571.

Gross, M., and Jaenicke, R. (1994). Proteins under pressure. The influence of high hydrostatic pressure on structure, function and assembly of proteins and protein complexes. *Eur. J. Biochem.* **221,** 617–630.

Harris, R. D., Jacobs, M., Long, M. M., and Urry, D. W. (1976). A high-pressure sample cell for circular dichroism studies. *Anal. Biochem.* **73,** 363–368.

Hawley, S. A. (1971). Reversible pressure-temperature denaturation of chymotrypsinogen. *Biochemistry* **10,** 2436–2442.

Hui Bon Hoa, G., Douzou, P., Dahan, N., and Balny, C. (1982). High-pressure spectrometry at subzero temperatures. *Anal. Biochem.* **120,** 125–135.

Ishimaru, D., Andrade, L. R., Teixeira, L. S., Quesado, P. A., Maiolino, L. M., Lopez, P. M., Cordeiro, Y., Costa, L. T., Heckl, W. M., Weissmuller, G., Foguel, D., and Silva, J. L. (2003). Fibrillar aggregates of the tumor suppressor p53 core domain. *Biochemistry* **42,** 9022–9027.

Jansen, R., Grudzielanek, S., Dzwolak, W., and Winter, R. (2004). High pressure promotes circularly shaped insulin amyloid. *J. Mol. Biol.* **338,** 203–206.

Kamatari, Y. O., Kitahara, R., Yamada, H., Yokoyama, S., and Akasaka, K. (2004). High-pressure NMR spectroscopy for characterizing folding intermediates and denatured states of proteins. *Methods* **34,** 133–143.

Kendrick, B. S., Chang, B. S., Arakawa, T., Peterson, B., Randolph, T. W., Manning, M. C., and Carpenter, J. F. (1997). Preferential exclusion of sucrose from recombinant interleukin-1 receptor antagonist: Role in restricted conformational mobility and compaction of native state. *Proc. Natl. Acad. Sci. USA* **94,** 11917–11922.

Kiefhaber, T., Rudolph, R., Kohler, H. H., and Buchner, J. (1991). Protein aggregation *in vitro* and *in vivo*: A quantitative model of the kinetic competition between folding and aggregation. *Biotechnology (NY)* **9,** 825–829.

Kim, Y., Wall, J. S., Meyer, J., Murphy, C., Randolph, T. W., Manning, M. C., Solomon, A., and Carpenter, J. F. (2000). Thermodynamic modulation of light chain amyloid fibril formation. *J. Biol. Chem.* **275,** 1570–1574.

Kim, Y. S., Randolph, T. W., Manning, M. C., Stevens, F. J., and Carpenter, J. F. (2003). Congo red populates partially unfolded states of an amyloidogenic protein to enhance aggregation and amyloid fibril formation. *J. Biol. Chem.* **278,** 10842–10850.

Kim, Y. S., Randolph, T. W., Stevens, F. J., and Carpenter, J. F. (2002). Kinetics and energetics of assembly, nucleation, and growth of aggregates and fibrils for an amyloidogenic protein. Insights into transition states from pressure, temperature, and co-solute studies. *J. Biol. Chem.* **277,** 27240–27246.

Kuwata, K., Li, H., Yamada, H., Legname, G., Prusiner, S. B., Akasaka, K., and James, T. L. (2002). Locally disordered conformer of the hamster prion protein: A crucial intermediate to PrPSc? *Biochemistry* **41,** 12277–12283.

Lange, R., and Balny, C. (2002). UV-visible derivative spectroscopy under high pressure. *Biochim. Biophys. Acta* **1595**, 80–93.

Lefebvre, B. G., Gage, M. J., and Robinson, A. S. (2004). Maximizing recovery of native protein from aggregates by optimizing pressure treatment. *Biotechnol. Prog.* **20**, 623–629.

Lefebvre, B. G., and Robinson, A. S. (2003). Pressure treatment of tailspike aggregates rapidly produces on-pathway folding intermediates. *Biotechnol. Bioeng.* **82**, 595–604.

Martins, S. M., Chapeaurouge, A., and Ferreira, S. T. (2003). Folding intermediates of the prion protein stabilized by hydrostatic pressure and low temperature. *J. Biol. Chem.* **278**, 50449–50455.

Mozhaev, V. V., Heremans, K., Frank, J., Masson, P., and Balny, C. (1996). High pressure effects on protein structure and function. *Proteins* **24**, 81–91.

Nash, D. P., and Jonas, J. (1997). Structure of pressure-assisted cold denatured lysozyme and comparison with lysozyme folding intermediates. *Biochemistry* **36**, 14375–14383.

Niraula, T. N., Konno, T., Li, H., Yamada, H., Akasaka, K., and Tachibana, H. (2004). Pressure-dissociable reversible assembly of intrinsically denatured lysozyme is a precursor for amyloid fibrils. *Proc. Natl. Acad. Sci. USA* **101**, 4089–4093.

Noudeh, G. D., Taulier, N., and Chalikian, T. V. (2003). Volumetric characterization of homopolymeric amino acids. *Biopolymers* **70**, 563–574.

Ohnishi, S., and Takano, K. (2004). Amyloid fibrils from the viewpoint of protein folding. *Cell Mol. Life Sci.* **61**, 511–524.

Panda, M., Gorovits, B. M., and Horowitz, P. M. (2000). Productive and nonproductive intermediates in the folding of denatured rhodanese. *J. Biol. Chem.* **275**, 63–70.

Panick, G., Malessa, R., and Winter, R. (1999). Differences between the pressure- and temperature-induced denaturation and aggregation of beta-lactoglobulin A, B, and AB monitored by FT-IR spectroscopy and small-angle X-ray scattering. *Biochemistry* **38**, 6512–6519.

Prusiner, S. B. (1998). Prions. *Proc. Natl. Acad. Sci. USA* **95**, 13363–13383.

Randolph, T. W., Seefeldt, M., and Carpenter, J. F. (2002). High hydrostatic pressure as a tool to study protein aggregation and amyloidosis. *Biochim. Biophys. Acta* **1595**, 224–234.

Richards, F. M. (1977). Areas, volumes, packing and protein structure. *Annu. Rev. Biophys. Bioeng.* **6**, 151–176.

Rochet, J. C., and Lansbury, P. T., Jr. (2000). Amyloid fibrillogenesis: Themes and variations. *Curr. Opin. Struct. Biol.* **10**, 60–68.

Ruan, K., and Weber, G. (1989). Hysteresis and conformational drift of pressure-dissociated glyceraldehydephosphate dehydrogenase. *Biochemistry* **28**, 2144–2153.

Ruan, K., Xu, C., Li, T., Li, J., Lange, R., and Balny, C. (2003). The thermodynamic analysis of protein stabilization by sucrose and glycerol against pressure-induced unfolding. *Eur. J. Biochem.* **270**, 1654–1661.

Sasahara, K., Sakurai, M., and Nitta, K. (1999). The volume and compressibility changes of lysozyme associated with guanidinium chloride and pressure-assisted unfolding. *J. Mol. Biol.* **291**, 693–701.

Seefeldt, M. B., Kim, Y. S., Tolley, K. P., Seely, J., Carpenter, J. F., and Randolph, T. W. (2005). High-pressure studies of aggregation of recombinant human interleukin-1 receptor antagonist: Thermodynamics, kinetics, and application to accelerated formulation studies. *Protein Sci.* **14**, 2258–2266.

Seefeldt, M. B., Ouyang, J., Froland, W. A., Carpenter, J. F., and Randolph, T. W. (2004). High-pressure refolding of bikunin: Efficacy and thermodynamics. *Protein Sci.* **13**, 2639–2650.

Silva, J. L., Foguel, D., and Royer, C. A. (2001). Pressure provides new insights into protein folding, dynamics and structure. *Trends Biochem. Sci.* **26**, 612–618.

Silva, J. L., Miles, E. W., and Weber, G. (1986). Pressure dissociation and conformational drift of the beta dimer of tryptophan synthase. *Biochemistry* **25,** 5780–5786.

Silva, J. L., and Weber, G. (1993). Pressure stability of proteins. *Annu. Rev. Phys. Chem.* **44,** 89–113.

Smeller, L., Rubens, P., and Heremans, K. (1999). Pressure effect on the temperature-induced unfolding and tendency to aggregate of myoglobin. *Biochemistry* **38,** 3816–3820.

St. John, R. J., Carpenter, J. F., Balny, C., and Randolph, T. W. (2001). High pressure refolding of recombinant human growth hormone from insoluble aggregates. Structural transformations, kinetic barriers, and energetics. *J. Biol. Chem.* **276,** 46856–46863.

St. John, R. J., Carpenter, J. F., and Randolph, T. W. (1999). High pressure fosters protein refolding from aggregates at high concentrations. *Proc. Natl. Acad. Sci. USA* **96,** 13029–13033.

St. John, R. J., Carpenter, J. F., and Randolph, T. W. (2002). High-pressure refolding of disulfide-cross-linked lysozyme aggregates: Thermodynamics and optimization. *Biotechnol. Prog.* **18,** 565–571.

Timasheff, S. N. (1998). Control of protein stability and reactions by weakly interacting cosolvents: The simplicity of the complicated. *Adv. Protein Chem.* **51,** 355–432.

Torrent, J., Alvarez-Martinez, M. T., Harricane, M. C., Heitz, F., Liautard, J. P., Balny, C., and Lange, R. (2004). High pressure induces scrapie-like prion protein misfolding and amyloid fibril formation. *Biochemistry* **43,** 7162–7170.

Torrent, J., Alvarez-Martinez, M. T., Heitz, F., Liautard, J. P., Balny, C., and Lange, R. (2003). Alternative prion structural changes revealed by high pressure. *Biochemistry* **42,** 1318–1325.

Tsumoto, K., Umetsu, M., Kumagai, I., Ejima, D., Philo, J. S., and Arakawa, T. (2004). Role of arginine in protein refolding, solubilization, and purification. *Biotechnol. Prog.* **20,** 1301–1308.

Van Eldik, R., Asano, T., and Le Noble, W. (1989). Activation and reaction volumes in solution. 2. *Chem. Rev.* **89,** 549–688.

Webb, J. N., Webb, S. D., Cleland, J. L., Carpenter, J. F., and Randolph, T. W. (2001). Partial molar volume, surface area, and hydration changes for equilibrium unfolding and formation of aggregation transition state: High-pressure and cosolute studies on recombinant human IFN-gamma. *Proc. Natl. Acad. Sci. USA* **98,** 7259–7264.

Zipp, A., and Kauzmann, W. (1973). Pressure denaturation of metmyoglobin. *Biochemistry* **12,** 4217–4228.

[14] Phage Display Screening for Peptides that Inhibit Polyglutamine Aggregation

By Daniel J. Kenan, Warren J. Strittmatter, and James R. Burke

Abstract

Proteins with expanded polyglutamine domains cause nine dominantly inherited, neurodegenerative diseases, including Huntington's disease. There are no therapies that inhibit disease onset or progression. To identify a novel therapeutic, we screened phage displayed peptide libraries for phage that bind preferentially to expanded polyglutamine repeats.

METHODS IN ENZYMOLOGY, VOL. 413
Copyright 2006, Elsevier Inc. All rights reserved.

0076-6879/06 $35.00
DOI: 10.1016/S0076-6879(06)13014-1

We identified a peptide motif that inhibits polyglutamine aggregation *in vitro* and inhibits death in cellular and *Drosophila* models of the polyglutamine repeat diseases. In this chapter, we describe in detail how to screen a peptide phage display library and highlight results demonstrating the success of this approach. A similar experimental approach could be used for other diseases caused by conformational change in disease proteins, including prion, Alzheimer's, and Parkinson's diseases.

Introduction

Proteins with expanded polyglutamine domains cause a number of dominantly inherited, neurodegenerative diseases, including Huntington's disease; dentatorubral-pallidoluysian atrophy (DRPLA); spinobulbar muscular atrophy; and spinocerebellar ataxia (SCA) types 1, 2, 3 (Machado-Joseph), 6, and 17 (Cummings and Zoghbi, 2000). The disease genes have no homology except for a CAG repeat domain that is enlarged in affected individuals and encodes a polyglutamine domain in the translated protein. A striking feature of these diseases is the sharp transition between the length of normal and pathological repeats. In Huntington's disease, for example, symptoms do not occur in individuals who express huntingtin protein containing 36 glutamines, whereas disease is virtually guaranteed in persons who express proteins containing 40 or more repeats (Rubinsztein *et al.*, 1996).

The mechanism underlying this glutamine length-dependent specificity is unknown, but it is likely attributable to conformational change in the disease protein (Walker and Levine, 2000). Structural data demonstrating a length-dependent conformational change in the polyglutamine domain are lacking, but identification of monoclonal antibodies that distinguish between proteins with pathological-length glutamine domains and the same protein containing a normal-length polyglutamine strongly supports a structural change. One monoclonal antibody, IC2, has been used to identify pathological-length polyglutamine domains in Huntington's disease; SCA 2, 3, and 7; and fusion proteins with Glutathione S-transferase (GST), thioredoxin, and Green Fluorescent Protein (GFP), demonstrating that it identifies the conformation of the polyglutamine domain itself rather than other conformational elements in the protein (Burk *et al.*, 1997; Evert *et al.*, 1999; Holmberg *et al.*, 1998; Nagai *et al.*, 2000; Stevanin *et al.*, 1996; Trottier *et al.*, 1995). Other monoclonal antibodies with similar specificity for pathological-length polyglutamine domains are also reported (Gutekunst *et al.*, 1999).

There are no therapies proven to ameliorate the pathogenesis of the polyglutamine repeat diseases. We hypothesized that a small molecule that preferentially binds to the pathological conformation of polyglutamine would prevent pathological protein interactions and aggregation and inter-rupt pathogenesis. To identify a lead compound, we screened M13 phage libraries expressing unique peptides at the amino terminus of the pIII protein for the ability to bind to proteins containing a pathological-length polyglutamine domain. This chapter describes the approach we used and a general method of phage screening suitable for use with other target proteins.

Why Use Phage Displayed Peptide Libraries?

Identification of molecules with the ability to bind a defined target is a common requirement in the development of therapeutics. Most drugs are small organic molecules discovered by screening of chemical libraries con-sisting of, at most, a few million entities. Peptide libraries, on the other hand, because they are composed of oligomers of 20 amino acids with a wide range of structural and chemical properties, can contain billions of unique constituents. It is not practical to synthesize large numbers of random peptides chemically to screen for binding to a target, but one can take advantage of biological synthesis and phage genetics to generate a large number of individual phage particles, with each expressing a unique peptide on its surface. Phage display was originally described by Smith (1985), who found that phage containing a fusion protein can be enriched more than 1000-fold over ordinary phage by affinity screening.

To generate a phage display library, segmentally randomized, defined-length oligonucleotides are recombinantly joined to a gene encoding a phage coat protein (Adey *et al.*, 1996). When the phage coat protein is expressed, the modified phage displays a unique peptide on its surface that is encoded by the foreign DNA. Surface displayed peptides are accessible for interaction with a target and behave similar to the isolated peptide in solution in most cases. Phage that bind to a target can be enriched in a process called "biopanning." Individual binding phage can be cloned and amplified by infection, and the interacting peptide can be identified by sequencing the DNA attached to the coat protein gene. (Construction of a peptide phage display library is beyond the scope of this chapter, which focuses on screening. For details on library construction, see Adey *et al.*, 1996.)

Library Selection

An early decision in phage display screening is library selection. Libraries are commonly constructed in phage M13, but λ and T7 phage libraries can also be used (Kay *et al.*, 1996). There are several capsid proteins in M13 that can be used to display peptides, but the most

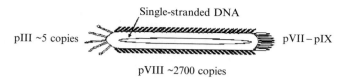

Single-stranded DNA

pIII ~5 copies   pVII – pIX

pVIII ~2700 copies

FIG. 1. Schematic of M13 phage. M13 are elongated cylindrical phage that are ≈6.5 nm in diameter and 930 nm in length. They contain single-stranded DNA that encodes 11 genes in 6407 nucleotides. The phage libraries described in this chapter contain a 33-nucleotide insert in the 5′-end of the pIII gene. Illustration by S. P. Burke.

commonly used is pIII, which exists in approximately five copies on one end of the phage (see Fig. 1 for schematic outline of M13 phage). To construct a library, segmentally degenerate double-stranded oligonucleotides are cloned into the 5′-end of the pIII capsid gene. Peptides from 5–50 amino acids in length have been inserted into the pIII protein by cloning of appropriate degenerate oligonucleotides. To reduce the incidence of stop codons in the inserted sequence, which would produce nonviable phage, oligonucleotides are typically synthesized using an NNK (K = G and T) or NNS (S = G and C) codon scheme that encodes all 20 amino acids but only one of the three stop codons (TAG). Readers interested in more details about library construction are referred to other publications (Kay *et al.*, 1996 or Chang *et al.*, 2003).

In our screening for peptides that bind to pathological-length polyglutamine domains, we used libraries with 11 amino acid inserts and a fixed central amino acid ($X_5FixedX_5$) at the amino terminus of the pIII protein. The presence of a fixed central amino acid imparts a constant structural context across the entire library. For example, a fixed central proline imposes a structural bend to all peptides in the library and further restricts the conformational degrees of freedom that may be explored by each library peptide. In contrast, a fixed central glycine would impose much more structural flexibility to the encoded peptides and would further enable exploration of unusual dihedral angles that may be essential for binding certain targets. In practice, an assortment of libraries representing different fixed residue and other structural constraints will provide the greatest likelihood of deriving high-affinity peptides specific for a given target.

Screening Phage Display Random Peptide Libraries

The purpose of phage display screening is to select peptides that selectively bind the native conformation of a target. Identification of binding peptides is usually performed by incubating phage containing random peptide inserts with the target protein immobilized on a solid support.

Numerous solid supports have been successfully used for phage library screening, including 96-well microtiter plates, immunotubes, and immobilization of target proteins on Sepharose or paramagnetic beads. Larger tubes allow more target protein binding and the addition of a greater number of phage, but 96-well plates are inexpensive and their standard format makes them convenient to use.

Binding of the target protein is a critical determinant for success in screening. If the protein is denatured or in a nonnative conformation, it is unlikely that isolated phage will recognize the desired target. It may be necessary, therefore, to modify binding conditions depending on the properties of the target protein. It is also essential to purify the target protein to the greatest extent possible, because impurities in the target protein may serve as binding sites just as well as the intended target. If possible, target fusion partners, affinity, and immunotags (e.g., GST, His-tags, myc) should be removed from the target protein before immobilization to reduce isolation of phage that bind to nondesired targets (Murthy et al., 1999).

Polyglutamine is a highly insoluble peptide, so a fusion partner to increase solubility was required. We chose GST because it can be expressed in bacteria at high concentrations and is easily purified. However, these advantages are partially offset because GST-binding phage can compete with the desired polyglutamine-binding phage. Phage binding to plastic can also be troublesome, even with efforts to block nonspecific interactions. An initial round of panning was performed against bovine serum albumin (BSA) and immobilized GST with a nonpathological-length polyglutamine domain to reduce the number of phage that bind to plastic, nonspecific proteins and GST. A summary of peptides that display nonspecific binding can be found in a recent article by Menendez and Scott (2005).

Reagents/Solutions

Anti-M13 Antibody Conjugated with Horseradish Peroxidase (Pharmacia)

　　Stock of TG-1 (Stratagene)
　　　　TG-1 genotype: supE thi-1 Δ(lac-proAB) Δ(mcrB-hsdSM)5 (rK− mK−) [F' *traD36 proAB lacIqZΔM15*]. (Genes listed signify mutant alleles. Genes on the F' episome, however, are wild-type unless indicated otherwise). TG1 cells contain the *lacIqZΔM15* gene on the F' episome, allowing blue-white screening for recombinant plasmids.
　　ABTS (2′,2′-azino-bis-3-ethylbenzthiazoline-6-sulfonic acid) solution
　　　　Dissolve 10.5 g citrate monohydrate in 1.0 L of sterile deionized water.
　　Adjust pH to 4 with approximately 6 ml of 10-*M* NaOH.

Add 220 mg ABTS; filter sterilize, and store at 4°.

40% glycerol

Dilute 4 ml glycerol with 6 ml of deionized water.

Sterilize by autoclaving for 20 min at 15 psi on liquid cycle.

1 M glycine pH 2.0

Dissolve 111.6 g glycine in approximately 800 ml of deionized water.

Adjust pH to 2 with concentrated HCl. Bring to a final volume of 1 L with deionized water; sterilize by autoclaving for 20 min at 15 psi on liquid cycle.

2% isopropylthio-β-d-galactoside (IPTG) (0.84 M, Formula weight [FW] = 283.3)

Dilute 20 mg IPTG with 1 ml of deionized water.

Filter sterilize and store at −20°.

2% X-gal (5-Bromom-4-chloro-3-indoyl-B-d-galactoside) (FW = 408.6)

Dissolve 20 mg X-gal in 1 ml dimethylformamide in a glass or polypropylene tube; filter sterilize, and limit light exposure by storing wrapped in foil at −20°.

100 mM NaHCO$_3$, pH 8.5

Dissolve 4.2 g NaHCO$_3$ in deionized water to 400 ml.

Adjust to pH 8.5 if necessary, and adjust volume to 500 ml; sterilize by autoclaving for 20 min at 15 psi on liquid cycle; store at room temperature. The solution is stable for months if not contaminated.

200 mM NaHPO$_4$, pH 8.5 (1000 ml)

NaH$_2$PO$_4$-7H$_2$O, 53.5 g

Deionized water, 800 ml

Dissolve solutes, and adjust pH to 8.5; bring volume to 1 L with deionized water.

Sterilize by autoclaving for 20 min at 15 psi on liquid cycle.

1× phosphate-buffered saline (PBS), pH 7.4

Deionized water, 800 ml

NaCl, 8 g

KCl, 0.2 g

Na$_2$HPO$_4$, 1.44 g

NaH$_2$PO$_4$, 0.24 g

Adjust to pH 7.4 with HCl; aliquot and sterilize by autoclaving for 20 min at 15 psi on liquid cycle; store at room temperature.

1× PBST, pH 7.4, with 0.1% Tween-20 (PBST)

Mix 1 ml Tween-20 with 999 ml 1× PBS

Store at room temperature. This solution is prone to contamination, so check frequently.

Acetylated BSA

Stock 20 mg/ml in water.

Ethanolamine (Aldrich 11,016–7)
   301.8 μl in 50 ml Diethylpyrocarbonate (DEPC)-treated water
   Store at room temperature.
2× YT medium (1000 ml)
   Bacto-Tryptone, 10 g
   Bacto-yeast extract, 10 g
   NaCl, 5 g
   Deionized water, 900 ml
   Stir to dissolve the solutes; adjust pH to 7.0 with 5 N NaOH, and
   adjust volume to 1 L with deionized water; sterilize by autoclaving
   for 20 min at 15 psi on liquid cycle.
2× YT plates (1000 ml)
   Bacto-Tryptone, 10 g
   Bacto-yeast extract, 10 g
   NaCl, 5 g
   Bacto-agar, 15 g
   Deionized water, 900 ml
   Stir to dissolve the solutes; adjust volume to 1 L with deionized
   water; sterilize autoclaving for 20 min at 15 psi on liquid cycle; let
   agar cool to ≈55°, and pour plates.
Top agar (soft agar (0.8%) in 2× YT)
   Bacto-Tryptone, 10 g
   Bacto-yeast extract, 10 g
   NaCl, 5 g
   Bacto-agar, 8 g
   Deionized water, 950 ml
   Stir to dissolve the solutes; adjust volume to 1 L with deionized
   water; sterilize autoclaving for 20 min at 15 psi on liquid cycle; let
   agar cool to ≈55°.

Methods

Day 1

Immobilize Target Protein in High-Binding Microtiter Plates. It is critically important to employ stringent phage hygiene procedures to prevent
inadvertent contamination of the experiment and the laboratory. Although
phage are not hazardous to humans, it is surprisingly commonplace for new
practitioners of phage display to lose weeks of work because of preventable
contamination. Efforts to minimize aerosol formation of phage solutions
must be observed at all times. Barrier tips must be employed. Any spills and
discarded phage cultures must be decontaminated with 10% bleach.

Gloves should be changed immediately if contamination is suspected. Bench tops and micropipettes should be decontaminated with bleach at least once per week and more frequently if contamination is suspected. Pay special attention when opening tubes and microtiter plate adhesive lids to prevent escape or cross-contamination of samples. It is probably also best to avoid the use of plate washers for phage display screening, because these are difficult to decontaminate.

1. Freshly prepare target protein (polyglutamine-GST with a polyglutamine domain containing 62 consecutive glutamines [Q62-GST]). Details of polyglutamine-GST vector construction are published (Onodera et al., 1996). Dilute Q62-GST to 2.5 μg/ml in 100 mM NaHCO$_3$, pH 8.5. For each phage library to be screened, you will need 300 μl of the protein solution. One hundred microliters of 2.5-μg/ml GST with a nonpathological polyglutamine domain (Q19-GST) should also be prepared in 100 mM NaHCO$_3$, pH 8.5 buffer to be used to preadsorb phage that bind the "normal" conformation of polyglutamine or GST. A positive control for which a known binding peptide has been identified should also be prepared. We used a fusion protein consisting of an Src SH3 domain fused to GST (300 μl of a 2.5-μg/ml GST-SrcSH3 in 100 mM NaHCO$_3$, pH 8.5) (Rickles et al., 1994).

2. Label four flat plates (Costar 3590 flat-bottom, high-binding, 96-well enzyme immunoassay [EIA]/radioimmunoassay [RIA] plate or equivalent) with numbers 1–4. Mask every other row with narrow laboratory tape, and use only every other well to prevent cross-contamination. Each library to be screened will occupy its own well, so label the well where each library will be added. For example, if you are planning to screen eight libraries, you would use wells A1, A3, A5, A7, C1, C3, C5, and C7. Typically, we screen the positive control against only a single library, so one well should also be labeled for the positive control. Use the same labeling pattern for all four plates.

The M13 phage display library (constructed by the Combinatorial Science Center at Duke University) used to identify peptides that bind to proteins with an expanded polyglutamine domain contained a random 11-amino acid peptide inserted at the amino terminus of the pIII capsid protein. The 11-mer peptide was not completely random, because a fixed amino acid was inserted in the sixth position of the peptide (X$_5$-fixed-X$_5$). A total of 2.5 $\times$ 10^{11} phage from each of the following fixed amino acid libraries was screened for binding: aspartate, phenylalanine, histidine, lysine, leucine, proline, and tryptophan.

3. To three of the plates, add 100 μl of the target protein solution (Q62-GST) to each well to be used in panning and 100 μl GST-SrcSH3 to the

control well. The fourth plate is coated with 100 μl immobilized Q19-GST to reduce selection of phage binding to GST. Seal the plate with an adhesive lid (USA Scientific #2920–000) to avoid evaporation, and incubate at 4°.

Bacterial Culture

The day before panning, start a fresh culture of TG-1 (or DH5aF') in 50 ml 2× YT from a single colony. Incubate overnight at 37°, with shaking at approximately 150–200 rpm. Note that shaking at higher rates will shear filamentous phage extruding from the bacterial surface and shear pili from bacteria, leading to poor phage yields.

Day 2

Bacterial Culture. A fresh log-phase culture should be available every day during phage display screening procedures. Log-phase cultures are used because bacteria display the greatest number of pili during rapid growth. Dilute the overnight culture of TG-1 (or DH5aF') 1:100 into 2× YT at 37°, and grow to log phase (Optical Density [OD]$_{600nm}$ = 0.5–1.0) to use in amplifying eluted phage from panning. Fresh log-phase cultures are important, especially in the first round of panning, to ensure efficient recovery of all the diverse phage that are selected. Prepare at least 1 ml of culture for each library to be panned. Incubate culture at 37°, with shaking for approximately 2 h before checking the OD. Most *Escherichia coli* strains will double approximately every 30 min.

It is optimal to start the culture at a time such that the culture will reach log phase at approximately the same time that the phage selection is completed and the phage are ready for infection. If the phage are not ready, the log-phase culture may be stored on ice for several hours, although efficiency of phage infectivity will drop with chilling or growth to higher density.

Preadsorption and First Round Panning

1. Remove the lid from plates 1 and 4, and add 150 μl 0.1% acetylated BSA in 100 mM NaHCO$_3$, pH 8.5 to each well to reduce nonspecific binding. Cover the plates with an adhesive seal, and incubate the plate at room temperature for 1 h. Leave plates 2 and 3 at 4°, and do not add BSA at this time.

WASH PROTOCOL

2. Wash the wells in plate 4 containing immobilized GST five times with PBST by inverting the plate and flicking the wash solution into the sink. Fill the wells, but avoid overfilling and cross-contamination. Remove residual liquid by slapping the dish against a clean stack of paper towels. Do not let the wells dry out completely.

3. Add 25 μl of peptide library phage (>10^{10} pfu) in 125 μl PBST to each appropriate well in plate 4. Plate 4 contains immobilized Q19-GST. Phage that bind nonpathological-length polyglutamine protein will be removed from subsequent panning by this step. Seal the plates with a fresh adhesive, and incubate for 1 h at room temperature. When working with phage, always use aerosol pipette tips and rinse contaminated materials in a bleach solution.

4. Collect nonbinding phage with a pipette, and transfer to a fresh microfuge tube. Plate 4 can be disposed after rinsing with bleach.

5. Wash the wells in plate 1 containing immobilized target protein (Q62-GST) and the positive control protein five times with PBST by inverting the plate and flicking the wash solution into the sink as described in the section on the wash protocol.

6. Add phage collected in step 4 to the appropriate well using a fresh aerosol-filtered pipette tip for each library to be screened. Seal the plates with a fresh adhesive covering, and incubate for 7 h at room temperature or overnight at 4°. Shorter incubation times may be appropriate for other targets.

7. Remove nonbinding phage by washing the wells five times with PBST as described in the section on the wash protocol. More washes may be appropriate for targets that are known to be "sticky" or that engage in multiple protein-protein interactions.

ELUTION. Bound phage are generally eluted by dramatic pH changes that disrupt the interaction between the phage and the target without destroying phage infectivity. A common approach is to elute bound phage with glycine pH 2.0, followed by a second elution with ethanolamine at pH 12.

8. Add 100 μl of 50-mM glycine-HCl (pH 2.0) prewarmed to 50° to each well to elute bound phage. Incubate the plate at room temperature for 15 min, and then use a fresh filtered pipette tip for each well to transfer the contents to a clean microfuge tube containing 200 μl of 200-mM NaPO$_4$, pH 8.5 buffer. It is essential to neutralize the solution containing the eluted phage, because the acidic elution conditions will destroy the phage.

9. Repeat the elution by adding 100 μl of 100-mM ethanolamine prewarmed to 50° to each well. Incubate for 15 min at room temperature. Remove the ethanolamine from each well using a clean tip for each well, and then combine with the glycine-eluted phage previously eluted from that same well. Keep each library separate, and avoid cross-contamination. Store at 4° until ready to amplify phage. Alternatively, phage can be frozen until ready to amplify. To freeze, gently mix the phage with an equal volume of sterile 40% glycerol in 1× PBS and then flash freeze on dry ice.

Amplification of Recovered Bound Phage

1. Label separate Falcon 4-ml sterile polypropylene culture tubes with the name of each library and the control protein. Add 1 ml of the log-phase culture of TG-1 or DH5aF′ begun earlier in the day and 400 μl of recovered phage from each screened library to the appropriate tube. Incubate at 37° without shaking for 10 min to allow infection, followed by incubation with shaking at 150–200 rpm.

Ideally, cultures should not grow for more than 8 h to minimize the chance of proteolytic degradation of the displayed peptide. The vast majority of peptides will not be rapidly degraded and remain intact after overnight incubation. If overnight cultures show significant signs of cell death (e.g., lack of opalescence in a swirled culture, evidence of particulates or clumps), discard and start with a fresh culture. Phage can also be stored after amplification at 4° for several days prior to use in binding assays.

Second and Third Rounds of Panning

1. Collect the amplified phage by centrifuging cells at 4000g for 10 min at 4°. Transfer the phage supernatant to a new 5-ml Falcon tube. Store amplified phage on ice.
2. Aliquot 200 μl of phage supernatant into a microfuge tube and store at 4° to use in enzyme-linked immunoassay (ELISA) of phage pools (described below).
3. Remove 100 μl of phage supernate for panning, and freeze the remaining phage in case there is a need to repan or analyze the pool/ isolates from this round.

FREEZING PHAGE. To freeze phage supernatant dilute 1:1 with sterile 40% glycerol in 1× PBS and prepare 200-μl aliquots. Flash freeze aliquots in a dry ice–ethanol bath, and store at −80°. Make sure the labeling on tube does not dissolve in ethanol.

4. Dilute the overnight culture of TG-1 (or DH5aF′) 1:100 into 2× YT, and grow to log phase (OD = 0.5–1.0) to amplify eluted phage. Prepare at least 1 ml of culture for each library to be panned.
5. Remove the lid from plate 2, and add 150 μl 0.1% acetylated BSA in 100 mM NaHCO₃, pH 8.5 to each well to reduce nonspecific binding. Cover the plate with an adhesive seal and incubate the plate at room temperature for 1 h.
6. Discard the target/BSA solution by inverting the plate and flicking the wash solution as described in the section on the wash protocol. Do not let the wells dry out completely.

7. After washing the plate, immediately add 100 μl of amplified phage from the previous round of panning and 100 μl PBST to the appropriate wells and incubate for 7 h (or overnight) as described for round 1 panning.
8. Wash, elute, and amplify as described for round 1 panning. Repeat for round 3 and 4 panning.

Measuring Binding Activity in Pooled Phage by ELISA. Measuring the binding activity of selected phage pools does not provide information on individual phage, but increased binding with successive rounds of panning is an indicator that enrichment for bound phage is occurring. ELISA signals from pools of a particular library should show a steady increase with subsequent rounds of panning. If enrichment is not observed after four rounds, there is a good chance that the target is denatured or inactive. Consider an alternative method of immobilization, such as neutral or acidic buffer or biotinylation of the target, and capture on streptavidin-coated plates or beads, which may permit retention of the target in a physiologic conformation (Scholle *et al.*, 2004).

Typically, pool ELISAs are performed on all pannings after the fourth round. Some targets may require an additional round of selection, but it is generally not advisable to go beyond four rounds because of selection pressure from rapid growing mutant phage that have deleted the foreign peptide sequence of interest.

1. Label the ELISA plate with one well for each panning of each library screened. Because each panning will occupy a well, the same labeling pattern used in the screening cannot be used in the pooled phage ELISA. Because the ELISA is analytic and not preparative, it is not critical to avoid trace cross-contamination; thus, adjacent wells can be used. Include an appropriate number of wells for the SrcSH3-positive control and a BSA-negative control for each pool being assayed.

2. Immobilize target protein, SrcSH3 (positive control), and BSA on ELISA plate as described above in the section on immobilization of protein in high-binding microtiter plates. Incubate the plate overnight at 4°.

3. Complete ELISA plate preparation as described in steps 1 and 2 in the section on preadsorption and first round panning.

4. Add 50 μl PBST to each well containing immobilized protein. Add 50 μl phage stock representing each pool to the appropriate well. Seal the plates, and incubate at room temperature for 1 h.

5. Remove nonbinding phage by washing the wells five times as described in the section on the wash protocol.

6. Dilute HRP-conjugated anti-M13 antibody 1:5000 in PBST. Add 100 μl of the diluted antibody to each well. Seal the wells with an adhesive

cover, and incubate the plate at room temperature for 1 h. Wash the wells as described in step 5 above.

7. Add 100 μl ABTS reagent containing 0.05% hydrogen peroxide to each well. Incubate the plate at room temperature for 10 min. Quantify the reaction by measuring the absorbance at 405 nm with a microtiter plate reader. Other detection systems are available for HRP-conjugated secondary antibodies.

Typically, the BSA-negative controls give absorbance values of less than 0.01 at 405 nm, whereas positive selected pools are typically greater than 0.5 in this ELISA format. In the example shown in Fig. 2, the selected pools showed a high degree of signal after only a single round of selection, whereas the BSA background binding remained low. As in any ELISA, it is advisable to conduct the assay over a range of dilutions to evaluate the binding behavior of the phage pools fully at various concentrations. Some selected pools will demonstrate significant background binding at very high concentrations but specific binding at lower concentrations.

Isolation and Propagation of Affinity Purified Phage Clones

If you have an increase in binding demonstrated in the pooled phage ELISA assay, isolate individual phage and screen for binding to your target. In our work, after the fourth round of panning, we isolated 50 phage from each of the seven libraries that we screened (for a total of 350 individual, putative, polyglutamine-binding phage). In general, it is best to isolate phage from the earliest possible pool that shows a significant positive signal. Otherwise, you risk focusing all your validation efforts characterizing sibling

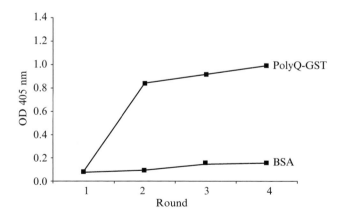

FIG. 2. Example of pooled enzyme-linked immunoassay results. Note the increase in signal between rounds 1 and 2 in wells coated with polyQ-GST compared with wells coated with bovine serum albumin (BSA), which is also in the wash buffer.

clones that happen to survive the applied selection method better than their library mates. These may or may not represent the best binders and, in any event, will not yield useful structure-activity data such as can be obtained from a diversity of clones representing both strong and weak binders.

Phage Isolation. Melt top agar in a microwave oven, and keep at 55° until ready for use. Top agar should be no warmer than 55° to prevent heat injury to the phage. Make sure that a fresh logarithmic-phase culture of TG-1 or DH5aF' is ready for use.

1. Perform a 10-fold serial dilution of the final round of panning from each library that demonstrated an increase in phage pool binding on ELISA.

Dilutions can be easily handled using a 96-well microtiter plate (non-ELISA plates with U-shaped bottoms) with each library assigned its own row. Add 180 μl PBS into each well with a multichannel pipette, and transfer 20 μl of appropriate selected phage pool into column 1 of the appropriate row. Mix by pipetting up and down. Transfer 20 μl of the phage from column 1 to column 2. Mix and continue the dilution series through column 12 (10^{-1} to 10^{-12} dilution). Use a fresh pipette tip for each dilution.

2. Prepare three sterile culture tubes for each library with the screened library's name and dilution. Typically, only dilutions in columns 7 to 9 (10^{-7}, 10^{-8}, and 10^{-9}) need to be plated, but all dilutions should be made because it is occasionally necessary to replate in cases of very low or very high titers. Add 100 μl of fresh bacterial culture (TG-1 or DH5aF') into each tube.

3. Transfer 100 μl of each phage dilution into the appropriate culture tube. Add 30 μl 2% IPTG, 30 μl 2% X-Gal, and 3 ml molten 0.8% top agar. Cap and mix gently by inversion, avoiding bubbles as much as possible, and then to pour the mixture rapidly and spread onto 2× YT plates that have been prewarmed at 37°. With practice, it is possible to plate the top agar without introducing bubbles, which can be confused with phage plaques to the untrained eye. It is often possible to remove bubbles by brief exposure to a Bunsen flame. Allow the plates to sit undisturbed until the top agar hardens (approximately 5 min). It is cautioned that top agar hotter than 55° will kill *E. coli.*

4. Invert the plates, and incubate at 37° for 8 h. Examine the plates for plaques. If the plaques are confluent at the highest dilution or no plaques are present at the lowest dilutions, additional dilutions will need to be plated. Plaques may be counted and multiplied by their dilution factor to determine the phage titer of the initial sample. Typically, phage titer is expressed as plaque-forming units per milliliter. Store plates at 4° until you are ready to pick the isolated phage clones.

5. Dilute an overnight culture of TG-1 or DH5aF' 1:100 into sterile 2× YT medium. For each plaque to be propagated, add 3 ml of bacterial culture into a 15-ml tube. At least 50 individual plaques should be picked for amplification from each library that gave positive pool ELISAs.

6. Pick blue-green plaques by touching the surface with a sterile wooden stick or toothpick. M13 phage contain a portion of the β-galactosidase gene, which complements the bacterial gene and restores galactosidase activity. TG-1 infected with phage that are grown on plates containing X-gal will cleave the dye moiety from the galactose, turning plaques blue green. TG-1 that are not infected with phage do not contain the full β-galactosidase gene, so they cannot cleave X-gal, which leaves plaques white. Incubate the tubes at 37° with vigorous agitation for 8–10 h. Pellet the bacterial cells by centrifuging at 4000g for 10 min at 4°. Transfer the supernatant to a new tube, and store it at 4°. Keep the bacterial pellet for replicative form (RF) plasmid purification of phage that display good binding characteristics.

Confirmation of Binding Activity of Affinity Purified Phage by ELISA. It is essential to confirm binding of the isolated phage to the intended target. To verify that phage bound to the "pathological" conformation of polyglutamine, we compared binding of individual phage to pathological-(Q62) and normal-(Q19) length polyglutamine domain GST fusion protein. Phage that demonstrated greater binding to the pathological-length repeat were further characterized by sequencing the DNA insert in the pIII gene.

1. For each phage clone to be tested, add 0.25 μg of target protein (Q62-GST), comparison protein (Q19-GST), and a negative control protein (BSA) in 100 mM NaHCO$_3$ (pH 8.5) into adjacent wells in a 96-well microtiter plate. It is not necessary to skip wells, but aerosol tips are required to prevent cross-contamination of phage.

2. Seal the plate with an adhesive lid to avoid evaporation, and incubate as described in the section on immobilization of target into high-binding microtiter plates.

3. The next morning, add 150 μl 0.1% BSA in 100 mM NaHCO$_3$, pH 8.5 to each well to reduce nonspecific binding. Seal the wells with an adhesive lid, and incubate for 1 h at room temperature.

4. Wash the wells three times with PBST by flicking the wash solution into the sink as described above in the section on the wash protocol. Do not let the wells dry out completely.

5. Add 50 μl PBST to each well containing immobilized protein. Add 50 μl of supernatant from each isolated phage to a set of wells containing target, comparison, and negative control proteins. Seal the plates, and

incubate at room temperature for 1 h. Keep the rest of the supernatant as phage stock at 4°. For long-term storage, add glycerol to a final concentration of 20% and store at −80° as described above.

6. Remove nonbinding phage by washing wells five times with PBST as described in the section on the wash protocol.

7. Dilute HRP-conjugated anti-phage antibody (Pharmacia) 1:5000 in PBST. Add 100 μl of the diluted antibody to each well. Seal the plate, and incubate the plate at room temperature for 1 h. Wash the wells five times to remove unbound antibody.

8. Add 100 μl ABTS reagent containing 0.05% hydrogen peroxide to each well. Incubate the plates at room temperature for 10 min. Quantify the signal by measuring the absorbance at 405 nm using a microplate reader. Most positive signals have OD values that range from 0.5–3.0 units, whereas negative values are typically in the range 0.05–0.1 units.

Phage that demonstrate the desired binding affinities following initial isolation should be replaque-purified, and binding should be confirmed by repeat ELISA.

Preparation of Plasmid from Positive Phage Clones for DNA Sequencing

1. Pick the cell pellet of confirmed positive binding phage to isolate double-stranded RF DNA for automated fluorescence sequencing.

2. Use the QIAprep Spin Plasmid kit (Qiagen) for isolation of RF DNA according to the manufacturer's instructions. Elute DNA with 50 μl of sterile water. Measure DNA concentration by reading the OD value at 260 nm (1 OD = 50 μg/ml DNA). Alternatively, phage DNA can be sequenced from the single-stranded genomic DNA contained within the phage particle by following instructions in the QIAprep Spin M13 Kit. Note that only the M13 PIII reverse primer can be used to sequence single-stranded genomic DNA, whereas either primer can be used to sequence the double-stranded RF DNA:

> M13 pIII forward oligo sequence: ATTCACCTCGAAAGCAAG
> CTG
> M13 pIII reverse oligo sequence: ACCCTCATAGTTAGCGTAACG

3. Add 0.7 μg DNA and 10 pmol M13 primer plus sterile water to bring volume to 20 μl in a 0.65-ml microcentrifuge tube. Sequence DNA.

Results of Screening for Phage That Binds Proteins Containing Pathological-Length Polyglutamine Domains

Six of the 350 isolated phage that we isolated displayed greater binding to Q62-GST compared with Q19-GST in the ELISA assay (greater binding was defined as binding ratios >1.2). We named these peptides

TABLE I

CHARACTERISTICS OF PHAGE PEPTIDES THAT PREFERENTIALLY BIND TO PATHOLOGICAL-LENGTH POLYGLUTAMINE PROTEINS

Sequence of isolated phage	Q62/Q19 ELISA binding ratio	Name
SNWKWWPGIFD	1.66	QBP1
HWWRSWYSDSV	1.31	QBP2
HEWHWWHQEAA	1.30	QBP3
WGLEHFAGNKR	1.27	QBP4
WWRWNWATPVD	1.25	QBP5
WHNYFHWWQDT	1.23	QBP6

Solutions containing 0.25 μg Q19-GST or Q62-GST are bound and washed as described in the section on immobilization of the target protein in high-binding microtiter plates. Fifty microliters PBST is added to each well containing immobilized protein, and the plates are sealed and incubated for 1 h at room temperature. The wells are washed five times with PBST, and 100 μl monoclonal antibody against M13 phage (Amersham Pharmacia Biotech; 1:5000 in PBST) is added to detect bound phage. The plates are then sealed and incubated for 1 h at room temperature. The plates are then washed five times in PBST, and 100 μl ABTS reagent containing 0.05% hydrogen peroxide is added to each well. The plates are incubated for 10 min at room temperature, and the reaction is quantified by measuring the ABTS signal at 405 nm using a microtiter plate reader. Binding ratios are calculated by determining the signal from wells in which phage was incubated with Q62-GST divided by the signal from the phage incubated with Q19-GST. Modified and reproduced with permission from Nagai *et al.* (2000).

glutamine (Q) binding (B) peptides (P) or QBP 1 through 6. Table I shows the results of individual phage binding to Q62-GST versus Q19-GST expressed as a ratio. None of the selected phage bound GST or BSA.

We sequenced the DNA encoding the insert in the pIII gene and found that peptides that preferentially bind Q62-GST have a conserved motif. Most of the peptides are tryptophan (W)-rich, with many containing three tryptophans, including a pair of adjacent tryptophans. The phage peptides containing tandem tryptophans also have an adjacent positively charged (basic) amino acid.

QBP1 Blocks Aggregation of Polyglutamine Proteins In Vitro *and in Cells*

In Vitro. To determine whether QBP1 interacts with other proteins containing a polyglutamine domain, we developed an *in vitro* aggregation assay by fusing a polyglutamine domain to thioredoxin. Pathological-length polyglutamine domains fused to thioredoxin demonstrate length-dependent aggregation with a similar cutoff to human polyglutamine disease proteins (i.e., thioredoxin fusion proteins containing less than 35 glutamines do not aggregate, whereas proteins with glutamine domains longer than 40 do aggregate; not shown) (Nagai *et al.*, 2000). Aggregation of the polyglutamine-thioredoxin fusion protein is quantified by measuring the turbidity of

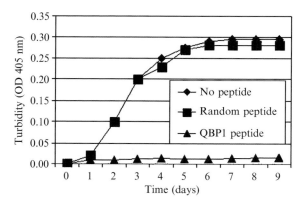

FIG. 3. Synthetic QBP1 peptide inhibits aggregation of polyglutamine. A 17-μM solution of 62 glutamines fused to thioredoxin Thio-Q62 becomes turbid in a pattern consistent with nucleation kinetics (there is an initial lag phase, followed by rapid polymerization and, finally, a plateau). The plateau is reached by 4 days of incubation. Results shown are from representative experiments. Variation between duplicate wells is less than 10%. OD, optical density.

the protein solution at 405 nm using a plate reader. Figure 3 shows that a random peptide has no effect on aggregation of Q62-thioredoxin, whereas QBP1 completely blocks aggregation (Nagai *et al.*, 2000; Ren *et al.*, 2001). A scrambled version of QBP1 has no effect on aggregation, confirming the importance of QBP1's amino acid sequence in inhibiting aggregation (not shown). We have used this simple screening assay to identify the amino acids required to inhibit polyglutamine protein aggregation (see Ren *et al.*, 2001 for an example using polyglutamine binding peptides) and suggest that this approach is a reasonable option to identify novel therapeutics for other disease proteins caused by a conformational change.

QBP1 Activity in Cells. Ultimately, a therapy for the polyglutamine repeat diseases must be effective in cells. To determine whether QBP1 blocks polyglutamine aggregation and toxicity in cells, we co-expressed QBP1 fused to cyan fluorescent protein (QBP1-CFP) and Q57 fused to yellow fluorescent protein (Q57-YFP) in COS cells. QBP1-CFP co-localizes with the fluorescent polyglutamine aggregate, but co-expression of a scrambled version of QBP1, or a random peptide, fused to CFP remains diffusely expressed (not shown) (Nagai *et al.*, 2000). Figure 4 demonstrates that QBP1-CFP inhibits aggregation and cell death induced by Q57 fused to YFP. A tandem repeat of the QBP1 sequence is more potent than a single copy in inhibiting aggregation and cell death. Higher level multimers of QBP1 were not more effective than a tandem repeat (data not shown).

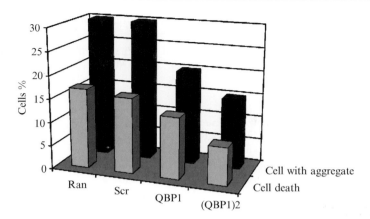

FIG. 4. QBP1 inhibits polyglutamine protein aggregation and cell death. COS-7 cells were examined 48 h after transfection with either random peptide (Ran)- or scrambled peptide (Scr) QBP1, or a tandem repeat of QBP1 fused to cyan fluorescent protein (CFP) and Q57-yellow fluorescent protein (YFP) using a Zeiss fluorescence microscope with the CFP/YFP filter set from Omega Optical. The percentage of dead cells was calculated by counting the number of transfected cells stained with ethidium homodimer and dividing by the total number of transfected cells multiplied by 100. The percentage of cells with aggregates was calculated by counting the number of transfected cells containing aggregate and dividing by the total number of transfected cells multiplied times 100. In each experiment, at least 200 transfected cells were counted. Experiments were repeated at least three times. (QBP1)2 is a tandem repeat of QBP1 (SNWKWWPGIFDSNWKWWPGIFD). QBP1 and (QBP1) 2 inhibition of aggregation and cell death is significantly different from cells expressing scrambled or random peptides ($P < 0.01$, Student t test).

QBP1 Activity in Drosophila. We also examined the effect of expressing QBP1 on polyglutamine pathogenesis in *Drosophila* models of disease. A tandem repeat of QBP1 fused to CFP reduces aggregate formation and normalizes eye development in a transgenic fly expressing a pathological-length polyglutamine protein with 92 glutamines fused to a FLAG epitope (Nagai *et al.*, 2003). QBP1 expression also dramatically prolonged the lifespan of *Drosophila* expressing a truncated portion of the Machado-Joseph protein, which causes SCA type 3. Expression of a tandem repeat of QBP1 in the Machado-Joseph fly increased the median lifespan from 5.5–52 days, suggesting that QBP1 can prevent or minimize pathogenesis (Nagai *et al.*, 2003).

Conclusion

Phage display screening is–a powerful tool that can be used to identify peptides that inhibit protein–protein interactions. We have demonstrated that QBP1 has activity *in vitro*, in cells, and in an animal model of polyglutamine repeat diseases. A similar screening approach could be used to identify peptides that inhibit pathogenesis in other diseases characterized by disease proteins that undergo a conformational change, such as Parkinson's, Alzheimer's, motor neuron, and prion diseases.

Acknowledgments

This work was supported by the Deane Laboratory, the Hereditary Disease Foundation, and National Institutes of Health grants RO1NS40540 and R41NS48700 to J. R. Burke and RO1 CA77042 and R21 CA81088 to D. J. Kenan.

References

Adey, N. B., Sparks, A. B., Beasley, J., and Kay, B. K. (1996). *In* "Construction of Random Peptide Libraries in Bacteriophage M13" (B. K. Kay, J. Winter, and J. McCafferty, eds.), pp. 66–78. Academic Press, San Diego.

Burk, K., Stevanin, G., Didierjean, O., Cancel, G., Trottier, Y., Skalej, M., Abele, M., Brice, A., Dichgans, J., and Klockgether, T. (1997). Clinical and genetic analysis of three German kindreds with autosomal dominant cerebellar ataxia type I linked to the SCA2 locus. *J. Neurol.* **244,** 256–261.

Chang, C. Y., Norris, J. D., Jansen, M., Huang, H. J., and McDonnell, D. P. (2003). Application of random peptide phage display to the study of nuclear hormone receptors. *Methods Enzymol.* **364,** 118–142.

Cummings, C. J., and Zoghbi, H. Y. (2000). Fourteen and counting: Unraveling trinucleotide repeat diseases. *Hum. Mol. Genet.* **9,** 909–916.

Evert, B. O., Wullner, U., Schulz, J. B., Weller, M., Groscurth, P., Trottier, Y., Brice, A., and Klockgether, T. (1999). High level expression of expanded full-length ataxin-3 in vitro causes cell death and formation of intranuclear inclusions in neuronal cells. *Hum. Mol. Genet.* **8,** 1169–1176.

Gutekunst, C. A., Li, S. H., Yi, H., Mulroy, J. S., Kuemmerle, S., Jones, R., Rye, D., Ferrante, R. J., Hersch, S. M., and Li, X. J. (1999). Nuclear and neuropil aggregates in Huntington's disease: Relationship to neuropathology. *J. Neurosci.* **19,** 2522–2534.

Holmberg, M., Duyckaerts, C., Durr, A., Cancel, G., Gourfinkel-An, I., Damier, P., Faucheux, B., Trottier, Y., Hirsch, E. C., Agid, Y., and Brice, A. (1998). Spinocerebellar ataxia type 7 (SCA7): A neurodegenerative disorder with neuronal intranuclear inclusions. *Hum. Mol. Genet.* **7,** 913–918.

Kay, B. K., Winter, J., and McCaffery, J. (ed.) (1996). "Phage Display of Peptides and Proteins: A Laboratory Manual." Academic Press, San Diego.

Menendez, A., and Scott, J. K. (2005). The nature of target-unrelated peptides recovered in the screening of phage-displayed random peptide libraries with antibodies. *Anal. Biochem.* **336,** 145–157.

Murthy, K. K., Ekiel, I., Shen, S. H., and Banville, D. (1999). Fusion proteins could generate false positives in peptide phage display. *Biotechniques* **26,** 142–149.

Nagai, Y., Tucker, T., Ren, H., Kenan, D. J., Henderson, B. S., Keene, J. D., Strittmatter, W. J., and Burke, J. R. (2000). Inhibition of polyglutamine protein aggregation and cell death by novel peptides identified by phage display screening. *J Biol. Chem.* **275,** 10437–10442.

Nagai, Y., Fujikake, N., Ohno, K., Higashiyama, H., Popiel, H. A., Rahadian, J., Yamaguchi, M., Strittmatter, W. J., Burke, J. R., and Toda, T. (2003). Prevention of polyglutamine oligomerization and neurodegeneration by the peptide inhibitor QBP1 in *Drosophila. Hum. Mol. Genet.* **12,** 1253–1259.

Onodera, O., Roses, A. D., Tsuji, S., Vance, J. M., Strittmatter, W. J., and Burke, J. R. (1996). Toxicity of expanded polyglutamine-domain proteins in *Escherichia coli. FEBS Lett.* **399,** 135–139.

Ren, H., Nagai, Y., Tucker, T., Strittmatter, W. J., and Burke, J. R. (2001). Amino acid sequence requirements of peptides that inhibit polyglutamine-protein aggregation and cell death. *Biochem. Biophys. Res. Commun.* **288,** 703–710.

Rickles, R. J., Botfield, M. C., Weng, Z., Taylor, J. A., Green, O. M., Brugge, J. S., and Zoller, M. J. (1994). Identification of Src, Fyn, Lyn, PI3K and Abl SH3 domain ligands using phage display libraries. *EMBO J.* **13,** 5598–5604.

Rubinsztein, D. C., Leggo, J., Coles, R., Almqvist, E., Biancalana, V., Cassiman, J. J., Chotai, K., Connarty, M., Crauford, D., Curtis, A., Curtis, D., Davidson, M. J., Differ, A. M., Dode, C., Dodge, A., Frontali, M., Ranen, N. G., Stine, O. C., Sherr, M., Abbott, M. H., Franz, M. L., Graham, C. A., Harper, P. S., Hedreen, J. C., and Hayden, M. R. (1996). Phenotypic characterization of individuals with 30-40 CAG repeats in the Huntington disease (HD) gene reveals HD cases with 36 repeats and apparently normal elderly individuals with 36-39 repeats. *Am. J. Hum. Genet.* **59,** 16–22.

Scholle, M. D., Collart, F. R., and Kay, B. K. (2004). *In vivo* biotinylated proteins as targets for phage-display selection experiments. *Protein Expr. Purif.* **37,** 243–252.

Smith, G. P. (1985). Filamentous fusion phage: Novel expression vectors that display cloned antigens on the virion surface. *Science* **228,** 1315–1317.

Stevanin, G., Trottier, Y., Cancel, G., Durr, A., David, G., Didierjean, O., Burk, K., Imbert, G., Saudou, F., Abada-Bendib, M., Gourfinkel-An, I., Benomar, A., Abbas, N., Klockgether, T., Grid, D., Agid, Y., Mandel, J. L., and Brice, A. (1996). Screening for proteins with polyglutamine expansions in autosomal dominant cerebellar ataxias. *Hum. Mol. Genet.* **5,** 1887–1892.

Trottier, Y., Lutz, Y., Stevanin, G., Imbert, G., Devys, D., Cancel, G., Saudou, F., Weber, C., and Tora, L. (1995). Polyglutamine expresion as a pathological epitope in Huntington's disease and four dominant cerebella atarias. *Nature* **378,** 403–406.

Walker, L. C., and LeVine, H. (2000). The cerebral proteopathies: Neurodegenerative disorders of protein conformation and assembly. *Mol. Neurobiol.* **21,** 83–95.

Further Reading

Gourfinkel-An, I., Cancel, G., Trottier, Y., Devys, D., Tora, L., Lutz, Y., Imbert, G., Saudou, F., Stevanin, G., Agid, Y., Brice, A., Mandel, J. L., and Hirsch, E. C. (1997). Differential distribution of the normal and mutated forms of huntingtin in the human brain. *Ann. Neurol.* **42,** 712–719.

[15]   Peptide-Based Inhibitors of Amyloid Assembly

By Kimberly L. Sciarretta, David J. Gordon, and
Stephen C. Meredith

Abstract

This review considers the design, synthesis, and mechanistic assessment of peptide-based fibrillogenesis inhibitors, mainly focusing on β-amyloid, but generalizable to other aggregating proteins and peptides. In spite of revision of the "amyloid hypothesis," the investigation and development of

METHODS IN ENZYMOLOGY, VOL. 413   0076-6879/06 $35.00
          DOI: 10.1016/S0076-6879(06)13015-3

fibrillogenesis inhibitors remain important scientific and therapeutic goals for at least three reasons. First, it is still premature to dismiss fibrils altogether as sources of cytotoxicity. Second, a "fibrillogenesis inhibitor" is typically identified experimentally as such, but these compounds may also bind to intermediates in the fibrillogenesis pathway and have hard-to-predict consequences, including improved clearance of more cytotoxic soluble oligomers. Third, inhibitors are valuable structural probes, as the entire field of enzymology attests. Screening procedures for selection of random inhibitory sequences are briefly considered, but the bulk of the review concentrates on rationally designed fibrillogenesis inhibitors. Among these are internal segments of fibril-forming peptides, amino acid substitutions and side chain modifications of fibrillogenic domains, insertion of prolines into or adjacent to fibrillogenic domains, modification of peptide termini, modification of peptide backbone atoms (including N-methylation), peptide cyclization, use of D-amino acids in fibrillogenic domains, and nonpeptidic β-sheet mimics. Finally, we consider methods of assaying fibrillogenesis inhibitors, including pitfalls in these assays. We consider binding of inhibitor peptides to their targets, but because this is a specific application of the more general and much larger problem of assessing protein–protein interactions, this topic is covered only briefly. Finally, we consider potential applications of inhibitor peptides to therapeutic strategies.

Rationale for Developing Peptide-Based Fibrillogenesis Inhibitors

β-sheet fibrils, or amyloids, are the hallmarks of many neurodegenerative diseases, including Alzheimer's, Huntington's, and Parkinson's diseases, as well as the systemic amyloidoses. Fibrillogenesis, as depicted schematically in Fig. 1, is complex and includes oligomers, both on- and off-pathway, for fibril formation and immediate precursors of fibrils, called protofibrils or protofilaments. Each of the steps shown schematically is assumed to be reversible to some degree, although in some steps, disaggregation is slow enough that the reaction can be considered irreversible.

Protein aggregates are firmly linked with disease, but the mechanisms by which these aggregates lead to cell death remain largely unknown. The study of inhibitors of fibril formation began with the naive assumption that fibrils were not only associated with disease but were etiologically implicated in disease. Whether or not this early assumption was justified, the inhibition of fibril formation remains a potential scientific and therapeutic goal for several reasons. First, despite much progress, the mechanisms by which protein aggregates lead to cell injury and death are poorly understood, and it remains imprudent to dismiss the fibrils as possible cytotoxins. Second, a "fibrillogenesis inhibitor" could act at multiple steps in the

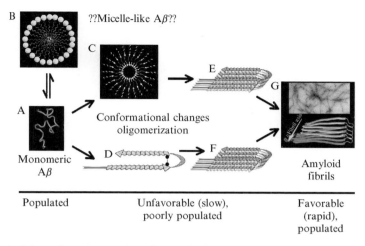

FIG. 1. Schematic representation of steps in fibrillogenesis. Some of the steps depict β-amyloid, but the figure is intended to be general. An unstructured, monomeric peptide (A) or a protein with an unstructured, flexible or highly dynamic loop (structure of β-amyloid from Massi *et al.*, 2001; also see Lee *et al.*, 1995; Zhang *et al.*, 2000) must undergo conformational changes (C to E and/or D to F) and self-associate into soluble oligomeric intermediates (C and F) in order to evolve into a fibril (G, structure from Petkova *et al.*, 2002). Self-association is often depicted as preceding conformational changes (C to E), but the order of these events is generally not known (for D to F, see Lazo *et al.*, 2005; Sciarretta *et al.*, 2005). Micelles (B) and other "off-pathway" species could also occur. In addition, the figure shows only two pathways toward fibril formation, but all steps could be multiple.

pathway and bind to multiple species, and this fact could have important but hard-to-predict consequences. Consider the *gedank* experiment depicted in Fig. 2. Assume, as many people now believe, that a soluble oligomer is more cytotoxic than fibrils. A worry about fibrillogenesis inhibitors is that they might drive fibril-forming peptides into the cytotoxic oligomer pool. This is not necessarily the case, however. What if there is more than one type of oligomer, as is widely believed? Consider the case in which one is cytotoxic (A) and one is relatively innocuous (B). Whether the fibrillogenesis inhibitor increases the concentration of the cytotoxic species will depend on many factors (e.g., relative degradation rates of A and B, whether the inhibitor binds to oligomers as well as fibrils). Suppose (Fig. 2) that an inhibitor favors formation of oligomer B, which is more readily degraded by the cell than oligomer A. In that case, both the concentration of the cytotoxin (A) and of the overall oligomer would decrease in the cell. This highly hypothetical example suggests that even given the minimalist pathway depicted in Fig. 1, the kinetics of protein aggregation and degradation are sufficiently complex in the cell that it is not obvious, *a priori*,

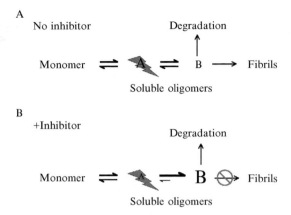

FIG. 2. A *gedank* experiment shows a possible effect of a fibrillogenesis inhibitor. A concern about fibrillogenesis inhibitors is that by preventing fibril formation, they could increase the cellular concentration of more toxic soluble oligomeric intermediates, but this is not necessarily the case, as depicted in the figure. Consider a fibrillogenesis inhibitor for a peptide that forms two oligomeric intermediates in a slow equilibrium with each other, a cytotoxic intermediate "A" and a less or nontoxic intermediate "B." Whether a fibrillogenesis inhibitor increases the concentration of the cytotoxic species will depend on many factors (e.g., the relative degradation rates of A and B by the cell, whether the inhibitor binds to oligomers made of the same protein that comprises the fibrils, whether the inhibitor stabilizes A or B). In the example depicted, the fibrillogenesis inhibitor binds to both fibrils and the less toxic soluble oligomers (B). If the cell can degrade B, the concentration of both A and B would decrease in the cell. Thus, a fibrillogenesis inhibitor might increase or decrease the concentration of a cytotoxin in a cell, depending on the details of the particular system.

which site is optimal for controlling cellular levels of a cytotoxic species, even if one knew with certainty what that cytotoxic species is.

Third, and perhaps most importantly, inhibitors can lead to invaluable structural and mechanistic insights, as the entire field of enzymology attests. An inhibitor of fibrillogenesis is likely to bind to and stabilize one or more of the intermediates in the pathway that would be otherwise difficult or impossible to capture for structural studies. In light of the well-known difficulty in crystallizing amyloids and the low water solubility of many of their precursors, inhibitors may be important aids in future structural investigations of intermediates.

Advances in our understanding of fibril sequence and structure have allowed the rational design of peptide-based inhibitors of fibrillogenesis. Peptide-based inhibitors include peptides themselves, peptidic polymers with modification of the main chain backbones (e.g., N-methyl groups, esters, reduced peptide bonds), modified peptides with adducts to either side chains and termini, and peptidomimetics (i.e., non-peptide-based compounds resembling β-strand structures). Peptide-based inhibitors complement small molecule fibrillogenesis inhibitors (see Chapter by Kocisko and Caughey in this volume), and each approach to inhibiting aggregation and fibril formation has its advantages and disadvantages. Small molecule inhibitors often can be readily modified for optimal pharmacokinetics and for chemical stability. The discovery of such compounds, however, is often fortuitous rather than rational, and in many cases, the mechanism of action is obscure and the likely sites and modes of action are multiple. Peptide-based inhibitors of fibrillogenesis also have their pros and cons. Although methods of peptide synthesis are well developed, some modified peptides are difficult to synthesize, especially on a large scale. Many peptides are subject to proteolysis and have unfavorable pharmacokinetics for use of these compounds therapeutically. The chief advantages of peptides, however, are the possibility of rational design, their specificity, the wealth of highly developed methods for analyzing their modes of action, and the possibility of building upon structural data from solid-state nuclear magnetic resonance (NMR) (see Chapter 6 by Tycko in this volume) and other techniques to target different aspects of the protein aggregation problem specifically. A recent model of the β-amyloid fibril, based on solid-state NMR and other structural data, holds that a single molecular unit of the fibril consists of a double-layered β-sheet structure with a hydrophobic core and one exterior face that is considerably more hydrophobic than the other exterior face (Petkova *et al.*, 2002). Although the details of this model remain to be refined, it illustrates that one could design fibrillogenesis inhibitors that target different "faces" of the β-sheet (e.g., the internal hydrophobic pocket, lateral aggregation site formed by the more hydrophobic exterior face of the single molecular layer).

Approaches to the Design of Peptide-Based Inhibitors

Random Selection of Sequences

Hughes *et al.* (1996) used a yeast two-hybrid system to demonstrate interaction of β-amyloid with itself, but such a system could easily be adapted to a wider screen for interacting partners, some of which could prove to be fibrillogenesis inhibitors. A striking example of this approach is that of Burke and co-workers (Nagai *et al.*, 2000; also see Chapter 14 by Kenan and colleagues in this volume), who screened a combinatorial peptide library expressed on M13 phage pIII to identify six tryptophan-rich peptides that bind to and interfere with pathologically expanded polyglutamine peptide aggregation and toxicity. The original QBP1 peptide, SNWKWWPGIFD, could be shortened to WKWWPGIF without loss of activity; the tryptophan repeat appeared to be necessary for this activity. In more recent studies, they also showed that a tandem repeat of this protein suppresses polyglutamine-induced neurodegeneration in the compound eye of a *Drosophila* model of polyglutamine expansion disease and rescues premature death of flies expressing an expanded polyglutamine protein in the nervous system. More recently, Zhou *et al.* (2004) performed a screen of 40 peptides from random sources and observed that a tetrapeptide, WMDF (Trp-Met-Asp-Phe) derived from cholecystokinin but unrelated to the serpins effectively blocked polymerization of the serpins antitrypsin and antithrombin. The initial approach in this article was a more or less random selection of peptides that bound to an aggregation site.

Random searches could, in theory, be supplemented by elaborate, large-scale screening approaches, such as phage display. The requirements for such an approach include (1) a rapid, sensitive, and simple screening test for blockage of polymerization; (2) the ability to generate a large number of peptides for screening; and (3) the ability to analyze the binding of inhibitor to fibril-forming peptide structurally to guide inhibitor optimization. Although some polymerizing proteins/peptides (e.g., serpins) are amenable to crystallographic studies, most amyloidogenic peptides (e.g., β-amyloid) are not thus far. In addition, any attempt to generate fibrillogenesis inhibitors through large-scale screening of random sequences will encounter many of the same problems as large-scale screening in other arenas. Nevertheless, such approaches could have promise in this field as they have in others as diverse as the generation of peptide ligands for antibodies and the development of angiogenesis inhibitors, oncogene inhibitors, spermicides, and vaccines.

Rational Design of Fibrillogenesis Inhibitors

Most fibrillogenesis inhibitors described to date were designed rationally and have been based on presumed molecular recognition elements, generally a presumed site of aggregation. In many cases, the "aggregation site" is not known with certainty. Indeed, the concept of a unique aggregation site in a small peptide like β-amyloid may be useful on a practical level, but it inevitably simplifies the complex interactions that lead to fibril formation. Often, identification of an aggregation site is based on mutational/deletional data that lead to little more than an educated guess as to the residues that should be targeted by fibrillogenesis inhibitors.

Because peptide-based fibrillogenesis is a process of self-association, most fibrillogenesis inhibitors are somewhat homologous to the presumed site at which self-association occurs; that is, they are self-similar in some way. An exception to this statement is the antibodies used in recent vaccinations against Alzheimer's β-amyloid (Morgan et al., 2000; Schenk et al., 1999). Whether the binding is complementary or self-similar,[1] a tacit or explicit assumption in designing fibrillogenesis inhibitors is that inhibitors bind to the fibril-forming peptide. This assumption, however, is sometimes extrapolated to the assumption that the best binder is also the best inhibitor, and this extrapolation is not valid. Good binding, as defined by ensemble measurements of affinity, is not sufficient to make a good fibrillogenesis inhibitor. Indeed, one should not even assume that there is a rough correlation between affinity of binding and efficacy as a fibrillogenesis inhibitor. A good fibrillogenesis inhibitor should bind well to some form of the fibril-forming protein or peptide but not necessarily to the fibril itself and not necessarily to the predominant (by mass) soluble form of the peptide. In addition, the efficacy of a fibrillogenesis inhibitor depends on properties other than, and not necessarily related to, its affinity for its target peptide. For example, an effective fibrillogenesis inhibitor should also be water soluble, resistant to proteases, and not form fibrils itself.

Thus, high affinity of the fibrillogenesis inhibitor for its target is not sufficient to make a good inhibitor, but one might also ask whether high affinity of the fibrillogenesis is necessary or even desirable. Consider the extreme case of an inhibitor that binds irreversibly to its target. Such an inhibitor would "cap" the growing fibrils and lead to the production of a greater number of shorter or smaller fibrils, or protofibrils, and prevent dissolution of possible cytotoxic species. Thus, an "ideal" fibrillogenesis inhibitor would have some affinity for the aggregated protein, but binding would be reversible and allow fibril dissolution. What is the ideal affinity? The answer to this question is not known. Thus, even leaving aside, for the moment, the important biological question of which is more toxic, fibrils or their precursors, the "right" affinity of inhibitor for target remains an open question.

Internal Segments of Fibril-Forming Peptides as Fibrillogenesis Inhibitors

Some of the first fibrillogenesis inhibitor peptides targeting β-amyloid aggregation were based on amyloidogenic sequences within the peptide, amyloid β $(A\beta)_{16-20}$, KLVFF (Tjernberg et al., 1996, 1997; Hetenyi et al., 2002). Although these peptides were effective at preventing the formation of

---

[1] The mechanism by which anti-$A\beta$ vaccines might reduce cerebral $A\beta$ load is beyond the scope of this chapter (for a review, see Bennett and Holtzmann, 2005; Gelinas et al., 2004). In either case, the present point is that the antibody must bind to its antigen, whether monomeric, oligomeric, or fibrillar $A\beta$, and that this binding is heterologous. On the other hand, aggregation of immunoglobulin light chains can be inhibited in vitro by a synthetic peptide identical to the presumed light chain binding site for the chaperone, BiP (Davis et al., 2000), where interaction between peptide and light chain is presumably homologous rather than heterologous.

thioflavin T-positive fibrils of $A\beta_{1-40}$, they themselves self-associated into fibrils, albeit thioflavin T-negative ones. A similar observation was made for peptides derived from a highly hydrophobic sequence from the prion protein (residues 113–120) (Chabry et al., 1998). These peptides could inhibit the conversion of "PrPsen" to "PrPres" (i.e., the protease sensitive to the protease resistant form, which is believed to model the conversion of the cellular form of PrP to its disease-associate or "Scrapie" form) but also formed β-sheet fibril aggregates themselves as shown by Fourier transform infrared (FTIR).

Amino Acid Substitutions and Side Chain Modifications of Fibrillogenic Domains. Although unmodified internal segments of fibrillogenic peptides can sometimes serve as fibrillogenesis inhibitors, there are advantages to modifying peptide backbones (discussed in another section below) or side chains. Modifications of side chains within the putative aggregation site could, in principle, increase the affinity of the inhibitor for the target. Because many aggregation sites consist of stretches of hydrophobic amino acids, increasing the hydrophobicity of an inhibitor could increase its affinity for its target—at the expense of water solubility, however. (A compromise strategy, discussed below, is the use of block co-polymers, in which one of the blocks is hydrophobic and mediates interaction with the target, whereas the other is polar and enhances water solubility of the inhibitor.) In addition, modification of side chains can decrease susceptibility of inhibitor peptides to proteases.

Many of the side chain substitutions as well as backbone modifications of β-amyloid have focused on the central hydrophobic core of this peptide, $-^{17}LVFF^{21}A-$ (Hilbich et al., 1992). Modification of the central Phe residues in $A\beta_{1-40}$ (e.g., replacing them with Thr) abolished fibril formation. Thus, Hughes et al. (1996) demonstrated that an octapeptide derived from this region ($A\beta_{16-23}$) binds to $A\beta_{1-40}$ but also showed by electron microscopy (EM) that the peptide containing the substitution of TT for FF (i.e., QKLVTTAE) was a fibrillogenesis inhibitor. Phenylalanines appear to be especially important in self-aggregation in other systems. Porat et al. (2004) showed that whereas the NFGAIL was a minimal fibril-forming fragment from human islet amyloid polypeptide (hIAPP), a core peptide from hIAPP with Tyr substituted for Phe (i.e., $^{22}NFGAILSS^{29}$ to $^{22}NYGAILSS^{29}$) did not form fibrils by itself and inhibited fibril formation by hIAPP. hIAPP undergoes a transition, observable by circular dichroism (CD), to a soluble β-sheet form that precedes fibril formation; $^{22}NYGAILSS^{29}$ also inhibited this conformational transition of hIAPP. The authors argue, furthermore, that inhibition stems from the geometrical constraints of the heteroaromatic benzene-phenol interaction, a notion supported by further observations that a small polyphenol, phenol red, also inhibits hIAPP fibril formation and decreases cytotoxicity of hIAPP toward βTC-tet rodent β cells.

Most of the fibril-forming peptides described in this article thus far contain large hydrophobic domains. Some amyloids, however, are formed from highly polar amino acids (e.g., huntingtin with polyGln expansions and short polyGln peptides) (Perutz *et al.*, 2002). Indeed, the hallmark study of homopolymers of amino acids by Greenfield and Fasman (1969) showed that poly-L-Lys could be heated and made to undergo transition from α-helix to β-sheet and that shortly after becoming β-sheet, the peptide precipitated from solution. Although the authors did not use the term *amyloid* at that time, it is likely, in view of more recent studies (e.g., Fändrich and Dobson, 2002), that these peptides were forming fibrils. Along similar lines, in a host-guest system, one peptide, Ad-EKEK-2α (Fig. 3) readily formed fibrils (Yamashita *et al.*, 2003). The polar residues of the peptide were then varied systematically. In general, the zwitterionic peptides readily formed fibrils, whereas fibril formation was inhibited in several homologous cationic peptides with Lys replacing Glu (Yamashita *et al.*, 2003).

Proline Substitutions and Additions. Proline residues are found uncommonly in β-sheets for at least two major reasons: their rigid structure restrains their φ and ψ angles outside of the β-sheet range, and they also lack the amide proton needed for the hydrogen bonding found in β-sheets. Because of the very low propensity of Pro residues to form or even to reside in β-sheets, early studies using "proline scanning" of $A\beta_{15-23}$ Q15K (Wood *et al.*, 1995) and hIAPP$_{20-29}$ (Moriarty and Raleigh, 1999) showed that substitutions of Pro at various positions reduced fibril-forming capacity to various extents; Pro residues appear to act by increasing the critical concentration (C_r ; see below for more discussion of C_r). In the latter study on hIAPP, FTIR and EM of hIAPP$_{20-29}$ peptides with proline substitutions confirmed the lack of β-sheets and fibrils, respectively. Indeed, although human IAPP forms fibrils, rodent islet amyloid polypeptide (IAPP) does not, probably because rodent IAPP contains Pro residues at key positions, especially Ser/Pro28 (Westermark *et al.*, 1990).

"β-sheet breaker" peptides utilize either single Pro residues or strings of Pro residues flanking amyloidogenic sequences in addition to Pro substitutions within an amyloidogenic sequence, such as the hydrophobic

Ad–βAla-ALEQKLAALEQKLA-βAla-C-NH$_2$

Ad–βAla-ALEQKLAALEQKLA-βAla-C-NH$_2$

Ad:

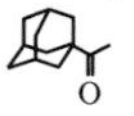

FIG. 3. Structure of the amyloid peptide Ad-EKEK-2α (Yamashita *et al.*, 2003) from which 15 analogous peptides, substituted at Glu (E) and Lys (K) residues, were synthesized. Inhibitors of amyloid formation of Ad-EKEK-2α were derived from Ad-KKKK-2α.

core domain of $A\beta_{17-20}$. One inhibitor, called $iA\beta5p$ (Ac-LPFFD-NH$_2$), was found to inhibit amyloid formation of $A\beta_{1-40}$ and $A\beta_{1-42}$ as quantified by ThT, sedimentation assays, Congo red binding, and EM (Soto *et al.*, 1998). These same investigators also added charged residues to increase the solubility of the inhibitor. Although these peptides do appear to inhibit fibril formation, the mechanism by which they bind to fibril growth sites is not well understood. One would expect, *a priori*, that a fibrillogenesis inhibitor would need to bind to a fibril growth site and, at the same time, block the binding of the additional amyloidogenic peptide molecules. Because Pro residues distort the β-strand geometry, it is not apparent how a short peptide containing one or more prolines could be structurally homologous to the fibril itself or to the incoming peptide/protein that binds to a fibril growth site. One inference is that the binding of "β-blocker" peptides to their targets is not actually homologous (i.e., it is heterologous), which leaves the mechanism unexplained.

Modification of the Peptide Termini. Another group of fibrillogenesis inhibitors consists of groups (peptide and nonpeptide) added to the termini of a short sequence consisting of the putative aggregation site. In general, the aggregation site segment is intended to mediate binding of the inhibitor to the growing fibril or its precursor, whereas the group added to the terminus or termini is intended to disrupt propagation of the fibril. Such compounds are really block co-polymers, and in the nomenclature used by Ghanta *et al.* (1996) to describe their inhibitors of β-amyloid toxicity, the first site is called a recognition site, whereas the second block is called the disrupting element. The disrupting element added to the termini can disrupt aggregation in several ways.[2] First, it can impede fibril formation or propagation by steric effects (e.g., a bulky group that prevents efficient packing of molecules at the growing ends of the fibril). Second, highly polar or charged groups added to the termini can increase the solubility of prefibrillar aggregates. Finally, adducts at the termini can protect inhibitors from exoproteases. In some cases, the mechanisms by which adducts inhibit fibril formation are not obvious, as discussed below.

C-Terminal Modifications

CHARGED BLOCKS ADDED TO RECOGNITION DOMAINS. Ghanta *et al.* (1996) devised the block co-polymer strategy for modifying β-amyloid aggregation, as described above. The first such inhibitor linked residues

---

[2] As discussed below, not all these groups disrupt fibril formation; indeed, some of them, including the ones described by Ghanta *et al.* (1996), actually accelerate fibril formation.

15–25 of β-amyloid, designed to function as the recognition element, to an oligolysine disrupting element. Comparing three peptides (R1 = VFFAEDVG, H1 = *KKKKK*GGQKLVFFAEDVG, and H2 = GQKLV FFAEDVGGa*KKKKK*), the authors found that all peptides increased hydrodynamic radius (dynamic light scattering) and had CD spectra similar to that of Aβ(1–39). However, H2 decreased thioflavin T fluorescence and reversed toxicity (MTT assays) of Aβ(1–39) to PC-12 cells, whereas R1 did not. In addition, the placement of the disruption domain was important; whereas H2 inhibited cytotoxicity of Aβ(1–39), H1 did not, and rather than inhibiting fibrillogenesis, it formed fibrils itself. In a later article (Pallitto *et al.*, 1999), the same group refined the recognition domain and found peptides containing the recognition domain KLVFF, (residues 16–20 of Aβ) to yield their effective inhibitor of Aβ cytotoxicity.

The effects on cytotoxicity observed by Ghanta *et al.* (1996) were disputed recently by Moss *et al.* (2003). The latter authors have attempted to distinguish between two modes of protofibril growth—elongation by monomer deposition and direct protofibril-protofibril association—through differential dependencies of these two processes on Aβ monomer and NaCl concentrations (see Nichols *et al.*, 2003). Although these authors found that larger aggregates formed in the presence of KLVFF-K6 from enhanced protofibril association, they did not find that promotion of Aβ protofibril association by KLVFF-K6 affected Aβ-induced cytotoxicity, as measured by decreases in cellular MTT reduction. In other words, their data did not support the proposal that insoluble fibrils formed with KLVFF-K6 are less toxic than soluble protofibrils. Resolution of this conflict must await additional data.

POLYHYDROXY AND OTHER UNCHARGED POLAR BLOCKS. Burkoth *et al.* (1998) described inhibition of fibrillogenesis of Aβ(10–35) by the addition of a C-terminal polyethyleneglycol (PEG) block (Fig. 4).

This peptide, although truncated at both ends, maintains the four-domain structure of full-length Aβ (i.e., from N- to C-terminus), a hydrophilic domain, followed by a hydrophobic "core domain," followed by a bend structure, and terminating in a hydrophobic C-terminal domain. The addition of a PEG-3000 block to the C-terminus resulted in concentration-dependent oligomerization to the β-sheet form, consistent with monomer to hexamer or heptamer. The Aβ-PEG3000 co-polymer formed fibrils but with much delayed kinetics. Furthermore, the fibrils were amyloid by

FIG. 4. Sequence of C-PEG$_{3000}$Aβ(10-35) (Burkoth *et al.*, 1998, 1999) synthesized by the addition of amyloid β(10–35) to a C-terminal polyethyleneglycol (PEG) block bound to a resin.

criteria like Congo red binding, but in contrast to unmodified $A\beta(10–35)$, they were strictly linear rather than twisted, and fibril formation could be reversed (e.g., by adjusting pH to ≈2.5). From neutron scattering experiments using contrast matching (through adjustment of the solvent deuteration level), the PEG block was proposed to be localized at the periphery of the fibrils. A similar design was used by Wantanabe et al. (2002), who incorporated a disrupting element to $A\beta(16–20)$, KLVFF, but, instead, utilized the hydrophilic, aminoethoxy ethoxy acetic acid (AEEA) with an ethylene glycol skeleton, varying the number of units added. In addition, two aspartic acids were added to increase hydrophilicity (Fig. 5).

They observed that these compounds, especially DDX3, were effective inhibitors of $A\beta$ fibrillogenesis by thioflavin T and Congo red binding assays, in contrast to KLVFF-K6 (Lowe et al., 2001). In addition, DDX3 and related compounds suppressed the cytotoxicity of $A\beta$ to IMR-32 neuroblastoma cells in vitro, as measured by an MTT assay.

N-Terminal Modifications. Findeis et al. (1999) examined numerous internal segments of $A\beta$ and performed a scan of 15-residue sections. This showed that the free amine acid of $A\beta(16–30)$ prevented $A\beta$ polymerization. A wide variety of N-terminal modifying reagents were used to explore the effects of changing polarity, charge, hydrophobicity, and size of the adducts on the inhibition of $A\beta(1–40)$ aggregation, which was measured by turbidity and thioflavin T fluorescence. The authors were also able to delimit further the recognition domain, and of the compounds tested,

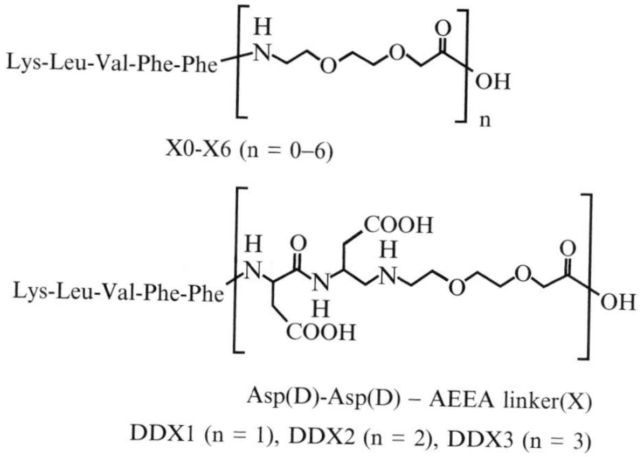

X0-X6 (n = 0–6)

Asp(D)-Asp(D) – AEEA linker(X)

DDX1 (n = 1), DDX2 (n = 2), DDX3 (n = 3)

FIG. 5. An amyloid β ($A\beta$) fibrillogenesis inhibitor, DDX3 (Wantanabe et al., 2002) formed by incorporating a hydrophilic, aminoethoxy ethoxy acetic acid (AEEA) into an ethylene glycol skeleton, and appended to $A\beta(16–20)$, KLVFF.

the most potent was cholyl-LVFFA-OH. This inhibitor, however, also had limited stability, presumably because of proteolytic degradation. The all-D-amino acyl analogue of this peptide retained inhibitory activity but had enhanced stability in monkey cerebrospinal fluid.

Adessi *et al.* (2003) also altered the above-mentioned β-sheet–breaker peptide derived from β-amyloid, LPFFD, with various N-terminal protections and found that acetylation enhanced stability of the peptide in human plasma. Gordon *et al.* (2002) added an anthranilic acid as a fluorescent probe for an Aβ(16–20) N-methyl peptide (see below) and found that this group also enhanced efficacy of the inhibitor.

N-terminal Arg residues have been used to solubilize fibrillogenesis inhibitors. In one instance, an inhibitor was designed to target the C-terminal hydrophobic region of Aβ, which has also been shown to play a role in aggregation (Fülöp *et al.*, 2004). Although the peptide NH_2-RIIGL-$CONH_2$ (IIGL = amino acids 31-34 of Aβ) did not form fibrils and inhibited fibril formation and cytotoxicity to differentiated SHSY-5Y neuroblastoma cells of Aβ(1–42), the same peptide with an N-terminal propionyl group formed fibrils and did not inhibit either fibril formation or cytotoxicity. Similarly, Arg was also used to solubilize an inhibitor targeting the central hydrophobic region of α-synuclein, residues 68–75 (El-Agnaf *et al.*, 2004); the resulting peptide inhibited amyloid formation as shown by thioflavin T fluorescence and EM.

Modification of Peptide Backbone Atoms. In an idealized β-sheet fibril with a "cross-β" architecture (Fig. 6), the longitudinal (meridional) axis of each β-sheet of the fibril is formed by stacks of peptide chains interacting with each other through hydrogen bonds. The β-sheets can also laminate through interactions between side chains, in an axis perpendicular (equatorial) to the long axis of the fibril. The inhibitors described in the previous sections are directed mainly toward side chain interactions. More recently, peptide inhibitors have been designed to interfere with β-sheet formation by preventing propagation of hydrogen bonds along the long axes of the β-sheets. These inhibitors include peptides with modification of or substitution for the amide nitrogen, or α-carbon atom. In most cases, modification to the amide bonds has been by alkylating the amide nitrogen (e.g., N-methylation) or substitution of the amide nitrogen (e.g., by oxygen in an ester bond).

N-Methylation. N-methyl amino acids can prevent protein aggregation in at least three ways: (1) the methyl groups in N-methyl amino acids replace the amide proton that otherwise could stabilize β-sheets through hydrogen-bonding interactions between individual β-strands; (2) the

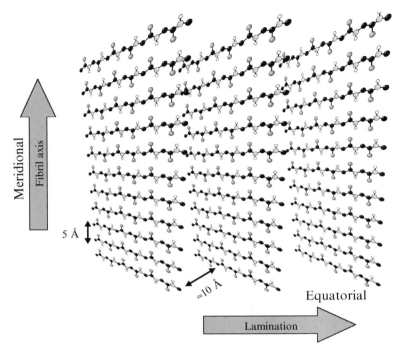

FIG. 6. A minimalist model of the amyloid fibril, in this case, depicted for an in-register, parallel β-sheet, as was developed for amyloid β (Aβ)(10–35). Along the longitudinal axis (meridional reflections at 4.7–5.0 Å in fiber X-ray diffraction patterns), β-strands interact through backbone hydrogen bonds to form a single β-sheet. The minor axes (width) of the fibril give rise to meridional reflections (≈10 Å) in fiber X-ray diffraction patterns. In addition, two or more β-sheets can laminate (three β-sheets are shown in the figure) through side chain interacts that can be predominantly hydrophobic (e.g., β-amyloid) or polar (e.g., polyglutamines). For β-amyloid, this picture is a simplification; in a recent structural model of Aβ(1–40), the molecules fold over through a bend structure, so that a single molecular layer of the fibril is composed of two β-sheet segments of the peptide (Petkova et al., 2002). In addition, fibrils of a given peptide can be polymorphic (Petkova et al., 2005).

methyl group is larger than the amide proton and could thus prevent the close approach of the peptide chains by steric hindrance; and (3) N-methyl amino acids form tertiary amides, and these strongly favor the trans conformation.[3] Thus, peptides incorporating N-methyl amino acids are

---

[3] Given the resemblance of peptide bonds of the form L-AA-NMeAA (i.e., an L-amino acid in amide linkage to an N-methyl amino acid) to X-Pro bonds, N-methyl amino acids might be expected to favor cis-peptide bond formation. Indeed, a number of examples of cis-

torsionally constrained to adopt the β-strand conformation, allowing inter-action with β-sheets in fibrils (Patel and Tonelli, 1976; Tonelli, 1971; Vitoux *et al.*, 1986).

N-methyl amino acids were first used to control aggregation of a non-fibrillar protein, interleukin-8, which was dimeric in both NMR and X-ray crystallographic studies. Replacement of Leu[25] by N-methyl-Leu resulted in disruption of the dimer interface at two sites, because the β-sheet forming that part of the dimer interface was antiparallel, but the structure and bioactivity of the peptide were otherwise unaltered (Rajarathnam *et al.*, 1994). N-methyl amino acids also were used to control the aggrega-tion of peptide nanotubes made of even numbers of alternating D- and L-amino acids. Without the incorporation of N-methyl amino acids, these disk-shaped molecules formed insoluble, infinitely long tubular aggregates, as shown in Fig. 7. However, when N-methyl amino acids were incorpor-ated into only one face of the peptide ring, the ring could only dimerize, giving rise to dimeric peptide nanocylinders (Clark *et al.*, 1998).

Doig (1977) designed a three-stranded β-sheet peptide called the β-meander, in which aggregation was prevented by using both N-methyl amino acids and Asp side chains on both faces of the sheet (Fig. 8).

Along similar lines, N-methyl amino acids were incorporated into pep-tides assembled on 2,3′-substituted, biphenyl-based, β-hairpin-like templates in order to prevent the aggregation of parallel β-sheet dimers extending from the biphenyl template (Chitnumsub *et al.*, 1997).

Doig and co-workers synthesized Aβ(25–35) congeners containing single N-methyl amino acids (Hughes *et al.*, 2000). In some cases, these peptides were found either to alter the morphology or to prevent aggrega-tion and neurotoxicity (as measured by MTT assays on PC12 cells) of a truncated form of Aβ, Aβ(25–35). In particular, NMe-Gly25-Aβ(25–35) had properties similar to the nonmethylated peptide,[4] but NMe-Gly33-Aβ (25–35) prevented fibril assembly and reduced the toxicity of preformed amyloid, whereas NMe-Leu34-Aβ(25–35) altered fibril morphology and

---

peptide bonds involving N-methyl amino acid-containing peptides have been found. In all cases, however, these occur in the context of highly unusual peptides (e.g., short cyclic depsipeptides (Elseviers *et al.*, 1988), linear depsipeptides (Bersch *et al.*, 1993), pseudopep-tides (Howell *et al.*, 1995), short highly constrained cyclic peptides (Higuchi *et al.*, 1983), and peptides containing mixtures of L- and D-amino acids or other amino acids not typically found in proteins (Penkler *et al.*, 1993; Vitoux *et al.*, 1981). In other words, there are no published examples of cis-peptide bonds occurring in linear peptides containing only L-amino acids and a small number (n = 3) of N-methyl amino acids. Nevertheless, the possibility that these amino acids could form cis-peptide bonds should be borne in mind.

[4] This peptide, however, contains N-methyl-Gly as an N-terminal amine rather than as part of a peptide bond.

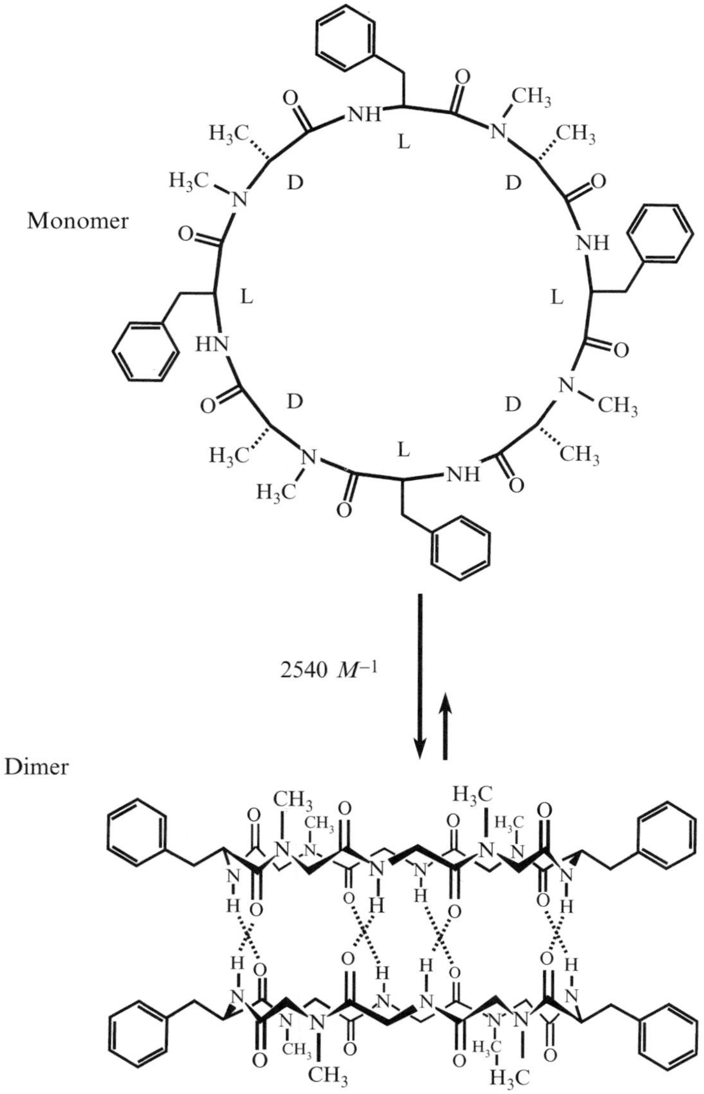

FIG. 7. Self-assembling nanotubes were constructed from a planar, cyclic D,L-peptide stacked as antiparallel β-sheets. If one face of the peptide ring is N-methylated, the construct gives rise to dimeric peptide nanocylinders. In the example illustrated here, *cyclo*[(-L-Phe-D-MeN-Ala)4-] self-assembles in anhydrous deuteriochloroform with an association constant of 2540 M^{-1}, giving rise to a single dimeric species detectable by ^{1}H nuclear magnetic resonance. For clarity, most side chains have been omitted. Reproduced with permission from Clark *et al.* (1998).

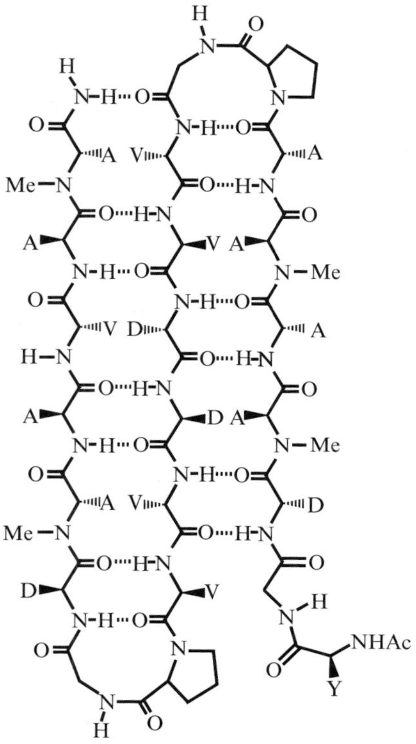

FIG. 8. A three-stranded β-sheet peptide, the β-meander, in which aggregation was prevented by using both N-methyl amino acids and Asp side chains on both faces of the sheet (Doig, 1997).

reduced toxicity. Thus, the location of the N-methyl groups was critical in determining the efficacy of inhibition of fibril formation and toxicity. In view of later structural models of Aβ, their two most effective inhibitor peptides had N-methyl amino acids in the C-terminal β-sheet domain, whereas the peptide that was not effective as an inhibitor, NMe-Gly25-Aβ(25–35), had the N-methyl amino acid in a non-β-sheet region of the Aβ peptide. Similar approaches have been used in the design of inhibitors of α-synuclein using the amyloidogenic region 68–78 (N-methyl Gly at position 73; Bodles *et al.*, 2004).

Further development of the N-methylation strategy was to incorporate N-methyl amino acids into alternate positions in the putative fibrillogenic region of the peptide, as shown in Fig. 9. Because the β-sheet is a "two-faced" or two-sided structure (with respect to both hydrogen bonds and side chains), placement of N-methyl groups at alternate positions likewise

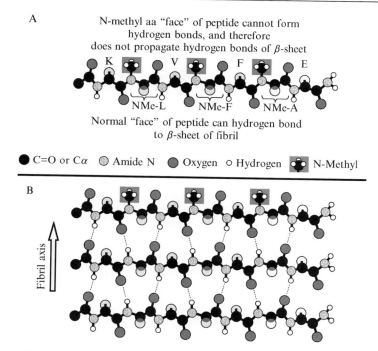

Fig. 9. Fibrillogenesis inhibitor with N-methyl amino acids at alternate residues. (A) Design of an inhibitor based on residues 16–22 of amyloid β. Placing the N-methyl amino acid at alternate positions forces the peptide chain into an extended conformation and creates a peptide with one normal hydrogen-bonding face and one face that does not form hydrogen bonds with peptides because of the lack of hydrogen bond donor amide hydrogens. (B) Possible capping of the growing fibril or oligomer edge by the hydrogen-bonding face of the inhibitor. Although not depicted in the figure, side chain interactions are clearly important for association of inhibitor to the fibril-forming peptide, because inhibition is highly sequence specific.

makes a two-faced inhibitor: one with normal hydrogen-bonding properties and one that cannot form hydrogen bonds, and thus could block the propagation of the β-sheet. Using this motif, Gordon *et al.* (2001, 2002) synthesized inhibitors targeted against Aβ as well as a peptide from the human prion protein (residues 106–126). A similar strategy was employed for IAPP, using the peptide SNNF-NMeG-A-NMeI-LSS and, more recently, [(N-Me)G24, (N-Me)I26-IAPP] (Yan *et al.*, 2006).

Two additional points emerged from these studies (Gordon *et al.*, 2002). First, the inhibitors are sequence specific: the inhibitor directed against Aβ did not cross-inhibit fibril formation by the prion peptide, and *vice versa*, clearly indicating that these peptides bind to their targets through side chain as well as hydrogen-bonding interactions. Second, in a comparison of a peptide with N-methyl amino acids in alternate positions (Ac-K-NMeL-V-NMeF-F-NMeA-E-CONH$_2$) and one with a greater number of N-methyl amino acids but in consecutive order (Ac-K-L-NMeV-NMeF-NMeF-NMeA-E-CONH$_2$), the peptide with N-methyl groups at alternate positions was the better inhibitor.

N-methyl amino acids have unusual physical properties that make them attractive as fibrillogenesis inhibitors. N-methylated peptides are highly water soluble. For example, although Ac-KLVFF-CONH$_2$ (residues 16–20 of Aβ) forms fibrils at concentrations above its critical concentration,[5] Ac-K-NMeL-V-NMeF-F-CONH$_2$ is water soluble at concentrations >30 mM. Even at very high concentrations, the peptide does not form fibrils; in fact, it remains monomeric (shown by CD, NMR, size exclusion chromatography, and analytical ultracentrifugation). Surprisingly, these peptides are also highly soluble in organic solvents, including alcohols, dimethyl sulfoxide (DMSO), diethyl ether, toluene, DMF, dichloromethane, and chloroform.[6] Thus, although the peptide is hydrophobic and can be solvated by organic solvents, it also retains sufficient polar groups to be solvated by water. CD spectra of these peptides are similar to a β-strand conformation but with an unusual, red-shifted, single minimum at 225–226 nm. Furthermore, the CD spectrum does not change with temperature from 0°–90°, with addition of urea up to 8 M, or with pH from 2–11. These findings indicate a high degree of structural constraint, which does not permit the molecules to self-associate. NMR studies also were consistent with a β-strand conformation. The unusual solubility and structural properties of these peptides suggest that they might be able to diffuse through phospholipid bilayers, as was shown to be the case (Fig. 10A, B). Ac-K-NMeL-V-NMeF-F-CONH$_2$ diffuses rapidly through single bilayer POPC vesicles (Gordon et al., 2002). In addition, a derivative of this peptide labeled with N-methyl-anthranilic acid, a fluor put in place of the acetyl group, was able to pass vectorially into COS cells (Fig. 10C); the directionality of transport into the cell probably results from the fact that this peptide has a single, positive charge. Finally, peptide bonds with N-methyl amino acids are highly resistant to proteolytic enzymes (e.g., chymotrypsin), an additional factor that could be of importance in the design of therapeutic agents based on this motif. Similarly, the

---

[5] The critical concentration has been studied rigorously by O'Nuallain et al. (2005) and is defined as the final concentration of unpolymerized monomeric peptide obtained after seeded fibril elongation. These authors obtained a value of 0.7–1.0 (μM for Aβ(1–40). The C$_r$ is sensitive to conditions used for fibrillogenesis; however, under a given set of conditions, it is a robust number.

[6] The state of aggregation of these peptides in organic solvents is not known. They might form dimers through hydrogen bonds between the nonmethylated faces of the peptide, leaving their methylated faces to be solvated by organic solvents.

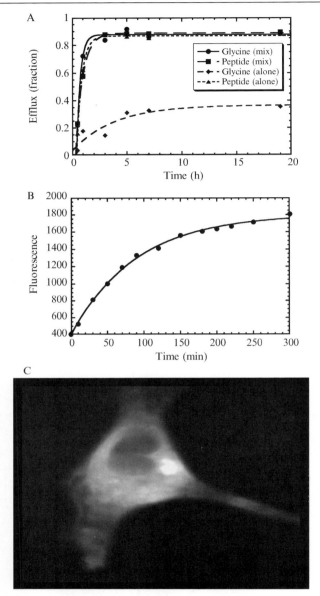

Fig. 10. Solubility and bilayer permeation properties of N-methylated peptides. Amyloid β (Aβ) 16–20 m (Ac-K-NMeL-V-NMeF-F-CONH2) passes through phospholipid bilayers. (A) Efflux of ^{14}C-Aβ16–20 m alone (▲),[3]H-glycine alone (H◆), and a mixture of ^{14}C-Aβ16–20 m (■) and ^{3}H-glyine (●) from phosphatidylcholine single bilayer vesicles prepared in the presence of these various compounds. (B) Efflux of calcein (a fluor that self-quenches at high concentrations in the vesicle) from phosphatidylcholine vesicles caused by Aβ16–20 m but not

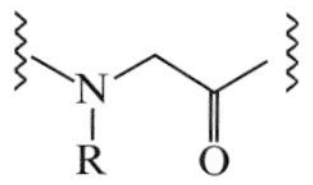

FIG. 11. The basic structure of peptoids, polymers of N-substituted-glycines; in most cases, the substitutions are α-chiral (i.e., they have side chain chiral carbon atoms adjacent to the backbone nitrogen).

fibrillogenesis inhibitor, iAβ5p-A1 (Ac-LP-NMeF-FD-CONH₂ [iAβ5p-A1]), showed increased stability *in vitro* (exposure to plasma or rat brain homogenates) and *in vivo* (10-fold longer half-life in plasma) compared with the nonmethylated counterpart, iAβ5p (Adessi *et al.*, 2003).

Other N-Alkylation (Peptoids). Peptoids are polymers of N-substituted glycines; in most cases, the substitutions are α-chiral, (i.e., they have side chain chiral carbon atoms adjacent to the backbone nitrogen). Methods are now available for efficient solid phase synthesis of these polymers. They lack backbone hydrogen bond donors but adopt a polyproline type I helical structure in solution and in the solid state (Wu *et al.*, 2003). Some peptoids, however, even those of only five residues, show CD spectra resembling those of α-helices (Kirshenbaum *et al.*, 1998) (Fig. 11).

Few attempts have been made to design fibrillogenesis inhibitors from peptoids. One related example is the incorporation of an N-butylated Ser residue at position 28 of a peptide based on residues 20–29 of hIAPP (NH₂-SNNFGAIL-nButS-S-CONH₂) (Rijkers *et al.*, 2002). Although this peptide-peptoid formed some aggregates, these were unstructured by EM and FTIR. Moreover, it inhibited gelation of IAPP, possibly by inhibiting hydrogen bonding or through the introduction of a bulky substituent on the backbone nitrogen that was unfavorable for β-sheet packing.

Ester Bonds. As discussed above, N-methylated and other alkylated peptides could act through a number of different mechanisms, alone or in combination. To investigate the importance specifically of hydrogen bonds in fibril formation, amides have been replaced by ester bonds, because esters lack a potential hydrogen bond donor present in amides, the amide proton (Beligere and Dawson, 2000; Bramson *et al.*, 1985; Lu *et al.*, 1997), but share many structural similarities with amides, including a transplanar conformation and similar bond lengths and angles on Ramachandran plots (Ingwall and

---

Aβ16–20. Aβ16–20 m does not cause vesicle fusion. (C) Fluorescence microscopy of COS cells incubated for 12 h with 40 μM Anth-Aβ16–20 m, an analogue of Aβ16–20 m with the fluorescent N-methyl-anthranilic acid in lieu of the acetyl group. Fluorescence microscopy (using a DAPI filter) shows net ingress of the peptide, which is cationic because of the single positive charge. Reproduced with permission from Gordon *et al.* (2002).

Goodman, 1974). Other than the absence of a hydrogen bond donor in esters, the main difference between esters and amides is that the carbonyl carbon is more electron deficient than that of the amide, rendering esters more base labile than amides. In addition, the ester carbonyl oxygen is less basic than that of the amide, and, as a consequence, the oxygen is a somewhat weaker hydrogen bond acceptor (Arnett *et al.*, 1973).

Gordon and Meredith (2003) synthesized fibrillogenesis inhibitors with essentially the same design as their earlier inhibitors containing N-methyl amino acids (i.e., with ester bonds at alternate residues within the $A\beta(16-20)$ peptide). Like the N-methylated peptides, the ester-containing peptides remained water soluble, did not form fibrils, were effective inhibitors of $A\beta$ fibrillogenesis (thioflavin T assays and EM), and yet also showed the same side chain specificity of the N-methyl inhibitor peptides. As expected, the ester bonds in these depsipeptides were base labile and were partially hydrolyzed over the course of ≈ 24 h even at slightly alkaline pH (e.g., pH 7.40). Nevertheless, the equal effectiveness of the ester-containing and N-methylated inhibitors suggests that both act mainly by disrupting hydrogen bonds rather than by steric interference with fibril packing. An ester-containing inhibitor of hIAPP aggregation based on residues 20–29 of IAPP and substituting a single ester bond for the amide at Ser 28 delayed aggregation and formation of β-sheets, as shown by FTIR and EM (Rijkers *et al.*, 2002); some aggregation did occur, however, possibly because the ester still allowed hydrophobic interactions that favored aggregation.

Modification of α-Carbon. Modifications of the peptide α-carbon cannot interfere with hydrogen bonding, but peptides containing such modifications could, through steric effects or restrictions of conformational freedom, disrupt fibril propagation. One set of fibrillogenesis inhibitors based on α-carbon modification were peptides substituting α-amino isobutyric acid, also called α-methyl alanine (αMeA), in place of alanine or leucine in sequences derived from hIAPP (Gilead and Gazit, 2004). In particular, there were three such peptides, αMeA-NF-αMeA-VHSS, αMeA-NF-αMeA-VH, and αMeA-NF-αMeA-V, based on the sequences ANFLVHSS, ANFLVH, and ANFLV (residues 13–17, 13–18, and 13–20, respectively). Ala and Leu are both α-helix formers, but the addition of the second methyl group to the α-carbon of Ala severely constrains this residue to φ and ψ angles in the α-helical range. However, because αMeA is achiral, this residue equally favors left- and right-handed α-helices (i.e., the Ramachandran plot for αMeA shows two small areas of permissible torsional angles at the center of those favored by both D- and L-Ala). The restrictions imposed on conformational freedom by αMeA are even more severe, and the permissible φ and ψ angles are further from those of β-sheets than those of proline.

Thus, peptides containing αMeA could be even more effective at disrupting β-sheets than proline-containing peptides.[7] The effect of these peptides, however, was not notably different from that of other inhibitors. Nevertheless, the use of αMeA represents another direction for future development of inhibitors.

Peptide Cyclization. In general, cyclization of peptides increases their stability, but in the case of amyloidogenic peptides, the stabilized form may be fibrillogenic. Thus, Sciarretta *et al.* (2005) recently showed that a homologue of Aβ containing a lactam cross-link between Asp23 and Lys28 (Aβ40-Lactam [D23/K28]) formed fibrils resembling those of unmodified Aβ but at a $\approx$1000-fold faster rate and without the usual lag period observed for the nucleation step(s) of fibril formation. At the lowest measurable concentrations, $\approx$1 μM, Aβ40-Lactam(D23/K28) formed aggregates that favored nucleation. This critical concentration of <1 μM compares with that of nonmodified Aβ, in the range of 50–100 μM.[8] In contrast to Aβ40-Lactam(D23/K28), Kapurniotu *et al.* (2003) synthesized cyclo[17, 21]-[Lys[17], Asp[21]] Aβ(1–28), a homologue of Aβ(1–28) with Lys substituted for Leu17, and incorporated it into the lactam-bridge with Asp21. Cyclo[17, 21]-[Lys[17], Asp[21]] Aβ(1–28) did form some amorphous precipitate but did not form fibrils, was able to inhibit fibril formation by Aβ(1–28), and interfered with Aβ(1–40) fibril formation. The difference between these lactam-containing peptides can be rationalized. Apart from differences in the Aβ chain length and solvent conditions, the lactam in cyclo[17, 21]-[Lys[17], Asp[21]]Aβ(1–28) is between the ith and i + 4th residues, which strongly favors the α-helical conformation. In contrast, the lactam in Aβ40-Lactam(D23/K28) is between the ith and i + 5th residues, which favors the formation of the bend structure found in Aβ(1–40) fibrils and not the α-helix.

One other attempt to use cyclization to produce a fibrillogenesis inhibitor peptide involved the incorporation of Cys residues at both the N- and C-termini of the "β-breaker peptide" LPFFD (Adessi *et al.*, 2003) and cyclization through the formation of a disulfide bond. This peptide did not inhibit Aβ(1–40) fibrillogenesis, however. A potential disadvantage of using disulfide cross-links is the possibility of disulfide exchange at alkaline or even neutral pH.

---

[7] Indeed, the ability of αMeA to disrupt β-sheets had been noted starting more than a decade ago (Karle *et al.*, 1990).

[8] Note that Sciarretta *et al.* (2005) used a different set of solvent conditions from those of O'Nuallain *et al.* (2005), which can account for the difference in apparent C_r.

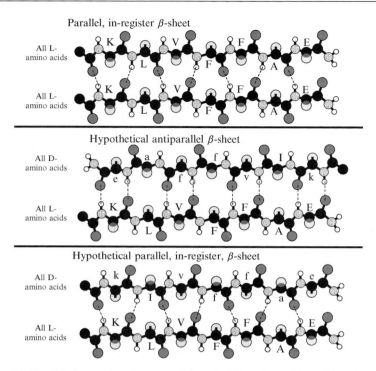

FIG. 12. Possible interactions between all-L and all-D amino acid peptides of the same sequence, in this case, from amyloid β (Aβ). (Top) Segment from Aβ aligned in an in-register and parallel orientation. (Middle and bottom) All-D peptide-forming hydrogen bonds with an all-L peptide in either the antiparallel (middle) or parallel in-register (lower) β-sheet orientation; side chains of adjacent peptide molecules point to opposite sides of the planes of the peptide bonds, however, rather than the same side, as if true for a single chirality. In the case of short peptides, such as KLVFFAE (Aβ[16–22]), the antiparallel orientation would yield shorter and straighter hydrogen bonds and relieve unfavorable charge interactions of placing residues of like charge (both Lys and Glu) close to one another in adjacent chains, but some of the favorable interactions between hydrophobic side chains (Leu, Val, and Phe) would also be lost. At present, the actual structure of these hypothetical β-sheets of mixed chirality is not known.

Use of D-Amino Acids. Several groups have incorporated D-amino acids in the design of their inhibitor peptides, probably for the same reason that bacteria incorporated them aeons ago into their cell walls—to inhibit proteolysis. In theory, it should be possible to design D-amino acid–containing inhibitor peptides of similar intrinsic potency as those containing only L-amino acids. As shown in Fig. 12, β-strand peptides containing all D-amino acids could be incorporated into β-sheets composed of all L-amino acid peptides. Indeed, mixtures of poly(D-lysine) and poly(L-lysine) form

β-sheet fibrils more readily than do peptides of a single chirality (Furhop *et al.*, 1987; Dzwolak *et al.*, 2004). Several investigators have reported that D-amino acid–containing inhibitor peptides were more potent than their L-amino acid–containing counterparts. Based on their earlier work showing that KLVFF (residues 16–20 of Aβ) and related peptides bind to Aβ(1–40) and disrupt fibril formation, Tjernberg *et al.* (1997) screened combinatorial pentapeptide libraries exclusively composed of D-amino acids and identified several ligands with a general motif containing phenylalanine in the second position and leucine in the third position. The issue of chirality in inhibiting fibril formation was addressed further by Chalifour *et al.* (2003), who found that D-enantiomers of five peptides based on Aβ(16–20) unexpectedly inhibited fibril formation (thioflavin T and EM) and prevented Aβ from adopting a β-sheet structure more effectively than L-KLVFFA, whereas L-KLVFFA was a more potent inhibitor of D-Aβ fibril formation than was D-KLVFFA. D-peptides were also more potent at reducing toxicity of Aβ peptides toward neuronal cells in culture. Similarly, cholyl-lvffa-OH was a more potent fibrillogenesis inhibitor than the L analogue (Findeis *et al.*, 1999). Similar results were observed for inhibitors of hIAPP aggregation (Rijkers *et al.*, 2002). Along the same lines, putrescine was added to the N-terminus of a β-breaker peptide, RDLPFYPVPID, to increase its permeability to the blood–brain barrier (Podluso *et al.*, 1999). The all-D-amino acid homologue partially inhibited Aβ fibril formation to approximately the same extent as the L-amino acid peptide.

The preceding results were somewhat at odds with those of Ban *et al.* (2004), who used total internal reflection fluorescence microscopy (TIRFM), combined with the binding of thioflavin T, to visualize seed-dependent fibril growth of Aβ(1–40) in real-time at the single fibril level. Their study focused on a fibril-forming and cytotoxic region of Aβ, Aβ(25–35), -GSNKGAIIGLM-. In experiments examining seeded fibril growth by TIRFM, D-Aβ(25–35) monomers did not extend L-Aβ(25–35) seeds, or *vice versa*, showing that heterologous extension between L- and D-enantiomers does not occur, and implying that enantiomeric peptides also would not prevent fibril extension on seeds heterologously, because heterologous binding did not occur. However, they also examined a series of chimeric peptides based on the same region of Aβ: GsnkGAIIGlm, GSNKGaaiGLM, and GsnkGAIIGLM. Their results can be summarized as follows. Binding of peptides required that the seed and monomer have the same stereochemistry (L- binding to L- or D- binding to D-), and the most effective binders were those that conserved stereochemistry of the hydrophobic domain (i.e., the order of effectiveness was GsnkGAIIGLM > GsnkGAIIGlm > GSNKGaaiGLM). Furthermore, three D-amino acids were sufficient for GsnkGAIIGLM to disrupt fibril growth by GSNKGAII GLM. None of these peptides, however, were able to inhibit growth of L-Aβ(1–40). It should be noted that the assay used by Ban *et al.* (2004) examines only fibril growth (extension) from single seed nuclei and not ensemble fibril nucleation and extension. Thus, in their assay, a β-sheet breaker peptide, as described above (Soto *et al.*, 1996), also did not inhibit fibril growth. This could mean that the β-sheet breaker peptide, and perhaps other inhibitor peptides examined by other investigators, has a greater effect on fibril nucleation than extension from preformed seeds.

In considering the preceding studies, it is important to recall that although Aβ(1–40) and long fragments of Aβ, such as Aβ(10–35), form parallel, in-register β-sheets, short, nonamphiphilic peptides, such as KLVFFAE (residues 16–22 of Aβ), form antiparallel fibrils (Balbach *et al.*, 2000). For Aβ peptides in general, the somewhat higher energy parallel β-sheet conformation is assumed by amphiphilic peptides, where the parallel conformation affords the opportunity to shield hydrophobic sections of a peptide like Aβ from the aqueous medium. However, in either antiparallel (middle panel, Fig. 12) or parallel (lower panel, Fig. 12) in-register β-sheets composed of alternating all-L- and all-D-amino acid peptides, side chains of adjacent β-strands would be placed on opposite sides of the planes of the peptide bonds rather than on the same side, as would be true for either parallel or antiparallel β-sheets of a single chirality. Thus, for short, nonamphiphilic peptides, such as all D-KLVFF (Tjernberg *et al.*, 1997), an antiparallel orientation of the D-amino acid inhibitor peptide with all L-Aβ would have energetically favorable short straight hydrogen bonds but would eliminate the unfavorable charge interactions of placing Lys side chains close to each other. On the other hand, screens of D-amino acid–containing peptides suggested a preference for sequences resembling a reversed KLVFF (e.g., yfllr), suggesting that the D-amino acid peptides could form a parallel β-sheet rather than the antiparallel β-sheet formed by KLVFF according to molecular modeling. This question remains unresolved at present, because there are currently no direct experiment data to show whether D-amino acid peptides bind to L-amino acid peptides (polylysines or others) as part of a parallel or antiparallel β-sheet. In addition, the periodicity of β-sheets is generally slightly greater than 2.0 amino acids per turn (i.e., most β-sheets are nonplanar), and it is not clear how the opposite "twists" of L- and D-amino acid β-strands would affect each other. Whatever the structure of such mixtures might be, such D-amino acid congeners of fibril-forming peptides could bind to fibril growth sites and inhibit the self-association of fibril-forming peptides.

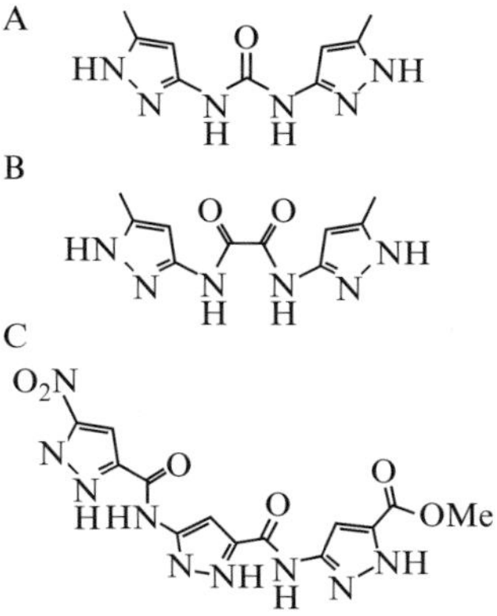

FIG. 13. Structures of the 3-aminopyrazole nucleus (A) used to form dimeric (B, Gly dimer) and trimeric (C, Gly trimer) β-sheet ligands. Adapted and reproduced with permission from Rzepecki *et al.* (2004).

Fibrillogenesis Inhibitors Designed as Nonpeptidic β-Sheet Mimics

Most nonpeptide inhibitors of protein aggregation are based on the structure of small, nonpeptidic molecules, such as melatonin and nicotine. One of the most important goals in the rational design of fibrillogenesis inhibitors, however, is to translate structural concepts from the study of β-sheets into the development of peptidomimetics with favorable structural and pharmacokinetic properties. Toward this goal, Rzepecki *et al.* (2004) designed nonpeptidic, oligomeric acylated aminopyrazole mimics of β-strands with a donor-acceptor-donor (DAD) hydrogen bond pattern complementary to that of a β-sheet (Fig. 13). From energy minimization calculations, they showed recognition of the hexapeptide, KKLVFF, by a trimeric aminopyrazole ligand, from which they devised new generations of dimeric and oligomeric ligands, retaining complementarity toward a β-sheet peptidic backbone. They then showed experimentally that such constructs bound to and capped the solvent-exposed β-sheet portions in Aβ (1–40), preventing propagation of β-sheets into insoluble protein aggregates. From ESI-mass spectrometry (MS) experiments, they demonstrated that their β-sheet ligands (Ampox or Trimer, Fig. 13B, C, respectively) allowed the formation of small oligomers with Aβ (and, after 5 days of incubation, higher molecular-weight oligomers) but prevented further aggregation of small oligomers in protofibrils and fibrils. For the trimeric aminopyrazole derivative (Fig. 13C), they were able to estimate a dissociation constant of $\approx 10^{-5} M$, a modest affinity, probably comparable to that of peptide inhibitors and approximately the affinity expected for efficient hydrogen bonding between this compound and the solvent-exposed edge of Aβ.

Methods of Inhibitor Peptide Synthesis and Assessment of Inhibition

Peptide Synthesis

Peptides Containing N-Methyl Amino Acids

Many N-methyl amino acids are now commercially available from several sources, such as Novabiochem, and can be purchased with either 9-fluorenmethoxycarbonyl (FMOC) or t-butoxycarbonyl (tBOC) α-amino protecting groups. In addition, a number of methods for synthesizing N-methyl amino acids have been published (Aurelio et al., 2004; Biron and Kessler, 2005; Di Gioia et al., 2003; White and Konopelski, 2005). Coupling yields during synthesis of peptides containing N-methyl amino acids are generally low for the addition of the amino acid following the N-methyl amino acid. There is not an obvious pattern as to which amino acid will be difficult to add to an N-methyl amino acid; although bulky, hydrophobic amino acids sometimes pose difficulties, in some cases, even sterically nonhindered amino acids can give poor yields. In general, the strong activating agent, HATU ([2-1H-9-azabenzotriazole-1-yl]-1,1,3,3-tetramethyluronium hexafluoridate) is recommended for coupling residues following N-methyl amino acids. In addition, prolonged coupling times are recommended, a minimum of 3–5 hours or even overnight. Although these couplings can be accomplished using automated synthesizers or manually, we generally perform these steps manually. For other residues in the inhibitor, more standard coupling conditions can be used (e.g., HOBt/HBTU [N-hydroxybenzotriazole/2-1H-benzotriazole-1-yl]-1,1,3,3-tetramethyluronium hexafluoridate).

Ester Inhibitors. As with N-methyl amino acids, α-hydroxy amino acid analogues are commercially available from several sources. In addition, methods are well established for synthesizing peptides containing ester bonds using both tBOC (Bramson et al., 1985; Lu et al., 1997) and FMOC (Gordon et al., 2003) chemistries.

A typical procedure for the coupling of an α-hydroxy amino acid analogue (e.g., leucic acid) and the following Boc-amino acid uses various slightly altered versions of 1-hydroxybenzotriazole/diisopropylcarbodiimide (NOBt/DIC) and 4-(dimethylamino)-pyridine/diisopropylcar bodiimide (DMAP/DIC) chemistry (e.g., Bramson et al., 1985; Lu et al., 1997). For example, in the synthesis of one peptide on the 0.2-mM scale, excess

L-leucic acid (or another α-hydroxy amino acid analogue) at 2.2 mM was coupled to the growing peptide chain in N,N-dimethylformamide/dichloromethane (1:1 DMF/DCM ratio) in the presence of 2.2 mM DIC, 2.5 mM HOBT, and 0.8 mM of NEM. The reaction proceeded to completion in 30 min as judged by standard ninhydrin assay. The coupling of Boc-Thr (2.2 mM) was carried out for 60 min in DMF/DCM (1:1) in the presence of 2.2 mM DIC, 0.8 mM N-ethylmorpholine (NEM), and a catalytic amount of 4-(dimethylamino)-pyridine (DMAP, less than 0.04 mM).

A typical procedure for FMOC chemistry is the following: for peptides synthesized on the 0.25-mM scale, 2.2 mM α-hydroxy acids is coupled to the resin-bound peptide using 2.2 mM DIC, 2.5 mM HOBT, and 0.8 mM NEM in DMF/DCM (1:1, v/v). Coupling under these conditions is rapid and is generally complete within 30 min. A typical procedure for the formation of the ester bond following coupling of the α-hydroxy acid (again, using the 0.25-mM scale) is to activate 2.2 mM FMOC amino acid with 2.2 mM DIC and couple for 3 h in the presence of 1.2 mM NEM and 0.04 mM DMAP.

Inhibition/Disassembly Assays: Some of the Pitfalls

Inhibition and Disassembly. Assays for potential fibrillogenesis inhibitors include assays for fibrillogenesis inhibition and for fibril disassembly. In the former case, a putative inhibitor is added to a solution of the peptide still in the monomeric (or soluble oligomeric) state, or at least in the state taken to be predominantly monomeric. In a fibril disassembly assay, the inhibitor is added to preformed fibrils. The latter assay assumes that fibril formation, in the absence of an inhibitor, is essentially irreversible, or at least so slowly reversible that fibril dissolution in the absence of inhibitor is negligible. This assumption, although infrequently tested directly, may not be strictly correct even in the case of a peptide like Aβ(1–40), where simple dilution of fibrils into a large excess of buffer can lead to varying degrees of solubilization, depending upon the solvent, variations of peptide sequence (Williams *et al.*, 2004), and mode of production of fibrils. Indeed, O'Nuallain *et al.* (2005) convincingly demonstrate that Aβ(1–40) monomers dissociate from the fibrils over the course of 1 day, reaching a plateau value of about 0.7 μM of soluble peptide (i.e., the C_r). For this reason, assays of fibril disassembly should always include a control sample in which solvent alone is added to the fibrils.

Our typical assay for assessing inhibitors of Aβ(1–40) fibrillogenesis is the following. The putative inhibitor peptide, stored as a lyophilized powder at −20°, is first dissolved in hexafluoroisopropyl alcohol (HFIP) or another volatile solvent as a concentrated stock solution (50 mg/ml). The solvent is evaporated under nitrogen and then dissolved in a buffer. As described below, the choice of a buffer is one of many variables to be considered in fibrillogenesis assays. An aliquot of Aβ(1–40) peptide in DMSO (see below) is then added to the solution, containing or not containing an inhibitor peptide. The mixtures are vortexed very briefly ($\approx$30 s) and then incubated at 37° for 5–7 days without shaking. The final concentration of Aβ(1–40) in the mixture is typically either 100 or 200 μM. At various times, aliquots are removed and the extent of fibril formation is assayed by thioflavin T fluorescence, Congo red binding, and EM.

A typical assay that we have used for assessing disassembly of Aβ(1–40) fibrils is the following. Aβ(1–40) is incubated alone for 7 days to allow fibrils to form, as described in the previous paragraph. An aliquot of the formed fibrils in buffer is then added to inhibitor peptide that has been dried from HFIP or another volatile solvent. The extent of fibrils remaining intact is assayed using thioflavin T fluorescence, Congo red binding, and EM.

Production of Fibrils. The kinetics of fibril formation are very complex, and even seemingly insignificant variations in the procedure for producing fibrils can lead to differences in the products on which fibrillogenesis inhibition and disassembly assays are performed. It is the common experience of investigators in the amyloid field that some proteins or peptides can be maddeningly variable in the kinetics with which they form fibrils. Furthermore, an inherent trait of fibril-forming proteins that is their conformationally plasticity (i.e., a single, pure protein or peptide in a single type of solvent at a single peptide concentration) can give different fibril products depending upon subtle variables that are not known in most cases. One recently described instance of this seemingly capricious behavior is that of Aβ(1–40), which forms fibrils of at least two distinct structures, as demonstrated by solid state NMR techniques, depending upon when the solution is "quiescent" or "agitated" during the formation of fibrils (Petkova *et al.*, 2005).

Thus, different investigators working with slightly different procedures for fibril formation could obtain strikingly different results in their inhibition assays on any given inhibitor. Any procedure for forming fibrils *in vitro* necessarily represents a reductionist approach to the complexity of fibrils as they exist biologically, and at this point, it is almost a philosophical question as to which fibrils formed *in vitro* most closely resemble fibrils formed *in vivo*. Nevertheless, one cannot overstate the necessity for absolutely rigorous consistency in the choice of a fibril-forming procedure.

In general, fibril-forming peptides, such as Aβ, need a preliminary "disaggregation" step, of which many have been reviewed by Zagorski *et al.*, (1999) as well as by Teplow in Chapter 2 and O'Nuallain and colleagues in Chapter 3 in this volume. We recently assessed multiple conditions for forming fibrils (Sciarretta *et al.*, 2005), and among the variables to be considered are the following:

- Solvents to induce disaggregation (e.g., neat HFIP, DMSO, or tri-fluoroacetic acid (TFA), 1 mM NaOH)
- Agitation (of various degrees) or quiescent conditions; among the former conditions are those that include sonication for various times
- Removal of the above disaggregation solvents by complete drying or dilution to a low (e.g., <2% v/v) concentration
- Buffers for resolubilizing disaggregated peptides
- Temperature at which fibrils are formed
- Size exclusion chromatography for isolating monomeric peptides, and if this is performed, procedures for reconcentrating peptide

The pros and cons of the various procedures are beyond the scope of this article, but, again, the main issue to be considered for fibrillogenesis inhibitor assays, given the current state of our knowledge, is consistency. A recent study, however, raised a concern about the use of HFIP (Nichols *et al.*, 2005). These authors observed two classes of soluble Aβ aggregates, one of which could be generated rapidly (<10 min) in buffered 2% HFIP. Such aggregates showed increased thioflavin fluorescence and were rich in β-sheet by CD spectroscopy but were unstable to dilution out of HFIP. Although the stability of these aggregates increased as they progressed to form fibers, they seeded elongation by Aβ monomers poorly. It is not known whether these aggregates resemble aggregates formed *in vivo*, although the same comment applies to all aggregates formed by all published procedures. Nevertheless, the type of fibrillar product formed using HFIP to solubilize Aβ appears to vary unacceptably with the detailed nuances of fibrillogenesis procedure (e.g., length of incubation after the samples develop thioflavin T binding and fluorescence, quantity of HFIP added or remaining after procedures to remove HFIP). Forming fibrils consistently reproducibly is difficult at best, and for the above reasons, we have avoided the use of HFIP in our recent studies (Sciarretta *et al.*, 2005).

Assays of Fibril Formation. As in the previous section, the pitfalls of all the simple procedures for monitoring fibril growth or disaggregation are common knowledge and beyond the scope of this chapter (for review, see Levine, 1999). Suffice it to say that the simplicity of performing widely used procedures, such as the thioflavin T or Congo red binding assay, belies the true complexity of the interactions involved in these measurements. For example, the fluorescent yield from the binding of thioflavin T to different fragments and variants of Aβ peptides is far from constant. Some of the shorter fragments of Aβ (e.g., Aβ[16–22]) do not cause thioflavin T fluorescence at all, even when fibrils can be easily demonstrated by Congo red binding assays or EM, whereas other peptides (e.g., Aβ40-Lactam [D23/K28] described above) bind thioflavin T but in a manner that gives consistently lower fluorescent yields than Aβ(1–40). Still other peptides cause high levels of thioflavin T fluorescence, even when fibrillar material cannot be demonstrated by Congo red binding assays or EM (unpublished observations). The above comments are not meant to "single out" thioflavin T fluorescence assays or to gainsay their use; similar comments can be adduced for other common assays of fibril formation (e.g., the difficulties in quantifying Congo red binding, the problems in sampling inherent in EM of fibril slurries).

Additional Caveats. Many of the physical techniques for following fibril formation and, hence, of fibril disassembly beg the question of which aggregates are on-pathway and which are off-pathway for fibril formation. Peptide aggregates can be detected by dynamic light scattering, for example, and may include some micelle-like aggregates that are not directly on the pathway toward fibril formation.

One additional issue to bear in mind is the need to test very high molar ratios of putative inhibitor to fibril former. There are few data on the affinity of inhibitors for their targets, however, and those data suggest modest affinity ($K_d \approx \mu M$). In any case, proteins and peptides that form fibrils have a high affinity for self. Thus, the development of potent inhibitors may require an iterative process, and one should not overlook compounds with even modest affinity.

Finally, because no assay of fibrillogenesis is entirely satisfactory, it is probably prudent not to rely exclusively on any single assay. For example, a fibrillogenesis inhibitor that leads to a reduction in thioflavin fluorescence might be a fibrillogenesis inhibitor in fact, or it might only compete with thioflavin T for the binding site on its target without actually preventing fibril formation. This is an issue, for example, in the case of short fragments of Aβ, such as Aβ(16–20), which themselves form fibrils that do not cause thioflavin T fluorescence (Gordon *et al.*, 2002). When mixed with Aβ(1–40), Aβ(16–20) leads to an overall reduction of thioflavin fluorescence, but does this result from inhibition of fibril formation by Aβ(1–40) or from the formation of hybrid fibrils (containing both Aβ[1–40] and Aβ[16–20]) that do not cause thioflavin T fluorescence? The answer to such a question is never straightforward but must begin by relying on multiple assays of fibril formation/disassembly.

Binding of Inhibitor Peptides

The implicit or explicit assumption in studying fibrillogenesis inhibitors is that an inhibitor must bind to the aggregating species; that is, the inhibitor must recognize, directly or indirectly, the domain of a protein or peptide responsible for self-aggregation. It is not strictly true, however, that a fibrillogenesis inhibitor must bind to the domain that aggregates. An inhibitor could stabilize the native structure of a protein, as in the case of small molecule inhibitors of transthyretin aggregation (Foss et al., 2005; Johnson et al., 2005). Nevertheless, for those inhibitors targeting putative aggregation sites, techniques for assessing binding of an inhibitor to its target are available and include the many techniques used to assess the broader aspects of protein–protein interactions.

Many of the assays for binding of fibrillogenesis inhibitors to their targets are, in essence, the same as assays to measure self-association of a fibril-forming peptide. An example of the latter type of assay is that of Esler et al. (1996), in which the authors measured deposition of radiolabeled $A\beta$ peptides at physiological concentrations onto neuritic plaques and other forms of cerebral β-amyloid. The corresponding fibrillogenesis inhibition assay would be to measure the abrogation of binding in the presence of a putative inhibitor. Another approach, described by Han et al. (1996), is to measure displacement of a radiolabeled known ligand by test compounds. These authors, for example, synthesized ^{99}Tc-labeled complexes of various derivatives of diphenyl-linked Congo red and other aromatic azo dyes, such as chrysamine G. Another related set of techniques involves the binding of putative inhibitors to a cellulose membrane, after which the binding of ^{125}I-labeled $A\beta$ is assessed by radiography (Tjernberg et al., 1996). Again, the implicit assumption is that inhibitors must bind to both the target peptide and cellulose membrane. This technique was used in an attempt to pinpoint the sites of interaction between $A\beta$ peptides. Other investigators have used fluorescent labels in place of radiolabels. For example, in one system, short peptide fragments of $A\beta$ (KLVFF plus a linker peptide) were constructed on the surface of the solid-phase, Mulltipin (Chiron Technologies Pty. Ltd), and binding of fluoresceinated $A\beta$ to these pins was measured (Akikusa et al., 2003). Such systems lend themselves to automation and large-scale screening procedures.

Surface plasmon resonance (SPR) has been used to demonstrate interaction between small peptides derived from $A\beta$ and full-length $A\beta$. In a study described above (Tjernberg et al., 1996), binding of [125]I-$A\beta$ to the solid-phase support was confirmed by SPR studies. In a similar manner, SPR has been used to screen small molecule inhibitors of $A\beta$ fibrillogenesis (Cairo et al., 2002) and could be used to measure binding of peptide inhibitors as well. In this study, $A\beta(10–35)$ was immobilized as the "bait" through the addition of a C-terminal Cys through an aminohexanoic acid spacer. This immobilized peptide was then used to measure the affinities of various small peptide analytes, mainly derived, in part, from an internal sequence of $A\beta$, residues 16–20 (KLVFF). The best of these peptides bound effectively to β-amyloid with $K_d \approx 40$ μM. In general, these authors found a correlation between high binding affinity of peptides for $A\beta(10\text{-}35)$ and the ability of those peptides to inhibit cytotoxic effects of $A\beta(10\text{-}35)$ to human neuroblastoma (SH-SY5Y) cells. Compounds with dissociation constants above ≈ 100 μM did not show significant activity in the cell toxicity assays.

SPR has been used in several assays of $A\beta$ fibril extension (Hasegawa *et al.*, 2002; Myszka *et al.*, 1999). In one study using $A\beta$ fibrils immobilized on a biosensor surface utilizing amine coupling chemistry (Cannon *et al.*, 2004), the authors found that sonicated fibrils supported elongation better than unsonicated fibrils, as expected for fibril extension. They also varied the fibril immobilization density and could use monomeric $A\beta$ as the immobilized bait.

More recently, solution NMR techniques have been employed to gain atomic resolution of interactions between fibril-forming peptides and inhibitors of fibrillogenesis. Chen and Rief (2004) used transferred residual dipolar couplings (trRDCs) to restrain the structure of peptide inhibitors that bind transiently to amyloid fibrils. This approach relies on the fact that amyloid fibrils become oriented within a magnetic field, most often with the fibril axis parallel to the external field (Worcester, 1978), which is a property based on the anisotropic diamagnetic susceptibility of the peptide bond. In their study, they used an internal fragment of $A\beta$ ($A\beta[14$–$23]$) that forms fibrils. They demonstrated that the peptide inhibitor iAβ5^{inv} (DPFFL, derived from iAβ5, LPFFD; Soto *et al.*, 1998) adopts a β-sheet conformation with the backbone N-H and C-H dipolar vectors aligned preferentially parallel and perpendicular, respectively, to the fibril axis. Transferred NOESY experiments also allowed them to make qualitative estimates of the affinity of this inhibitor for the fibril.

Application of Inhibitor Peptides to Therapeutic Strategies

The design and synthesis of peptide inhibitors of fibrillogenesis have two major goals: obtaining structural information about fibril-forming peptides (and the diseases in which fibrils are prominent) and developing potential therapeutic agents. Many claims have been made for the potential usefulness of fibrillogenesis inhibitors in therapy of neurodegenerative and other diseases. In attempting to translate these inhibitors to the clinical arena, there will, of course, be the usual hurdles of trying to use peptides as drugs: issues of bioavailability and pharmacokinetics; the susceptibility of peptides to degradation by proteases, oxidants, and immune responses; and their limited capacity to cross cell membranes and the blood–brain barrier. Beyond these generic concerns is the even greater problem that the mechanisms by which aggregating peptides and proteins lead to cellular dysfunction and death and, ultimately, to disease remain hotly contested and largely unknown. Much debate, for example, has gone into the question of whether fibrils or an earlier oligomeric intermediate en route to the fibril is more cytotoxic. Our attempts to understand the neurotoxicity of $A\beta$, for example, remain hampered by the rather crude metrics of cell death and the lack of realistic cell culture systems for neurons. Most of this is beyond the purview of this chapter. Suffice it to say that it will be wise, in evaluating these and future peptide and nonpeptide inhibitors of fibrillogenesis, to temper enthusiasm with prudence.

References

Adessi, C., Frossard, M., Boissard, C., Fraga, S., Bieler, S., Ruckle, T., Vilbois, F., Robinson, S. M., Mutter, M., Banks, W. A., and Soto, C. (2003). Pharmacological profiles of peptide drug candidates for the treatment of Alzheimer's disease. *J. Biol. Chem.* **278**, 13905–13911.

Akikusa, S., Nakamura, K., Watanabe, K.-I., Horikawa, E., Konakahara, T., Kodaka, M., and Okuno, H. (2003). Practical assay and molecular mechanism of aggregation inhibitors of β-amyloid. *J. Pept. Res.* **61**, 1–6.

Arnett, E. M., Mitchell, E. J., and Murty, T. S. S. R. (1973). Basicity. Comparison of hydrogen bonding and proton transfer to some Lewis bases. *J. Am. Chem. Soc.* **96**, 3875–3891.

Aurelio, L., Brownlee, R. T., and Hughes, A. B. (2004). Synthetic preparation of N-methyl-alpha-amino acids. *Chem. Rev.* **104**, 5823–5846.

Balbach, J. J., Ishii, Y., Antzutkin, O. N., Leapman, R. D., Rizzo, N. W., Dyda, F., Reed, J., and Tycko, R. (2000). Amyloid fibril formation by $A\beta$16-22, a seven-residue fragment of the Alzheimer's β-amyloid peptide, and structural characterization by solid state NMR. *Biochemistry* **39**, 13748–13759.

Ban, T., Hoshino, M., Takahashi, S., Hamada, D., Hasegawa, K., Naiki, H., and Goto, Y (2004). Direct observation of $A\beta$ amyloid fibril growth and inhibition. *J. Mol. Biol.* **344**, 757–767.

Beligere, G. S., and Dawson, P. E. (2000). Design, synthesis and characterization of 4-ester CI2, a model for backbone hydrogen bonding in protein α-helices. *J. Am. Chem. Soc.* **122**, 12079–12082.

Bennett, D. A., and Holtzmann, D. M. (2005). Immunization therapy for Alzheimer disease? *Neurology* **64**, 10–12.

Bersch, B., Koehl, P., Nakatani, Y., Ourisson, G., and Milon, A. (1993). ^{1}H nuclear magnetic resonance determination of the membrane-bound conformation of senktide, a highly selective neurokinin B agonist. *J. Biomol. NMR* **3**, 443–461.

Biron, E., and Kessler, H. (2005). Convenient synthesis of N-methylamino acids compatible with Fmoc solid-phase peptide synthesis. *J. Org. Chem.* **70**, 5183–5189.

Bodles, A. M., El-Agnaf, O. M. A., Greer, B., Guthrie, D. J. S., and Irvine, G. B. (2004). Inhibition of fibril formation and toxicity of a fragment of α-synuclein by an N-methylated peptide analogue. *Neurosci. Lett.* **359**, 89–93.

Bramson, H. N., Thomas, N. E., and Kaiser, E. T. (1985). The use of N-methylated peptides and depsipeptides to probe the binding of heptapeptide substrates to cAMP-dependent protein kinase. *J. Biol. Chem.* **260**, 15452–15457.

Burkoth, T. S., Benzinger, T. L. S., Jones, D. N. M., Hallenga, K., Meredith, S. C., and Lynn, D. G. (1998). C-terminal PEG blocks the irreversible step in β-amyloid(10-35) fibrillogenesis. *J. Am Chem. Soc.* **120**, 7655–7655.

Cairo, C. W., Strzelec, A., Murphy, R. M., and Kiessling, L. L. (2002). Affinity based inhibition of β-amyloid toxicity. *Biochemistry* **41**, 8620–8629.

Cannon, M. J., Williams, A. D., Wetzel, R., and Myszka, D. G. (2004). Kinetic analysis of β-amyloid fibril elongation. *Analyt. Biochem.* **328,** 67–75.

Chabry, J., Caughey, B., and Chesebro, B. (1998). Specific inhibition of *in vitro* formation of protease-resistant prion protein by synthetic peptides. *J. Biol. Chem.* **273,** 13203–13207.

Chalifour, R. J., McLaughlin, R. W., Lavoie, L., Morissette, C., Tremblay, N., Boule, M., Sarazin, P., Stea, D., Lacombe, D., Tremblay, P., and Gervais, F. (2003). Stereoselective interactions of peptide inhibitors with the β-amyloid peptide. *J. Biol. Chem.* **278,** 34874–34881.

Chen, Z., and Reif, B. (2004). Measurements of residual dipolar couplings in peptide inhibitors weakly aligned by transient binding to peptide amyloid fibrils. *J. Biomol. NMR* **29,** 525–530.

Clark, T. D., Buriak, J. M., Kobayashi, K., Isler, M. P., McRee, D. E., and Ghadiri, M. R. (1998). Cylindrical β-sheet peptide assemblies. *J. Am. Chem. Soc.* **120,** 8949–8962.

Davis, P. D., Raffen, R., Dul, L. J., Vogen, M. S., Williamson, K. E., Stevens, J. F., and Argon, Y. (2000). Inhibition of amyloid fiber assembly by both BiP and its target peptide. *Immunity* **13,** 433–442.

Di Gioia, M. L., Leggio, A., Le Pera, A., Liguori, A., Napoli, A., Siciliano, C., and Sindona, G. (2003). "One-pot" methylation of N-nosyl-alpha-amino acid methyl esters with diazomethane and their coupling to prepare N-methyl dipeptides. *J. Org. Chem.* **68,** 7416–7421.

Doig, A. J. (1997). A three stranded β-sheet peptide in aqueous solution containing N-methyl amino acids to prevent aggregation. *J. Chem. Soc., Chem. Commun.* **22,** 2153–2154.

El-Agnaf, O. M., Paleologou, K. E., Greer, B., Abogrein, A. M., King, J. E., Salem, S. A., Fullwood, N. J., Benson, F. E., Hewitt, R., Ford, K. J., Martin, F. L., Harriott, P., Cookson, M. R., and Allsop, D. (2004). A strategy for designing inhibitors of α-synuclein aggregation and toxicity as a novel treatment for Parkinson's disease and related disorders. *FASEB J.* **18,** 1315–1317.

Elseviers, M., Van der Auwera, L., Pepermans, H., Tourwe, D., and Van Binst, G. (1988). Evidence for the bioactive conformation in a cyclic hexapeptide analogue of somatostatin containing a cis-peptide bond mimic. *Biochem. Biophys. Res. Commun.* **154,** 515–521.

Fändrich, M., and Dobson, C. M. (2002). The behaviour of polyamino acids reveals an inverse side chain effect in amyloid structure formation. *EMBO J.* **21,** 5682–5690.

Findeis, M. A., Musso, G. M., Arico-Muendel, C. C., Benjamin, H. W., Hundal, A. M., Lee, J. J., Chin, J., Kelley, M., Wakefield, J., Hayward, N. J., and Molineaux, S. M. (1999). Modified-peptide inhibitors of amyloid β-peptide polymerization. *Biochemistry* **38,** 6791–6800.

Foss, T. R., Kelker, M. S., Wiseman, R. L., Wilson, I. A., and Kelly, J. W. (2005). Kinetic stabilization of the native state by protein engineering: Implications for inhibition of transthyretin amyloidogenesis. *J. Mol. Biol.* **347,** 841–854.

Fülöp, L., Zarandi, M., Datki, Z., Soos, K., and Penke, B. (2004). β-amyloid-derived pentapeptide RIIGL$_a$ inhibits Aβ(1-42) aggregation and toxicity. *Biochem. Biophys. Res. Commun.* **324,** 64–69.

Gelinas, D. S., DaSilva, K., Fenili, D., St. George-Hyslop, P., and McLaurin, J. (2004). Immunotherapy for Alzheimer's disease. *Proc. Natl. Acad. Sci. USA* **101**(Suppl. 2), 14657–14662.

Ghanta, J., Chen, C.-L., Kiessling, L. L., and Murphy, R. M. (1996). A strategy for designing inhibitors of β-amyloid toxicity. *J. Biol. Chem.* **271,** 29525–29538.

Gilead, S., and Gazit, E. (2004). Inhibition of amyloid fibril formation by peptide analogues modified with α-aminoisobutyric acid. *Angew. Chem. Int. Ed. Engl.* **43,** 4041–4044.

Gordon, D. J., and Meredith, S. C. (2003). Probing the role of backbone hydrogen bonding in β-amyloid fibrils with inhibitor peptides containing ester bonds at alternate positions. *Biochemistry* **42,** 475–485.

Gordon, D. J., Sciarretta, K. L., and Meredith, S. C. (2001). Inhibition of β-amyloid(40) fibrillogenesis and disassembly of β-amyloid(40) fibrils by short β-amyloid congeners containing N-methyl amino acids at alternate residues. *Biochemistry* **40,** 8237–8245.

Gordon, D. J., Tappe, R., and Meredith, S. C. (2002). Design and characterization of a membrane permeable N-methyl amino acid containing peptide that inhibits Aβ(1-40) fibrillogenesis. *J. Pept. Res.* **60,** 37–55.

Greenfield, N., and Fasman, G. D. (1969). Computed circular dichroism spectra for the evaluation of protein conformation. *Biochemistry* **8,** 4108–4116.

Hasegawa, K., Ono, K., Yamada, M., and Naiki, H. (2002). Kinetic modeling and determination of reaction constants of Alzheimer's β-amyloid fibril extension and dissociation using surface plasmon resonance. *Biochemistry* **41,** 13489–13498.

Hetenyi, C., Szabo, Z., Klement, E., Datki, Z., Kortvelyesi, T., Zarandi, M., and Penke, B. (2002). Pentapeptide amides interfere with disaggregation of β-amyloid peptide of Alzheimer's disease. *Biochem. Biophys. Res. Commun.* **292,** 931–936.

Higuchi, N., Kyogoku, Y., Shin, M., and Inouye, K. (1983). Origin of slow conformer conversion of triostin A and interaction ability with nucleic acid bases. *Int. J. Pept. Protein Res.* **21,** 541–545.

Hilbich, C., Kisters-Woike, B., Reed, J., Masters, C. L., and Beyreuther, K. (1992). Substitutions of hydrophobic amino acids reduce the amyloidogenicity of Alzheimer's disease βA4 peptides. *J. Mol. Biol.* **228,** 460–473.

Howell, P. L., Pangborn, W. A., Marshall, G. R., Zabrocki, J., and Smith, G. D. (1995). A thyrotropin-releasing hormone analogue: pGlu-Phe-D-Pro-psi [CN4]-NMe at 293 and 107 K. *Acta Crystallographica C* **51,** 2575–2579.

Hughes, E., Burke, R. M., and Doig, A. J. (2000). Inhibition of toxicity in the β-amyloid peptide fragment β-(25-35) using N-methylated derivatives—a general strategy to prevent amyloid formation. *J. Biol. Chem.* **275,** 25109–25115.

Hughes, S. R., Goyal, S., Sun, J. E., Gonzalez-Dewhitt, P., Fortes, M., Riedel, N. G., and Sahasrabudhe, S. R. (1996). Two hybrid system as a model to study the interaction of β-amyloid peptide monomers. *Proc. Natl. Acad. Sci. USA* **93,** 2065–2070.

Ingwall, R. T., and Goodman, M. (1974). Polydepsipeptides. III. Theoretical conformational analysis of randomly coiling and ordered depsipeptide chains. *Macromolecules* **7,** 598–605.

Johnson, S. M., Petrassi, H. M., Palaninathan, S. K., Mohamedmohaideen, N. N., Purkey, H. E., Nichols, C., Chiang, K. P., Walkup, T., Sacchettini, J. C., Sharpless, K. B., and Kelly, J. W. (2005). Bisaryloxime ethers as potent inhibitors of transthyretin amyloid fibril formation. *J. Med. Chem.* **48,** 1576–1587.

Karle, I. L., and Balaram, P. (1990). Structural characteristics of α-helical peptide molecules containing Aib residues. *Biochemistry* **29,** 6747–6756.

Kirshenbaum, K., Barron, A. E., Goldsmith, R. A., Armand, P., Bradley, E. K., Truong, K. T. V., Dill, K. A., Cohen, F. E., and Zuckermann, R. N. (1998). Sequence-specific polypeptoids: A diverse family of heteropolymers with stable secondary structure. *Proc. Natl. Acad. Sci. USA* **95,** 4303–4308.

Lazo, N. D., Grant, M. A., Condron, M. C., Rigby, A. C., and Teplow, D. B. (2005). On the nucleation of amyloid β-protein monomer folding. *Protein Sci.* **14,** 1581–1596.

Lee, J. P., Stimson, E. R., Ghilardi, J. R., Mantyh, P. W., Lu, Y. A., Felix, A. M., Llanos, W., Behbin, A., Cummings, M., Van Criekinge, M., Timms, W., and Maggio, J. E. (1995). ^{1}H NMR of Aβ amyloid peptide congeners in water solution. Conformational changes correlate with plaque competence. *Biochemistry* **34,** 5191–5200.

LeVine, H., 3rd. (1999). Quantification of beta-sheet amyloid fibril structures with thioflavin T. *Methods Enzymol.* **309,** 274–284.

Lowe, T. L., Strzelec, A., Kiessling, L. L., and Murphy, R. M. (2001). Structure-function relationships for inhibitors of β-amyloid toxicity containing the recognition sequence KLVFF. *Biochemistry* **40**, 7882–7889.

Lu, W., Qasim, M. A., Laskowski, M. J., and Kent, S. B. H. (1997). Probing intermolecular main chain hydrogen bonding in serine proteinase-protein inhibitor complexes: Chemical synthesis of backbone-engineered turkey ovomucoid third domain. *Biochemistry* **36**, 673–679.

Massi, F., Peng, J. W., Lee, J. P., and Straub, J. E. (2001). Simulation study of the structure and dynamics of the Alzheimer's amyloid peptide congener in solution. *Biophys. J.* **80**, 31–44.

Morgan, D., Diamond, D. M., Gottschall, P. E., Ugen, K. E., Dickey, C., Hardy, J., Duff, K., Jantzen, P., Dicarlo, G., Wilcock, D., Connor, K., Hatcher, J., Hope, C., Gordon, M., and Arendash, G. W. (2000). Aβ peptide vaccination prevents memory loss in an animal model of Alzheimer's disease. *Nature* **408**, 982–985.

Moriarty, D. F., and Raleigh, D. P. (1999). Effects of sequential proline substitutions on amyloid formation by human amylin$_{20-29}$. *Biochemistry* **38**, 1811–1818.

Moss, M. A., Nichols, M. R., Reed, D. K., Hoh, J. H., and Rosenberry, T. L. (2003). The peptide KLVFF-K(6) promotes β-amyloid(1-40) protofibril growth by association but does not alter protofibril effects on cellular reduction of 3-(4,5-dimethylthiazol-2-yl)-2,5-diphenyltetrazolium bromide (MTT). *Mol. Pharmacol.* **64**, 1160–1168.

Myszka, D. G., Wood, S. J., and Biere, A. L. (1999). Analysis of fibril elongation assays using surface plasmon resonance biosensors. *Methods Enzymol.* **309**, 386–402.

Nagai, Y., Tucker, T., Ren, H., Kenan, D. J., Henderson, B. S., Keene, J. D., Strittmatter, W. J., and Burke, J. R. (2000). Inhibition of polyglutamine protein aggregation and cell death by novel peptides identified by phage display screening. *J. Biol. Chem.* **275**, 10437–10442.

Nichols, M. R., Moss, M. A., Reed, D. K., Cratic-McDaniel, S., Hoh, J. H., and Rosenberry, T. L. (2005). Amyloid-β protofibrils differ from amyloid-β aggregates induced in dilute hexafluoroisopropanol in stability and morphology. *J. Biol. Chem.* **280**, 2471–2480.

Pallitto, M. M., Ghanta, J., Heinzelman, P., Kiessling, L. L., and Murphy, R. M. (1999). Recognition sequence design for peptidyl modulators of β-amyloid aggregation and toxicity. *Biochemistry* **38**, 3570–3578.

Patel, D. J., and Tonelli, A. E. (1976). N-methylleucine gramicidin-S and (di-N-methylleucine) gramicidin-S conformations with Cis L-Orn-L-N-MeLeu peptide nonds. *Biopolymers* **15**, 1623–1635.

Penkler, L. J., Van Rooyen, P. H., and Wessels, P. L. (1993). Conformational analysis of mu-selective [D-Ala2,MePhe4]enkephalins. *Int. J. Pept. Protein Res.* **41**, 261–274.

Perutz, M. F., Pope, B. J., Owen, D., Wanker, E. E., and Scherzinger, E. (2002). Aggregation of proteins with expanded glutamine and alanine repeats of the glutamine-rich and asparagine-rich domains of Sup35 and of the amyloid beta-peptide of amyloid plaques. *Proc. Natl. Acad. Sci. USA* **99**, 5596–5600.

Petkova, A. T., Ishii, Y., Balbach, J. J., Antzutkin, O. N., Leapman, R. D., Delaglio, F., and Tycko, R. (2002). A structural model for Alzheimer's β-amyloid fibrils based on experimental constraints from solid state NMR. *Proc. Natl. Acad. Sci. USA* **99**, 16742–16747.

Petkova, A. T., Leapman, R. D., Guo, Z., Yau, W. M., Mattson, M. P., and Tycko, R. (2005). Self-propagating, molecular-level polymorphism in Alzheimer's β-amyloid fibrils. *Science* **307**, 262–265.

Porat, Y., Mazor, Y., Efrat, S., and Gazit, E. (2004). Inhibition of islet amyloid polypeptide fibril formation: A potential role for heteroaromatic interactions. *Biochemistry* **43**, 14454–14462.

Rajarathnam, K., Sykes, B. D., Kay, C. M., Dewald, B., Geiser, T., Baggiolini, M, and Clark-Lewis, I. (1994). Neutrophil activation by monomeric interleukin-8. *Science* **264**, 90–92.

Rijkers, D. T., Hoppener, J. W., Posthuma, G., Lips, C. J., and Liskamp, R. M. (2002). Inhibition of amyloid fibril formation of human amylin by N-alkylated amino acid and α-hydroxy acid residue containing peptides. *Chem. Eur. J.* **8,** 4285–4291.

Rzepecki, P., Nagel-Steger, L., Feuerstein, S., Linne, U., Molt, O., Zadmard, R., Aschermann, K., Wehner, M., Schrader, T., and Riesner, D. (2004). Prevention of Alzheimer's disease associated Aβ aggregation by rationally designed nonpeptidic β-sheet ligands. *J. Biol. Chem.* **279,** 47497–47505.

Schenk, D., Barbour, R., Dunn, W., Gordon, G., Grajeda, H., Guido, T., Hu, K., Huang, J., Johnson-Wood, K., Khan, K., Kholodenko, D., Lee, M., Liao, Z., Lieberburg, I., Motter, R., Mutter, L., Soriano, F., Shopp, G., Vasquez, N., Vandervert, C., Walker, S., Wogulis, M., Yednock, T., Games, D., and Seubert, P. (1999). Immunization with amyloid-β attenuates Alzheimer-disease-like pathology in the PDAPP mouse. *Nature* **400,** 173–177.

Sciarretta, K. L., Gordon, D. L., Petkova, A. T., Tycko, R., and Meredith, S. C. (2005). Aβ40-lactam(D23/K28) models a conformation highly favorable for nucleation of amyloid. *Biochemistry* **44,** 6003–6014.

Soto, C., Sigurdsson, E. M., Morelli, L., Kumar, R. A., Castano, E. M., and Frangione, B. (1998). β-sheet breaker peptides inhibit fibrillogenesis in a rat brain model of amyloidosis: Implications for Alzheimer's therapy. *Nat. Med.* **4,** 822–826.

Tjernberg, L. O., Naslund, J., Lindqvist, F., Johansson, J., Karlstrom, A. R., Thyberg, J., Terenius, L., and Nordstedt, C. (1996). Arrest of β-amyloid fibril formation by a pentapeptide ligand. *J. Biol. Chem.* **271,** 8545–8548.

Tjernberg, L. O., Lilliiehook, C., Callaway, D. J. E., Naslund, J., Hahne, S., Thyberg, J., Terenius, L., and Nordstedt, C. (1997). Controlling amyloid β-peptide fibril formation with protease stable ligands. *J. Biol. Chem.* **272,** 12601–12605.

Tonelli, A. E. (1971). Stability of cis and trans amide bond conformations in polypeptides. *J. Am. Chem. Soc.* **93,** 7153–7155.

Vitoux, B., Aubry, A., Cung, M. T., Boussard, G., and Marraud, M. (1981). N-methyl peptides. III. Solution conformational study and crystal structure of N-pivaloyl-L-prolyl-N-methyl-N'-isopropyl-L-alaninamide. *Int. J. Pept. Protein Res.* **17,** 469–7.

Westermark, P., Engstrom, U., Johnson, K. H., Westermark, G. T., and Betsholtz, C. (1990). Islet amyloid polypeptide: Pinpointing amino acid residues linked to amyloid fibril formation. *Proc. Natl. Acad. Sci. USA* **87,** 5036–5040.

White, K. N., and Konopelski, J. P. (2005). Facile synthesis of highly functionalized N-methyl amino acid esters without side-chain protection. *Org. Lett.* **7,** 4111–4112.

Williams, A. D., Portelius, E., Kheterpal, I., Guo, J. T., Cook, K. D., Xu, Y., and Wetzel, R. (2004). Mapping Aβ amyloid fibril secondary structure using scanning proline mutagenesis. *J. Mol. Biol.* **335,** 833–842.

Wood, S. J., Wetzel, R., Martin, J. D., and Hurle, M. R. (1995). Prolines and amyloidogenicity in fragments of the Alzheimer's peptide β/A4. *Biochemistry* **34,** 724–730.

Wu, C. W., Kirshenbaum, K., Sanborn, T. J., Patch, J. A., Huang, K., Dill, K. A., Zuckermann, R. N., and Barron, A. E. (2003). Structural and spectroscopic studies of peptoid oligomers with α-chiral aliphatic side chains. *J. Am. Chem. Soc.* **125,** 13525–13530.

Yamashita, T., Takahashi, Y., Takahashi, T., and Mihara, H. (2003). Inhibition of peptide amyloid formation by cationic peptides with homologous sequences. *Bioorg. Med. Chem. Lett.* **13,** 4051–4054.

Yan, L.-M., Tatrek-Nossol, M., Velkova, A., Kazantzis, A., and Kapurniotu, A. (2006). Design of a mimic of nonamyloidogenic and bioactive human islet amyloid polypeptide (IAPP) as nanomolar inhibitor of IAPP cytotoxic fibrillogenesis. *Proc. Natl. Acad. Sci. USA* **103,** 2046–2051.

Zagorski, M. G., Yang, J., Shao, H., Ma, K., Zeng, H., and Hong, A. (1999). MethodologicalIs "nonds" correct in article title in reference by Patel and Tonelli? and chemical factors affecting amyloid beta peptide amyloidogenicity. *Methods Enzymol.* **309,** 189–204.

Zhang, S., Iwata, K., Lachenmann, M. J., Peng, J. W., Li, S., Stimson, E. R., Lu, Y., Felix, A. M., Maggio, J. E., and Lee, J. P. (2000). The Alzheimer's peptide Aβ adopts a collapsed coil structure in water. *J. Struct. Biol.* **130,** 130–141.

Zhou, A., Stein, P. E., Huntington, J. A., Sivasothy, P., Lomas, D. A., and Carrell, R. W. (2004). How small peptides block and reverse serpin polymerization. *J. Mol. Biol.* **342,** 931–941.

Further Reading

Chitnumsub, P., Fiori, W. R., Lashuel, H. A., Diaz, H., and Kelly, J. W. (1999). The nucleation of monomeric parallel β-sheet-like structures and their self-assembly in aqueous solution. *Bioorg. Med. Chem.* **7,** 39–59.

Cruz, M., Tusell, J. M., Grillo-Bosch, D., Albericio, F., Serratosa, J., Rabanal, F., and Giralt, E. (2004). Inhibition of β-amyloid toxicity by short peptides containing N-methyl amino acids. *J. Pept. Res.* **63,** 324–328.

Doig, A. J., Hughes, E., Burke, R. M., Su, T. J., Heenan, R. K., and Lu, J. (2002). Inhibition of toxicity and protofibril formation in the amyloid-β peptide β(25-35) using N-methylated derivatives. *Biochem. Soc. Trans.* **30,** 537–542.

Formaggio, F., Bettio, A., Moretto, V., Crisma, M., Toniolo, C., and Broxterman, Q. B. (2003). Disruption of the β-sheet structure of a protected pentapeptide, related to the β-amyloid sequence 17-21, induced by a single, helicogenic Cα-tetrasubstituted α-amino acid. *J. Pept. Sci.* **9,** 461–466.

Gazit, E. (2002). Mechanistic studies of the process of amyloid fibrils formation by the use of peptide fragments and analogues: Implications for the design of fibrillization inhibitors. *Curr. Med. Chem.* **9,** 1725–1735.

Kumar, N. G., Izumiya, N., Miyoshi, M., Sugano, H., and Urry, D. (1975). Conformational and spectral analysis of polypeptide antibiotic N-methylleucine gramicidin-S dihydrochloride by nuclear magnetic-resonance. *Biochemistry* **14,** 2197–2207.

Kuner, P., Bohrmann, B., Tjernberg, L. O., Naslund, J., Huber, G., Celenk, S., Gruninger-Leitch, F., Richards, J. G., Jakob-Roetne, R., Kemp, J. A., and Nordstedt, C. (2000). Controlling polymerization of β-amyloid and prion-derived peptides with synthetic small molecule ligands. *J. Biol. Chem.* **275,** 1673–1678.

Mason, J. M., Kokkoni, N., Stott, K., and Doig, A. J. (2003). Design strategies for anti-amyloid agents. *Curr. Opin. Struct. Biol.* **13,** 526–532.

Nichols, M. R., Moss, M. A., Reed, D. K., Lin, W.-L., Mukhopadhyay, R., Hoh, J. H., and Rosenberry, T. L. (2002). Growth of β-amyloid(1-40) protofibrils by monomer elongation and lateral association. Characterization of distinct products by light scattering and atomic force microscopy. *Biochemistry* **41,** 6115–6127.

O'Nuallain, B., Shivaprasad, S., Kheterpal, I., and Wetzel, R. (2004). Thermodynamics of Aβ (1-40) amyloid fibril elongation. *Biochemistry* **44,** 12709–12718.

Poduslo, J. F., Curran, G. L., Kumar, A., Frangione, B., and Soto, C. (1999). β-sheet breaker peptide inhibitor of Alzheimer's amyloidogenesis with increased blood-brain barrier permeability and resistance to proteolytic degradation in plasma. *J. Neurobiol.* **39,** 371–382.

Walsh, D. M., Townsend, M., Podiisny, M. B., Shankar, G. M., Fadeeva, J. V., Agnaf, O. E., Hartley, D. M., and Selkoe, D. J. (2005). Certain inhibitors of synthetic amyloid β-peptide (Aβ) fibrillogenesis block oligomerization of natural Aβ and thereby rescue long-term potentiation. *J. Neurosci.* **25,** 2455–2462.

[16]   Screening for Modulators of Aggregation with a Microplate Elongation Assay

By VALERIE BERTHELIER and RONALD WETZEL

Abstract

Many protein misfolding or conformational diseases, a number of which are neurodegenerative, are associated with the presence of proteinaceous deposits in the form of amyloid/amyloid-like fibrils/aggregates in tissues. Little is known about the exact mechanisms by which fibrillar aggregates are formed and can impair cellular functions leading to cell death. Small molecules that can modulate aggregate formation and/or structure can be powerful tools for studying the aggregate assembly mechanism and toxicity and may also prove to be therapeutic. We describe here a microplate-based high-throughput screening assay for identification of such molecules. The assay is based on the ability of microplate-coated aggregates to grow by incorporating additional monomers. Compounds that influence the elongation reaction are selected as hits and are tested in dose–response experiments. We also discuss some additional experiments that can be used to characterize the modes of action of these aggregation modulators further.

Introduction

Formation and accumulation of protein aggregates are hallmarks of many neurodegenerative disorders, such as amyloid β (Aβ) fibrils in Alzheimer's disease and polyglutamine (polyGln) aggregates in CAG repeat disorders/Huntington's disease. However, the exact mechanisms by which abnormal, aggregation-prone proteins lead to neuronal dysfunction and death have yet to be elucidated. A variety of mechanisms have been proposed, including the ability of aggregates to impair the ubiquitin-proteasome system (Bence et al., 2001); inhibit specific neuronal functions, such as axonal transport and maintenance of synaptic integrity (Bayer et al., 2001; Gunawardena et al., 2003); and sequester critical cellular factors (Nucifora et al., 2001). The structural nature of the toxic aggregates is equally obscure.

Nonnative protein aggregates can take a variety of forms distinguished by their size, stability, morphology, and degree of order within the aggregate (Wetzel, 2006). Aggregates can exist as soluble low-molecular-weight oligomers (Bitan et al., 2003), spheroidal oligomers (Poirier et al., 2002),

METHODS IN ENZYMOLOGY, VOL. 413   0076-6879/06 $35.00

annular and filamentous protofibrils (Lashuel *et al.*, 2002a), and large inclusions whose compositions have not generally been well characterized and whose structures might range from amorphous assemblies to highly ordered, paracrystalline amyloid fibrils (Temussi *et al.*, 2003). Recent studies show that aggregates in their intermediate oligomeric state can exhibit significant toxicity that may rival or surpass that of large fibrils (Demuro *et al.*, 2005; Walsh and Selkoe, 2004). Some large inclusions might even prove to play cytoprotective roles (Arrasate *et al.*, 2004; Caughey and Lansbury, 2003).

One approach for tackling the role of aggregates in disease is to find tools that allow modulation of aggregation mechanisms, including inhibition of fibril formation, elongation, and alteration of intermediate/fibril structure. Development of antibodies, β-sheet breaker peptides, and other peptides that inhibit aggregation has been reported (Heiser *et al.*, 2000; McLaurin *et al.*, 2002; Soto *et al.*, 1998; see Chapter 15 by Sciarretta and colleagues in this volume). The advantage of some of these reagents is that they can be obtained by rational design. However, polypeptide-based therapeutics can present special problems, such as protease sensitivity, poor blood–brain barrier/cell membrane penetrance, low circulating lifetimes, and immunogenicity. An alternative strategy is to focus on small molecules (generally with a molecular weight lower than 500 Da) as inhibitors (Heiser *et al.*, 2002; Lashuel *et al.*, 2002b) or modulators (Williams *et al.*, 2005) of aggregation. In principle, such molecules can be discovered by an *in silico* structure-based design approach, but this is normally challenging and is particularly difficult when little structural information is available (as is the present situation with amyloid and other nonnative aggregates) (Guo, 2005). An alternative strategy is to apply systematic high-throughput screening on a compound library. This allows for the rapid identification of many hits, which can subsequently be sorted and validated by secondary assays.

A variety of high-throughput screening assays have been described (Stockwell, 2002), including microtiter plate–based *in vitro* and cell-based aggregation screens that focus on aggregation itself (Hughes *et al.*, 2001) or its toxic consequences (Apostol *et al.*, 2003). A major challenge in developing a screening assay is to be cognizant of the type of aggregation reaction one is targeting and then to be able to replicate this reaction under tight control. Another challenge is to keep the concentrations of the target protein assembly low, so that one maximizes the possibility of discovering a molecule with a reasonably low K_i (Wood *et al.*, 1996). In the aggregation screen described here, we focused on the elongation step in the growth of highly ordered amyloid/amyloid-like fibrils. Although spontaneous (nucleation-dependent) aggregation of amyloidogenic peptides *in vitro* often requires significant peptide concentrations and long incubation times, seeded amyloid elongation reactions occur without a lag time even at very low concentrations of monomer. If one has a high level of control over assay reproducibility and a very sensitive way of monitoring elongation, the reaction serves as an excellent basis for a high-throughput screening assay. Other assembly intermediates of importance, such as oligomers and pro-tofibrils, grow little or not at all by seeded elongation; thus, an assay like that described here is not advisable.

The assay was originally developed to allow measurement of the elongation kinetics of amyloid and amyloid-like fibrils (Berthelier and Wetzel, 2003; Berthelier et al., 2001). Here, we describe the modification of this assay for high-throughput screening of fibril elongation modulators. This assay is reproducible, sensitive, fast, and economical and allows the screening of thousands of compounds per month.

Methods

The screening assay that we describe here has been validated for elongation of both polyGln aggregates and $A\beta(1-40)$ fibrils and has been optimized for a 96-well format.

Preparation of Working Compound Plates

Libraries of compounds are acquired formatted in 96-well plates with compounds solvated in dimethyl sulfoxide (DMSO). Compounds are ali-quoted from their original 96-well plates to new 96-well working plates preserving their initial locations and are adjusted to a concentration of 1 mM in phosphate-buffered saline (PBS) containing 50% DMSO. Wells in the first and last columns, which do not contain any compounds, are filled with 50% DMSO in PBS and serve as controls. The working plates and lids used are chemical-resistant and low DMSO extractable (Becton Dickinson Labware, San Jose, CA). Once the transfer is complete, the plates are sealed with an aluminium adhesive overlay (Nalge Nunc International) to protect the compounds from any light damage, covered, and stored at $-20°$.

Preparation of Assay Plates

Aggregates and Biotinylated Peptides. Peptides are obtained by custom synthesis from the Keck Biotechnology Center at Yale University (available at: http://info.med.yale.edu/wmkeck/). PolyGln peptides with a nominal repeat length of 30 in the sequence context $K_2Q_{30}K_2$ are purified by reverse-phase high-performance liquid chromatography (RP-HPLC); typical product after purification has a weight average repeat length of

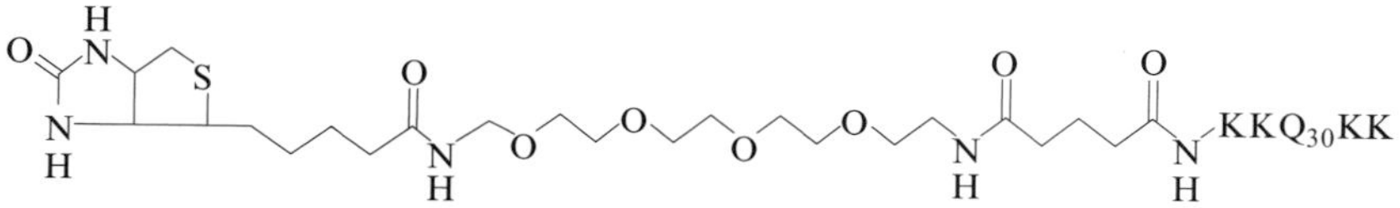

FIG. 1. Structure of the biotin-polyethylene glycol-Q30 peptide.

29 Gln as determined by mass spectrometry. Peptides are obtained with and without the addition of a biotin-polyethylene glycol (PEG) fixed to the N-terminus of the polyGln through the side chain of a modified Gln residue (biotin-PEG; Fig. 1). The PEG-based spacer enhances the signal for an aggregate elongation reaction compared with that of a biotin attached directly to the N-terminus of an analogous peptide (V. Berthelier and S. Shivaprasad, unpublished observations). Perhaps this is attributable to a reduction in steric hindrance that improves streptavidin binding to the recruited biotin molecules. $A\beta(1-40)$ is obtained by the large-scale synthesis option from the Keck Center, whereas biotin-PEG $A\beta(1-40)$ is obtained using the small-scale option and purified by RP-HPLC.

Methods for generation of pure, disaggregated monomeric peptides; their aggregation into elongation-competent aggregates/fibrils; and the routine characterization, handling, and storage of these species have been described (Wetzel, 2005; Williams, 2004; see Chapter 3 by O'Nuallain and colleagues in this volume). Briefly, purified polyGln peptides with and without biotin-PEG tag are dissolved in a 1:1 mixture of trifluoroacetic acid (TFA; Pierce, Rockford, IL) and 1,1,1,3,3-hexafluoro-2-propanol (HFIP; Fisher, Pittsburgh, PA). After complete dissolution, a stream of Argon is applied, the polyGln peptide residue is dissolved in water adjusted to pH 3 with TFA, and the resulting solution is ultracentrifuged to remove any presence of microaggregates. Determination of the exact polyGln peptide concentration of the supernatant is performed by RP-HPLC (Wetzel, 2005; see Chapter 3 by O'Nuallain and colleagues in this volume). Biotin-PEG polyGln peptides are then diluted into extension buffer (PBS and 0.01% Tween 20) with 5% DMSO to a concentration of 100 nM, snap-frozen, and stored at $-80°$. These stock solutions remain aggregate-free for a period up to 4 months. The eventual development of small amounts of small aggregates in the stored monomer solution leads to substantial increases in the fluorescent background signal of the assay, from values in the range of 5000–7000 counts to values of 18,000 counts or more.

Aggregates are prepared from peptides lacking the biotin-PEG tag. Thus, unmodified-polyGln peptides are adjusted to a 10 μM concentration in PBS, incubated for 24 h at $37°$, and then transferred to $-20°$ and incubated an additional 24 h. The resulting aggregates are collected by centrifugation, resuspended in extension buffer at 10 μM (monomer equivalent), aliquoted, snap-frozen, and stored at $-80°$.

$A\beta(1-40)$ peptides, with and without the biotin-PEG tag, are disaggregated by dissolving the peptide in TFA, drying under Argon, dissolving in HFIP, drying under Argon, and then removing residual organic solvent under vacuum. The peptide film is gently dissolved in 2 mM NaOH, and one-half volume of $2\times$ PBS is then added and gently mixed. The solution is subjected to ultracentrifugation at 350,000g for 3 h or overnight (Williams *et al.*, 2004). After determination of the concentration by RP-HPLC, 50 μM of the untagged peptide is incubated for 7 days at $37°$ to make amyloid fibrils. The resulting fibrils are snap-frozen and stored at $-80°$ (see Chapter 3 by O'Nuallain and colleagues in this volume). Biotin-PEG-$A\beta(1-40)$ monomers are stored at $-80°$ at a concentration of 100 nM and are stable (i.e., remain aggregate-free) for at least 1 year.

Preparation of Aggregate Coated Plate. By means of passive adsorption, 40 ng/well of polyGln aggregates or 100 ng/well of $A\beta(1-40)$ fibrils are fixed to an activated enzyme-linked immunosorbent assay (ELISA) microtiter plate (EIA/RIA plates; Costar, Atlanta, GA). Thus, 100 μl of an aggregate suspension at 0.4 μg/ml (for polyGln) or 1 μg/ml (for $A\beta$) in PBS is dispensed in all wells from column 2 through column 12; the eight wells of the first column are only filled with buffer and serve to measure the background signal as well as to test the stability of monomers, as mentioned above. Aggregates are allowed to dry by incubating the microplate uncovered for 17 h at $37°$. After incubation, the wells are washed three times with 300 μl of extension buffer and blocked for 1 h at $37°$ with 0.3% gelatin (Bio-Rad). For this and the following incubation steps, the plate is sealed with an adhesive overlay. After washing, the plates can be used immediately for screening or stored at $4°$ in PBS (100 μl/well) for 1 week without any loss in activity.

Screening Assay

For each working compound plate, screening assays are conducted in triplicate by processing three identical plates in parallel. Screening plates are set up such that the first column contains only monomers (background control); the last column contains aggregates and monomers (reaction control); and the other columns (2 through 11) contain aggregates, monomers, and compounds. In each assay well, 80 μl of extension buffer (PBS and 0.01% Tween 20) as well as 10 μl of compound from a working compound plate are delivered. After 3 min, 10 μl biotin-PEG monomer at 100 nM is added. Thus, the aqueous mixtures in all the wells contain 5%

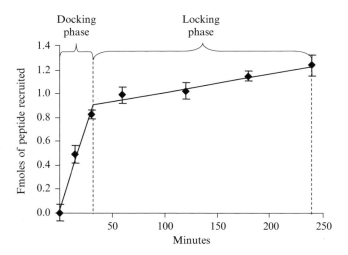

FIG. 2. Typical kinetics of polyglutamine (polyGln) aggregate elongating by recruiting polyGln monomer. K2Q30K2 aggregates at a rate of 40 ng/well were immobilized onto plastic and exposed for various periods to 10 nM biotin-polyethylene glycol-K2Q30K2 peptide. The fast phase of binding ending at about 45 min has been attributed to the initial binding phase of the multistep elongation cycle. Technically, the kinetics were obtained by setting up the individual reaction wells in reverse temporal order, so that the entire plate could be processed at once (see Berthelier *et al.*, 2001).

DMSO. The assay plate is then sealed, shaken gently for 5 s to ensure homogeneity of the different reactants, and placed at 37° for 45 min. (Figure 2 shows that the rapid binding phase of biotin-PEG polyGln to the polyGln aggregate is over after about 45 min; thus, a 45-min incubation gives a large signal that does not grow significantly larger until much longer incubation. This rapid phase appears to correspond to the binding of one polyGln to each growing end, after which slow rearrangement steps are required before more peptide can bind [Berthelier *et al.*, 2001; Esler *et al.*, 2000]). At the end of the 45-min incubation time, the screening plate is washed three times with extension buffer and incubated for 1 h in the dark with 100 μl/well of a europium-streptavidin (Perkin Elmer, Boston, MA) solution prepared at 1 ng/ml in extension buffer containing 0.5% bovine serum albumin (BSA). Three final washings are carried out, and after incubation for 5 min with 100 μl of enhancing solution (Perkin Elmer), europium fluorescence is measured by time-resolved fluorometry in a Victor2 counter (Perkin Elmer). Time-resolved fluorescence maximizes the signal-to-noise ratio by integrating over the long emission decay time of europium while, at the same time, discounting the short lifetime fluorescence signals associated with background from the plastic plate and compounds (Hemmila *et al.*, 1984). The fluorescent counts detected by the microtiter plate reader are then converted to femtomoles using a standard curve established using a calibrated europium solution (Perkin-Elmer). Femtomoles of europium are converted into femtomoles of biotin-PEG monomer recruited into aggregates using the manufacturer-determined seven Eu^{3+} ions per streptavidin molecule.

Data Analysis

Screening data are processed and analyzed. For each compound plate, three assay plates will have been run. An average value of the first column from each plate is calculated that corresponds to the signal background. Reproducibly, the magnitude of the signal background should be ≈ 0.2 fmol. Similarly, for the reaction control, an average value is calculated from the signals from the wells of column 12 in each of the three plates. A signal value of 4 and 10 times higher than the background is usually determined for the $A\beta$ and polyGln recruitment reactions, respectively. As mentioned above, the assays have been run in triplicate. Therefore, for each compound, an average signal from the three plates is calculated and compared with the signal of the reaction control in order to determine percentage inhibition or enhancement. Compounds with an effect greater than or equal to 50% (Fig. 3) are considered hits.

Remarks

In order to process the maximum number of compounds efficiently, the screening assays are run with the help of a Microlab 4000 MPH-48, a high-throughput liquid handling workstation (Hamilton, NV). When using a robot, especially for the preparation of the aggregate substrate plates, some precautions must be taken.

Preparing screening plates requires buffer without aggregates to be dispensed into column one (background reaction). This step must be performed before adding aggregates to the rest of the plate. Otherwise, these wells might be contaminated with aggregates. Even after a standard wash-cycle, aggregate deposition on the 48 separate stainless-steel probes can occur. Once the aggregate-coated plate is prepared and placed in the incubator for drying, a deep cleaning of the probe head is immediately performed: the probes are soaked in a soap solution (Liquinox) for 10 min, rinsed three times with water, soaked in ethanol for an additional 10 min, and rinsed again three times with water. Between each step of the screening phase, including delivery of buffer, compounds, and peptide, the probe head is rinsed three times with water.

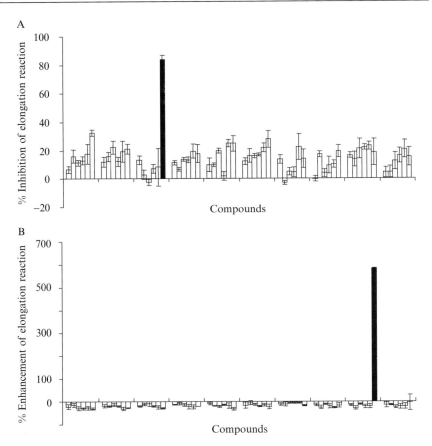

Fig. 3. Example of screening results revealing modulators of elongation. (A) Compound with 84% inhibition. (B) Compound capable of enhancing aggregate formation with an activity of 575%.

Finally, care should be taken in handling biotin-PEG-polyGln peptide. Once the peptide has been thawed at room temperature, which takes about 1 h, the screening assay should be conducted immediately in order to avoid aggregate formation in the biotin peptide stock solution, which can lead to high background in elongation reaction.

Results

Table I shows screening results of the National Institute of Neurological and Communicative Disorders and Stroke (NINCDS) library of 1041

TABLE I
COMPOUND INHIBITORS OF POLYGLN ELONGATION THAT HAVE BEEN IDENTIFIED FROM
HIGH-THROUGHPUT SCREENING OF THE NINCDS LIBRARY

Name	Inhibition (%)	IC$_{50}$ (μM)	Structure
Moxalactam disodium	91	1	
Guaiazulene	73	53	
Gossypol acetic acid	68	25	
Merbromin	62	38	
Protoporphyrin IX	59	114	
Ethynyl estradiol	59	123	
Methacycline	60	123	

compounds (library assembled for the NINCDS Drug Screening Project Assay Consortium in collaboration with the Hereditary Disease Foundation). Inhibitory activities of the hits are confirmed by conducting dose–response curve experiments. As described above, polyGln aggregates are immobilized onto wells and incubated for 3 min with various concentrations of a compound identified as an inhibitor, with the highest concentration starting at 400 μM down to at least 3.025 μM. Following the 3-min incubation, 10 μl of 100 nM biotinylated monomers are added for 45 min, and the rest of the protocol is applied as described above. As shown in Fig. 4, the acquisition of the dose–response curve allows us to determine the IC_{50} value, the concentration of the compound for which 50% of the inhibition effect is observed. If dealing with an enhancer of aggregation, a similar dose–response curve assay can be conducted and an EC_{50} value, the concentration of the compound for which 50% of the effective effect is observed, determined. IC_{50} and EC_{50} values are parameters that can be used to conduct structure–activity relation studies and to determine what chemical functions are required for a compound to be more active than its various homologues.

The high-throughput screening assay described here is a powerful method for identifying compounds that influence aggregate elongation. As described above, the assay will identify compounds that either inhibit or favor additional recruitment of monomer onto preformed aggregates. However, as with all screening assays, secondary screening assays are

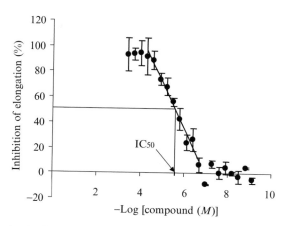

FIG. 4. Example of a dose–response curve allowing the determination of the concentration of the compound for which 50% of the inhibition effect is observed of an identified compound. K2Q30K2 aggregates were exposed at a rate of 40 ng/well of polyglutamine (polyGln) for 45 min to 10 nM biotin-polyethylene glycol-K2Q30K2 peptide plus different concentrations, ranging from 763 nM–400 μM, of a compound identified from screening.

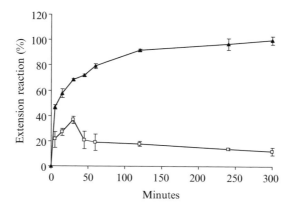

Fig. 5. Wash-out experiment indicating that this compound is acting on the aggregates. Aggregates were pre-exposed for 1 h to (□) compound or buffer (▲), washed, and incubated for various time with biotin-polyethylene glycol-K2Q30K2 as described in the legend for Fig. 2.

required to provide further confirmation, exclude artifacts, and categorize the observed activity. Determination of a dose–response curve serves as an independent confirmation of the result in addition to providing an approximate IC_{50} or EC_{50}. In addition, in order to determine if a compound is acting on aggregates or on monomers, one can conduct a "wash-out" experiment in which aggregates are exposed to the compound and then washed with buffer to remove the unbound compound before the elongation kinetics are measured using biotinylated monomer (Fig. 5). Another important backup assay is to conduct aggregation reactions in the presence of test compound in suspension phase with and without seeding. This eliminates possible artifacts arising from the use of a plastic surface and also tests the possible role of the test compound in the nucleation phase of aggregation (Williams *et al.*, 2005). Aggregates grown in the presence of test compound in suspension phase can be further characterized, for example, for thioflavin T (ThT) binding and compared with aggregates grown in the absence of compound (Williams *et al.*, 2005). Ultimately, cell or animal models can be used to test the compound's effect on aggregation and cytotoxicity *in vivo*.

Acknowledgment

The authors thank Erin Carney, Tina Richey, and Angela D. Williams for their assistance in the screening process as well as the Hereditary Disease Foundation for financial support.

References

Apostol, B. L., Kazantsev, A., Raffioni, S., Illes, K., Pallos, J., Bodai, L., Slepko, N., Bear, J. E., Gertler, F. B., Hersch, S., Housman, D. E., Marsh, J. L., and Thompson, L. M. (2003). A cell-based assay for aggregation inhibitors as therapeutics of polyglutamine-repeat disease and validation in *Drosophila*. *Proc. Natl. Acad. Sci. USA* **100**, 5950–5955.

Arrasate, M., Mitra, S., Schweitzer, E. S., Segal, M. R., and Finkbeiner, S. (2004). Inclusion body formation reduces levels of mutant huntingtin and the risk of neuronal death. *Nature* **431**, 805–810.

Bayer, T. A., Wirths, O., Majtenyi, K., Hartmann, T., Multhaup, G., Beyreuther, K., and Czech, C. (2001). Key factors in Alzheimer's disease: Beta-amyloid precursor protein processing, metabolism and intraneuronal transport. *Brain Pathol.* **11**, 1–11.

Bence, N. F., Sampat, R. M., and Kopito, R. R. (2001). Impairment of the ubiquitin-proteasome system by protein aggregation. *Science* **292**, 1552–1555.

Berthelier, V., Hamilton, J. B., Chen, S., and Wetzel, R. (2001). A microtiter plate assay for polyglutamine aggregate extension. *Anal. Biochem.* **295**, 227–236.

Berthelier, V., and Wetzel, R. (2003). An assay for characterizing *in vitro* the kinetics of polyglutamine aggregation. *Methods Mol. Biol.* **217**, 295–303.

Bitan, G., Kirkitadze, M. D., Lomakin, A., Vollers, S. S., Benedek, G. B., and Teplow, D. B. (2003). Amyloid beta-protein (Abeta) assembly: Abeta 40 and Abeta 42 oligomerize through distinct pathways. *Proc. Natl. Acad. Sci. USA* **100**, 330–335.

Caughey, B., and Lansbury, P. T. (2003). Protofibrils, pores, fibrils, and neurodegeneration: Separating the responsible protein aggregates from the innocent bystanders. *Annu. Rev. Neurosci.* **26**, 267–298.

Demuro, A., Mina, E., Kayed, R., Milton, S. C., Parker, I., and Glabe, C. G. (2005). Calcium dysregulation and membrane disruption as a ubiquitous neurotoxic mechanism of soluble amyloid oligomers. *J. Biol. Chem.* **280**, 17294–17300.

Esler, W. P., Stimson, E. R., Jennings, J. M., Vinters, H. V., Ghilardi, J. R., Lee, J. P., Mantyh, P. W., and Maggio, J. E. (2000). Alzheimer's disease amyloid propagation by a template-dependent dock-lock mechanism. *Biochemistry* **39**, 6288–6295.

Gunawardena, S., Her, L. S., Brusch, R. G., Laymon, R. A., Niesman, I. R., Gordesky-Gold, B., Sintasath, L., Bonini, N. M., and Goldstein, L. S. (2003). Disruption of axonal transport by loss of huntingtin or expression of pathogenic polyQ proteins in *Drosophila*. *Neuron* **40**, 25–40.

Guo, J.-T., Hall, C. K., Xu, Y, and Wetzel, R. (2005). Modeling Protein Aggregate Assembly and Structure. *In* "Computational Methods for Protein Structure Prediction and Modeling" (R. Xu, Y. Xu, and D. Liang, eds.) Springer Press.

Heiser, V., Engemann, S., Brocker, W., Dunkel, I., Boeddrich, A., Waelter, S., Nordhoff, E., Lurz, R., Schugardt, N., Rautenberg, S., Herhaus, C., Barnickel, G., Bottcher, H., Lehrach, H., and Wanker, E. E. (2002). Identification of benzothiazoles as potential polyglutamine aggregation inhibitors of Huntington's disease by using an automated filter retardation assay. *Proc. Natl. Acad. Sci. USA* **99**(Suppl. 4), 16400–16406.

Heiser, V., Scherzinger, E., Boeddrich, A., Nordhoff, E., Lurz, R., Schugardt, N., Lehrach, H., and Wanker, E. E. (2000). Inhibition of huntingtin fibrillogenesis by specific antibodies and small molecules: Implications for Huntington's disease therapy. *Proc. Natl. Acad. Sci. USA* **97**, 6739–6744.

Hemmila, I., Dakubu, S., Mukkala, V. M., Siitari, H., and Lovgren, T. (1984). Europium as a label in time-resolved immunofluorometric assays. *Anal. Biochem.* **137**, 335–343.

Hughes, R. E., Lo, R. S., Davis, C., Strand, A. D., Neal, C. L., Olson, J. M., and Fields, S. (2001). Altered transcription in yeast expressing expanded polyglutamine. *Proc. Natl. Acad. Sci. USA* **98**, 13201–13206.

Lashuel, H. A., Hartley, D., Petre, B. M., Walz, T., and Lansbury, P. T., Jr. (2002a). Neurodegenerative disease: Amyloid pores from pathogenic mutations. *Nature* **418,** 291.

Lashuel, H. A., Hartley, D. M., Balakhaneh, D., Aggarwal, A., Teichberg, S., and Callaway, D. J. (2002b). New class of inhibitors of amyloid-beta fibril formation. Implications for the mechanism of pathogenesis in Alzheimer's disease. *J. Biol. Chem.* **277,** 42881–42890.

McLaurin, J., Cecal, R., Kierstead, M. E., Tian, X., Phinney, A. L., Manea, M., French, J. E., Lambermon, M. H., Darabie, A. A., Brown, M. E., Janus, C., Chishti, M. A., Horne, P., Westaway, D., Fraser, P. E., Mount, H. T., Przybylski, M., and St. George-Hyslop, P. (2002). Therapeutically effective antibodies against amyloid-beta peptide target amyloid-beta residues 4-10 and inhibit cytotoxicity and fibrillogenesis. *Nat. Med.* **8,** 1263–1269.

Nucifora, F. C., Jr., Sasaki, M., Peters, M. F., Huang, H., Cooper, J. K., Yamada, M., Takahashi, H., Tsuji, S., Troncoso, J., Dawson, V. L., Dawson, T. M., and Ross, C. A. (2001). Interference by huntingtin and atrophin-1 with cbp-mediated transcription leading to cellular toxicity. *Science* **291,** 2423–2428.

Poirier, M. A., Li, H., Macosko, J., Cai, S., Amzel, M., and Ross, C. A. (2002). Huntingtin spheroids and protofibrils as precursors in polyglutamine fibrilization. *J. Biol. Chem.* **277,** 41032–41037.

Soto, C., Sigurdsson, E. M., Morelli, L., Kumar, R. A., Castano, E. M., and Frangione, B. (1998). Beta-sheet breaker peptides inhibit fibrillogenesis in a rat brain model of amyloidosis: Implications for Alzheimer's therapy. *Nat. Med.* **4,** 822–826.

Stockwell, B. R. (2002). Chemical genetic screening approaches to neurobiology. *Neuron* **36,** 559–562.

Temussi, P. A., Masino, L., and Pastore, A. (2003). From Alzheimer to Huntington: Why is a structural understanding so difficult? *EMBO J.* **22,** 355–361.

Walsh, D. M., and Selkoe, D. J. (2004). Oligomers on the brain: The emerging role of soluble protein aggregates in neurodegeneration. *Protein Pept. Lett.* **11,** 213–228.

Wetzel, R. (2005). Protein folding and aggregation in the expanded polyglutamine repeat diseases. *In* "The Protein Folding Handbook" (R. Buchner and J. Kiefhaber, eds.). Wiley. Wiley press, N.Y.

Wetzel, R. (2006). Chemical and physical properties of polyglutamine repeat sequences. *In* "Genetic Instabilities and Neurological Diseases" (R. Wells and R. Ashizawa, eds.). Elsevier, San Diego. Elsevier-Academic press, San Diego.

Williams, A. D., Portelius, E., Kheterpal, I., Guo, J. T., Cook, K. D., Xu, Y., and Wetzel, R. (2004). Mapping abeta amyloid fibril secondary structure using scanning proline mutagenesis. *J. Mol. Biol.* **335,** 833–842.

Williams, A. D., Sega, M., Chen, M., Kheterpal, I., Geva, M., Berthelier, V., Kaleta, D. T., Cook, K. D., and Wetzel, R. (2005). Structural properties of Abeta protofibrils stabilized by a small molecule. *Proc. Natl. Acad. Sci. USA* **102,** 7115–7120.

Wood, S. J., MacKenzie, L., Maleeff, B., Hurle, M. R., and Wetzel, R. (1996). Selective inhibition of βA4 fibril formation. *J. Biol. Chem.* 1996. **271,** 4086–4092.

[17] Conformation-Dependent Anti-Amyloid Oligomer Antibodies

By RAKEZ KAYED and CHARLES G. GLABE

Abstract

Although abundant evidence suggests that amyloid accumulation plays a significant role in the pathogenesis of degenerative disease, the mechanism of amyloid formation and toxicity remains elusive. Early hypotheses for disease pathogenesis proposed that large amyloid deposits, which are composed primarily of 6–10-nm mature amyloid fibrils, were the primary causative agent in pathogenesis, but this hypothesis required modification to consider the central role of oligomers or aggregation intermediates, because the accumulation of these large aggregates does not correlate well with pathogenesis. Recent evidence supports the hypothesis that small soluble aggregates representing intermediates in the fibril assembly process may represent the primary culprits in a variety of amyloid-related degenerative diseases. Investigating the role of soluble amyloid oligomers in pathogenesis presents a problem for distinguishing these aggregates from the mature fibrils, soluble monomer, and natively folded precursor proteins, especially *in vivo* and in complex mixtures. Recently, we generated a conformation-specific antibody that recognizes soluble oligomers from many types of amyloid proteins, regardless of sequence. These results indicate that soluble oligomers have a common, generic structure that is distinct from both fibrils and low-molecular-weight soluble monomer/dimer. Conformation-dependent, oligomer-specific antibodies represent powerful tools for understanding the role of oligomers in pathogenesis. The purpose of this chapter is to review the methods for the production, characterization, and application of this antibody to understanding the contribution of amyloid oligomers to the disease process.

Common Pathways of Amyloid Formation

In order to produce and characterize conformational antibodies, it is critical to have homogeneous and relatively stable populations of each amyloid species or conformer. This can be difficult because each species can undergo conformational transitions over time that lead to the formation of new species and result in mixed populations of conformers. This requires the establishment of conditions that favor the formation and stability of a particular conformation. The approach to this problem is

METHODS IN ENZYMOLOGY, VOL. 413
0076-6879/06 $35.00
DOI: 10.1016/S0076-6879(06)13017-7

necessarily empirical, analogous to finding the optimum conditions for protein renaturation or crystallization. Because amyloids arise from many different protein sequences, these conditions might be expected to be considerably different for different amyloids. However, recent evidence indicates that amyloid formation is governed largely by the generic polymer properties of the polypeptide backbone and that many proteins can form amyloids if they are partially or completely unfolded (Dobson, 1999). The specific amino acid sequence also plays a role, primarily by favoring the formation of stable native structures that are nonamyloidogenic (Chiti *et al.*, 2003; DuBay *et al.*, 2004). The understanding that amyloid formation is a generic polymer property of polypeptides suggests that there is a common pathway for amyloid aggregation and fibril formation and that the problem of establishing conditions for the formation and stabilization of different conformations may not be hopelessly complex.

It is remarkable that so many different amyloidogenic proteins form not only fibrils but small oligomers that appear to represent intermediates in the fibril formation process. Ultrastructural studies of fibril formation *in vitro* revealed a number of soluble intermediates, including spherical oligomers, protofibrils, and annular protofibrils, which appear as ring-like pore structures (Harper *et al.*, 1997; Lashuel *et al.*, 2003). Spherical oligomers appear at early times of incubation and then disappear as mature fibrils appear (Anguiano *et al.*, 2002; Harper *et al.*, 1997; Lashuel *et al.*, 2002). Although spherical oligomers are kinetic intermediates, it is not yet clear whether they coalesce directly to form fibrils or whether they buffer the amyloidogenic monomer concentration. These soluble intermediates have been referred to as micelles, protofibrils, prefibrillar aggregates, and Aβ-derived diffusable ligands (ADDLs) (Bucciantini *et al.*, 2002; Harper *et al.*, 1997; Lambert *et al.*, 1998; Lomakin *et al.*, 1997; Soreghan *et al.*, 1994; Walsh *et al.*, 1997). At later times of incubation, curvilinear structures form, which appear to be strings of spherical subunits that have a beaded appearance. These structures have also been commonly called "protofibrils" (Harper *et al.*, 1997). Spherical oligomeric intermediates have been identified in many other types of amyloids, suggesting that they represent a common state of assembly and aggregation for many different amyloids (Dobson, 2001). Additionally, it has been reported that some disease-related mutations accelerate and enhance the formation of oligomers and other nonfibrillar aggregates (Cardoso *et al.*, 2002; Conway *et al.*, 2000; Helms and Wetzel, 1996; Matsubara *et al.*, 2005; Nilsberth *et al.*, 2001; Pepys *et al.*, 1993). These mutations did not alter the normally folded structure of the proteins so as to facilitate fibril formation directly (Kelly, 1996). Once the amyloid fibril lattice has been established, it can grow by the addition of monomer onto the ends of the fibrils (Collins *et al.*, 2004; Tseng *et al.*, 1999).

An Antibody That Recognizes a Generic Epitope Common to
  Amyloid Oligomers

Although oligomers have been observed for many different types of
amyloids, the relations between the different types of oligomeric intermediates, their roles in fibril formation and growth, and their contributions
to disease pathogenesis are only now being clarified. The idea that amyloid
oligomers represent the primary toxic species of amyloids rather than the
mature fibrils arose as a consequence of the failure of fibril accumulation to
account for pathogenesis as noted above. A number of studies provided
evidence that amyloid β (Aβ) oligomers are, in fact, more toxic to cells
than mature fibrils (Klein et al., 2001). Increasing evidence indicates that
soluble amyloid oligomers are generically toxic whether they are derived
from disease-related proteins or not (Bucciantini et al., 2002). The idea that
soluble Aβ oligomers are the primary toxic species is an attractive explanation that can reconcile the seemingly inconsistent evidence that soluble
amyloid oligomers are causally related to disease, whereas the large fibrillar aggregates are not. However, the evidence in support of this hypothesis
is largely correlative.

It is difficult to assess the role of soluble oligomers in disease pathogenesis in vivo directly because the oligomers only differ from the native
protein, soluble monomer, and mature fibrils in their conformation or
aggregation state. To study these differences in vivo, we would need some
way of distinguishing these conformations. Conformation-dependent antibodies that distinguish between the different forms would be very useful.
We produced a conformation-dependent antibody that is specific for soluble oligomers and does not recognize natively folded proteins, monomers,
or fibrils (Kayed et al., 2003). Surprisingly, this antibody also recognizes
soluble oligomers from a wide variety of amyloid-forming peptides and
proteins equally well (Kayed et al., 2003). This suggests that the antibody
recognizes a peptide backbone epitope that is common to amyloid oligomers of different sequences and is independent of the amino acid side
chains. Similar antibodies specifically recognizing "generic" epitopes that
are common to fibrils from several different types of amyloids have been
previously described (Hrncic et al., 2000; O'Nuallain and Wetzel, 2002).
Recent studies indicate that the antifibril antibody WO-1 also recognizes
protofibrils, indicating that they may have a structure related to fibrils
(Williams et al., 2005). Because the anti-oligomer antibody does not recognize mature fibrils, this suggests that the prefibrillar and fibrillar generic
conformational epitopes are distinct. An interesting potential explanation for this unusual specificity is the possibility that the anti-oligomer
antibody recognizes a unique "alpha extended chain" pleated sheet

structure. This has been suggested based on molecular dynamics simulations indicating that the small soluble oligomers may preferentially form α-sheet aggregates (Armen et al., 2004).

The availability of an antibody that can specifically recognize the soluble oligomeric state and distinguish it from the native conformation, random coil monomer, and fibrillar states facilitates an examination of the localization of oligomers in disease and in cellular and animal models of pathogenesis. It also provides a practical tool for quantifying oligomer formation and identifying the presence of amyloid oligomers in diseases that are not known to be associated with amyloid. The anti-oligomer antibody detects a unique subset of amyloid deposits in Alzheimer's disease (AD) brain but not in nondemented control brain. Importantly, the oligomeric deposits do not co-localize thioflavin-S–positive fibrillar amyloid plaques (Kayed et al., 2003). The staining is predominantly extracellular and does not appear to be preferentially associated with any cell type. It remains to be established whether their presence is a good predictor of disease onset or progression.

The fact that many different types of oligomers react with one antibody argues for common sequence-independent conformation, which is a generic and fundamental property of polypeptide chains (Fandrich and Dobson, 2002; MacPhee and Dobson, 2000). This is an important insight because it suggests that the mechanism of toxicity is common to all amyloid diseases. Thus, a generic conformation inducing a similar mechanism of toxicity suggests that an anti-oligomer antibody may be a viable strategy for preventing amyloid toxicity.

Preparation of Homogeneous Populations of Stable Oligomers

Reasonably homogeneous populations of conformationally pure peptides are critical for producing and characterizing conformation-dependent antibodies. Pure and homogenous stocks are important not only for producing a specific immune response but for characterizing the specificity of antibodies and purifying them by affinity chromatography. The approach to this problem is necessarily empirical, involving a lot of systematic work, but several common themes that facilitate the preparation of pure solutions of oligomers have emerged from this work.

One of the most important factors is to start with completely solvated stock solutions that lack fibril seeds (see Chapter 2 by Teplow and Chapter 3 by O'Nuallain and colleagues in this volume). The presence of even tiny amounts of fibril seeds shifts the equilibrium toward the production of mature fibrils by favoring the addition of amyloidogenic monomer onto the ends of the fibril lattice rather than forming oligomers. Thus, the presence of small amounts of fibril seeds can drastically reduce the yield and stability of the oligomer preparation. A variety of initial solvation conditions have been shown to be effective for completely eliminating fibril seeds, including trifluoroacetic acid (Zagorski et al., 1999), NaOH (Fezoui et al., 2000), and fluorinated alcohols (Burdick et al., 1992; Dahlgren et al., 2002). Low pH (pH 2.5–4) has proven to be a useful condition for oligomerization of a variety of different amyloidogenic proteins (Bocharova et al., 2005; Kayed et al., 2003; Khurana et al., 2001; Lomakin et al., 1997). Low ionic strength is also generally useful in producing homogeneous samples of oligomers (Kayed et al., 2003). The reason for the general utility of these conditions is not clear, but our experience is that salt often causes the formation of larger aggregates, clusters of oligomers, or bundles of fibrils rather than discrete, individual structures. Stirring is important because it accelerates aggregation and makes the kinetics more consistent (Jarrett et al., 1993).

One common problem with amyloid oligomers is that they are unstable and disappear as mature fibrils form in solution (Harper et al., 1997; Kayed et al., 2003). Regardless of the method used for preparation, intermediates were detected at some time point when prepared from a seedless sample. Therefore, it is extremely important to establish the time course of oligomerization. In optimizing the conditions for oligomer formation, kinetics of amyloid formation are very sensitive to solvent conditions. Freezing and thawing of stock solutions should be avoided, because the samples tend to accumulate large clumps or bundles even if the samples do not change conformation. Storage at 4° or at room temperature in the presence of 0.02% NaN_3 as an antimicrobial agent is the preferred method of storage. Tubes should be sealed to avoid drying of the sample on the sides of the tube. One general means of stabilizing the oligomers after they have been formed is to add NH_4OH to a final concentration of 0.1%, (final pH 9.5–10.5). This stabilization method works for all types of amyloid oligomers that we have examined (Kayed et al., 2003). Solutions stabilized in this manner have remained stable populations of pure spherical oligomers for several months based on both morphological and antibody-binding criteria. Brief sonication of the stock solutions for 30 s can serve to disperse and homogenize the samples before use. For toxicity or in vivo assays, samples can be dialyzed (1–10-Kd cutoff) against 10 mM Na_2CO_3, pH 9–10, immediately before use to remove the azide or residual solvent.

Conformationally pure and homogeneous solutions of mature fibrils are also important because they are conformationally distinct from oligomers by antibody binding and morphology; thus, they represent a negative control for oligomer-specific conformation-dependent antibodies. Fibrils are typically prepared under the same solvent conditions as oligomers, except they are incubated with stirring for a longer period until oligomers are no longer detected either morphologically or by antibody binding (Kayed *et al.*, 2003). Fibrils may be prepared under many different conditions, but those prepared at low pH and low ionic strength resulted in more homogeneous populations containing fewer larger aggregates or fiber bundles. Using other conditions, the samples frequently contained large aggregates or fiber bundles and sometimes were contaminated with oligomers, as determined by electron microscopy (EM) analysis and anti-oligomer antibody.

Novel Nanoparticle Molecular Mimic of the Spherical Amyloid Oligomers

Coupling of nanoparticles and biomolecules has many applications in both nanotechnology research and biomolecule research, such as using biomolecules for nanoparticle organization, assembly, and stabilization (Niemeyer, 2001). Nanoparticles are used in protein-based recognition systems, DNA oligomerization and templating, gene-gun technology biomimetic systems, biotemplating, biolabeling, and the construction of model systems (Niemeyer, 2001). Previous studies indicated that Aβ oligomerization coincides with the formation of a hydrophobic environment into which hydrophobic fluorescent dyes partition (Soreghan *et al.*, 1994). It was also demonstrated that the hydrophobic carboxyl terminus of soluble oligomeric Aβ is much more shielded from the aqueous environment than it is in the fibrillar state, suggesting that it is buried in the interior of the oligomers (Garzon-Rodriguez *et al.*, 2000). Using the information about the parallel, in-register organization of Aβ in fibrils (Benzinger *et al.*, 1998) and soluble oligomers (Garzon-Rodriguez *et al.*, 2000), we were able to mimic the oligomeric structure by modifying the carboxyl terminus of Aβ to a thioester and covalently coupling it via the thiol moiety to colloidal gold nanoparticles (Fig. 1A).

The synthesis of Aβ40 C-terminal thioester analogues (Box I) was carried out according to the methods described (Ingenito *et al.*, 1999). The first amino acid was manually coupled to the sulfamylbutyry-AM-PEGA (aminomethyl-polyethylene-glycol-dimethylacrylamide Co-polymer) resin (Novabiochem, San Diego, CA), and 1 g (0.28 mM) of resin in 10 ml dichloromethane (DCM) and 5 eq of the first amino acid were added (Fmoc-Ala-OH for Aβ42, Fmoc-Val-OH for Aβ40). To this, 10 eq diisopropyl ethylamide (DIEA) was added, stirred for 20 min at room temperature, and then cooled using ice and salt to $-10°$ to $-20°$. Next, 4.7 eq (Benzatriazole-1-gl-oxy-tris-pyrolidino-phosphonium hexafluorophosphate) (PyBop) was added and stirred for 8–9 h at $-10°$ to $-0°$. The coupling efficiency was checked using the Kaiser test, and the substitution level was found to be around 0.18–0.20 mM/g, as determined using the Fmoc cleavage method. Acetylation was then performed using acetic anhydride. The sequence was elongated by fluoren-9-ylmethoxy carbonyl chemistry using a continuous flow semiautomatic instrument.

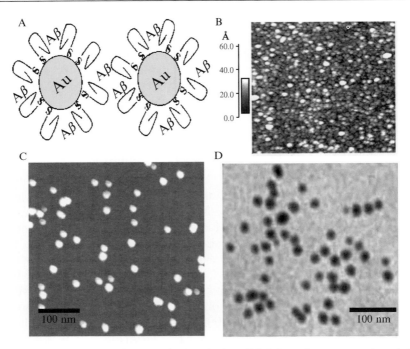

FIG. 1. Antigen preparation and characterization. (A) Schematic representation of the antigen. (B) Atomic force microscopy (AFM) image of Aβ40 spherical oligomers. (C) AFM image of the antigen. (D) Electron microscopy image of the antigen.

Box I

Aβ40
   H₂N-DAEFR⁵HDSGY¹⁰EVHHQ¹⁵KLVFF²⁰AEDVG²⁵SNK-GA³⁰IIGLM³⁵VGGVV-CO-OH

Aβ40 Carboxy Terminal Thioester
   H₂N-DAEFR⁵HDSGY¹⁰EVHHQ¹⁵KLVFF²⁰AEDVG²⁵SNK-GA³⁰IIGLM³⁵VGGVV-CO-SR

Activation of the Resin and Displacement by Thiols

The fully protected peptide on the resin (100 mg of resin, 0.18 m*M*/g) was placed in peptide synthesis vessel and washed five times with N-methyl-2-pyrrolidone (NMP). It was then treated with 5.0 ml NMP,

185 μl i-Pr$_2$EtN (1.1 mM), and 400 μl iodoacetonitrile (previously filtered through an alumina basic plug in the dark). The reaction mixture was shaken for 24 h in the dark on a rotary plate. The resin was then washed five times with NMP, five times with dimethyl-formamide (DMF), and five times with CH$_2$Cl$_2$ and then dried. The mixture was then activated with trimethylsilyldiazomethane resin (100 mg of resin, 0.18 mM/g), washed five times with tetrahydra furon (THF), and then treated with 2.7 ml THF and 2.7 ml TMS-CH$_2$N$_2$ (50%, v/v, in hexane). After stirring for 2 h, the resin was washed five times with 5 ml THF and five times with 5 ml DMF and dried under vacuum ready for the displacement reaction.

Displacement Reaction

Ethyl-3-mercaptopropionate (120 μl) and (800 μl) in CH$_2$Cl$_2$ were added to the dry activated resin. The mixture was shaken on a rotary plate for 24 h. The resin was filtered and then washed with 3 ml × 3 ml DMF. The filtrate and washes were collected, and the rotary was evaporated at 34°. The yield was approximately 60%. Using 1.1 mg sodium thiophenate as a catalyst improved the yield to more than 75%. The resulting fully protected peptide was deprotected using standard methods and purified by reverse phase high-performance liquid chromatography (RP-HPLC). The purity was checked by analytical RP-HPLC and electrospray mass spectrometry and was determined to be approximately 95%.

Preparation and Characterization of the Antigen

Gold colloids (5.3-nm mean diameter) were obtained from Ted Pella, Inc. The colloid was collected at 40,000g and washed two times with distilled water to remove the preservatives. The washed colloidal gold particles were mixed with a freshly prepared solution of 0.2 mg/ml Aβ40 thioester, (pH 5.0–5.5, 25 ml Aβ solution to 20 ml of washed gold colloids in water (approximately 1.0 × 10^{14} particles per milliliter of water). The initial peptide solution was in a monomeric or dimeric state, as determined by size exclusion chromatography. After 3 h of incubation at room temperature, the pH was adjusted to 7.4 with 100 mM Tris pH 8.0 (0.02% sodium azide). After incubation for 6 h at room temperature, the antigen was collected by centrifugation at 30,000g at 4° for 30 min, washed three times with phosphate-buffered saline (PBS), pH 7.6 to remove any unincorporated Aβ, and then redispersed in distilled H$_2$O (0.02% sodium azide). Before use, the antigen was washed three times with PBS, pH 7.4 and resuspended in PBS at the desired concentration.

To characterize the antigen, samples were examined by atomic force microscopy (AFM), EM (Fig. 1), and ultraviolet (UV)/visible spectroscopy (data not shown). We compared the structure organization of the

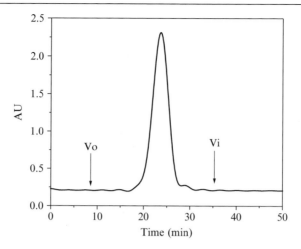

FIG. 2. Purity of amyloid β (Aβ)40 oligomers. Size exclusion chromatogram of pure Aβ40 oligomers. Oligomers were prepared as described in the methods section at a concentration of 0.3 mg/ml in distilled H_2O (pH < 5). Oligomers were applied to a Superdex 75-gel filtration column, with a flow rate of 0.4 ml/min and eluting buffer of 10 mM Tris, 50 mM NaCl (pH 7.4) containing 0.02% sodium azide. The absorbance was monitored at 280 nm. Void Volume (Vo); Included volume Vi.

gold-coupled peptide with that of naturally occurring Aβ oligomers (Fig. 1B–D) and with samples of natural Aβ spherical oligomers prepared by size exclusion chromatography (Fig. 2). The AFM and EM images show that the antigen is uniform and homogeneous, free of any fibril-like structures. Using circular dichroism (CD) (Fig. 3), thioflavin T fluorescence (Fig. 4), and B is-Anilino-Sulfonic Acid (Bis-ANS) (Fig. 5), the properties of the gold-conjugated Aβ peptide are remarkably similar to those of the naturally occurring Aβ oligomers. The CD spectrum gives a minimum at 218 nm, which is characteristic for β-sheet structure (the dominant secondary structure in amyloid protein aggregates) and very similar to the naturally occurring spherical oligomers (Fig. 3). The antigen binds thioflavin T and Bis-ANS, producing spectra that are very similar to authentic oligomers.

Antibody Production and Characterization of Conformational Specificity

Rabbits were immunized with the antigen, and the serum was tested by dot blot and enzyme-linked immunosorbent assay (ELISA), using soluble Aβ, Aβ oligomeric species, and fibrils (Kayed *et al.*, 2003). In order to characterize a conformation-dependent antibody, it is necessary to have homogeneous, conformationally pure solutions with which to characterize

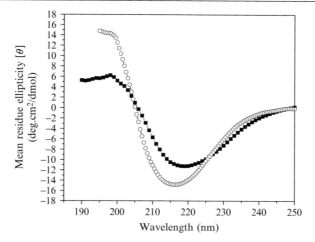

FIG. 3. Circular dichroism (CD) spectra of the antigen compared with natural oligomers (■). Antigen CD spectrum at a concentration of 0.25 mg/ml in phosphate-buffered saline (PBS) (○). Amyloid β (Aβ)42 oligomer CD spectrum at a concentration of 5 μM in PBS. Oligomers were prepared in H₂O at 0.3 mg/ml.

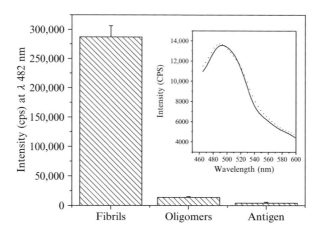

FIG. 4. Thioflavin T binding to amyloid β (Aβ)40 fibrils, Aβ40 natural oligomers, and antigen. Ten microliters of 58-μM solution was added to 130 μl of 3 μM thioflavin T in 10 mM sodium phosphate. (Insert) Thioflavin T binding to the antigen and natural oligomers (solid line); 20 μl of antigen (0.5 mg/ml in phosphate-buffered saline) was added to 130 μl of 3 μM thioflavin T in 10 mM sodium phosphate buffer, pH 6.5 (dotted line). Ten microliters of natural Aβ40 oligomers (0.3 mg/ml in H₂O) was added to 130 μl of 3 μM thioflavin T in 10 mM sodium phosphate buffer, pH 6.5. Excitation of 442 nm, emission of 465–600 nm.

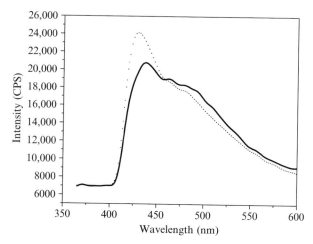

Fig. 5. Bis-ANS binding to the antigen compared with natural oligomers. (Solid line) Twenty microliters of antigen (0.5 mg/ml in phosphate-buffered saline) was added to 130 μl of 1 μm bis-ANS in 30 mM citrate, pH 2.5. (Dotted line) Ten microliters of natural Aβ40 oligomers (0.3 mg/ml in H$_2$O) was added to 130 μl of 1 μm bis-ANS in 30 mM citrate, pH 2.5. Excitation of 355 nm, emission of 365–600 nm.

the antibody specificity. All rabbits injected with the antigen showed the same immune response and produced the same antibody oligomer-specific antibody.

Both the serum and the affinity-purified antibody have no reactivity toward soluble Aβ or other soluble proteins tested. Moreover, the antibody does not react with fibrils immobilized on microplates. Rather, this antibody recognizes a conformation-dependent (three-dimensional) epitope found only on the spherical oligomer aggregates and protofibrils that is not displayed on the soluble Aβ monomer/dimer or amyloid fibrils. An advantage of covalently coupling the peptide antigen to colloidal gold is that its structure is stabilized. The peptide cannot aggregate further to form fibrils and cannot dissociate to form monomer because it is covalently constrained. This limits the conformational heterogeneity of the sample so that the exposure of multiple, different epitopes is minimized. Covalently bonding the peptide to colloidal gold also seems to restrict the processing of the antigen and to make the immune response more specific. Repeated boosting has not given rise to the appearance of substantial amounts of traditional, sequence-specific anti-Aβ antibodies that are conformation insensitive. When animals are immunized against Aβ peptides, a broad range of antibody specificities are obtained, corresponding to epitopes at both ends and the middle.

The colloidal gold-conjugated antigen also provides a facile means of affinity purifying the oligomer-specific antibodies from serum. The IgG fraction of serum is purified on a protein G Sepharose column and then incubated with the oligomer molecular mimic at 4° overnight. The gold particles containing bound antibody are washed twice in PBS by centrifugation at 10,000g for 10 min, and the antibody is eluted by the addition of 1 ml of 0.2 M glycine, pH 2.8. The eluted antibodies are separated by sedimenting the gold particles, and the antibody is neutralized by the addition of 50 mM Tris, pH 7.4. The antibodies are dialyzed against PBS containing 0.02% sodium azide as a preservative, and aliquots are stored frozen at $-20°$. Thawed aliquots are stored at 4° and not refrozen. The colloidal gold Aβ particles can be regenerated by washing them in 0.2 M glycine, pH 2.8, followed by washing with PBS containing 0.02% sodium azide; they are then stored at 4° and reused for antibody purification.

Spherical soluble oligomers have been observed for many different types of amyloids. When soluble oligomers formed from a wide variety of amyloidogenic proteins were tested for reactivity with the anti-oligomer antibody, all reacted equally well, regardless of sequence. No reactivity was observed with either the soluble low-molecular-weight species or the fibrils by ELISA and dot blots for any of the amyloids tested (Kayed et $al.$, 2003).

The specificity of the anti-oligomer antibody was also examined by screening lysates of SH-SY5Y cells for any cellular proteins that might react with the antibody by dot blot and ELISA assays. No detectable reactivity was observed for 2.8 μg of total cell protein on dot blots (Kayed et $al.$, 2003) or with 20 μg of total cell protein by ELISA assay, although approximately 1 ng of soluble Aβ oligomers is readily detected when mixed with total cell lysate. The detection of the added soluble oligomers when mixed with cell lysate depends on the presence of protease inhibitors, suggesting that soluble oligomers are sensitive to proteolysis (Kayed et $al.$, 2003).

Applications of Oligomer-Specific Conformation-Dependent Antibodies

One of the most significant issues facing scientists working on degenerative diseases is the role of amyloid accumulation in pathogenesis, and particularly whether amyloid oligomers or prefibrillar aggregates represent the primary species responsible for cell toxicity and pathogenesis. The availability of conformation-dependent antibodies that specifically recognize amyloid oligomers or prefibrillar aggregates provides a facile means of localizing and quantifying oligomer levels in tissues and complex biological samples. The ability to specifically distinguish oligomers from fibrils, monomer, and natively folded precursor proteins is a powerful tool for assessing the significance of oligomers in pathogenesis and identifying other diseases in which oligomers may play a role. Conformation-dependent antibodies are useful for distinguishing the specific misfolded conformation from a vast excess of native and other folded structures in complex mixtures or tissues. The anti-oligomer antibody has been utilized to localize oligomers in human AD brain and transgenic animal models of AD, using immuno-histochemistry (Crowther et al., 2005; Kayed et al., 2003; Kokubo et al., 2005; Oddo et al., 2003; Yang et al., 2005). Because of its unusual specificity for a generic epitope associated with many different types of amyloid oligomers, it should be useful for examining the localization and patho-genic significance of oligomers for other types of amyloid-related degen-erative diseases. Indeed, it may be possible to discover new diseases that were not previously known to be associated with amyloid oligomers on the basis of staining with anti-oligomer antibody. Recently, the α-B-crystallin–containing deposits in desmin-related cardiomyopathy were shown to stain intensely with anti-oligomer antibody in both human disease tissue and transgenic animal models, suggesting that the pathogenesis of this disease is also related to the accumulation of oligomeric forms of amyloid (Sanbe et al., 2004). The anti-oligomer antibody was used to study age-related macular degeneration (AMD), where it was found to stain diffuse deposits in the center of the Druzen that accumulates in the disease (Luibl et al., 2004). These recent findings suggest there may be more diseases that remain to be recognized as being related to the accumulation of amyloid oligomers.

One of the potential problems in utilizing conformation-dependent antibodies in immunohistochemical and immunofluorescence staining is that conditions required for fixation and tissue processing may denature or unfold the epitope recognized by the antibody. Because the epitope is sensitive to denaturing conditions, it would be best to use fixation and staining conditions that minimize the potential for protein denaturation. Lack of specific staining is difficult to interpret in the absence of positive controls for the particular type of amyloid oligomer to be examined. In order to trouble-shoot problems with lack of staining systematically, a positive control of purified oligomers is valuable. With such a positive control, conditions for fixation, antigen retrieval, and staining can be sys-tematically evaluated for their compatibility with antibody detection. Amyloid oligomers can be formed from many different proteins if they are partially or completely unfolded (Chiti et al., 1999). This presents a potential problem in artifactually creating amyloid oligomer immunoreac-tivity as a consequence of protein denaturation under sample preparation and antigen retrieval conditions that are often denaturing. Using the natively folded protein as a negative control should help to rule out this possibility. The generic oligomer specificity also presents some ambiguity as to what specific type of amyloid oligomer is being detected, but this can be clarified by double labeling with a second, sequence-specific antibody (Kayed *et al.*, 2003).

Some amyloids are derived from large proteins, such as tau and huntingtin, where only a small region of the polypeptide may be involved in forming the amyloid lattice. This may present problems for detection of amyloid oligomers if the polypeptide segments outside of this lattice mask the epitope or interfere with antibody binding. Amyloid oligomers formed from large proteins may be particularly challenging for detection with conformation-dependent antibodies, because conditions must be carefully chosen that reveal the epitope without denaturing it.

Another important application of anti-oligomer antibodies is quantifying the amounts of oligomeric species by ELISA assay or dot blots (Kayed *et al.*, 2003). The ability to evaluate oligomer levels quantitatively in biological samples is very important for determining their contribution to disease pathogenesis and evaluating the effect of treatments that target oligomer levels. One of the problems with using the generic anti-oligomer antibody for quantitative assays is that it does not discriminate between different types of amyloid oligomers. A simple way of overcoming this issue would be to employ a sandwich ELISA, using a sequence-specific antibody to either capture or read-out a particular type of amyloid oligomer. The anti-oligomer antibody also provides an easy readout of oligomer formation during fibril assembly and a means of screening for drugs that specifically inhibit oligomer formation. Shorter and Lindquist (2004) used the anti-oligomer antibody to clarify the enigmatic role of Hsp104 in both formation and elimination of self-replicating Sup35 [PSI+] yeast prions and the role of soluble oligomers in nucleating amyloid fibril formation.

Anti-oligomer antibody has also been used to examine the effect of potential drugs on oligomer levels in animal models (Yang *et al.*, 2005). Antibodies can be used for medium- to high-throughput screens for compounds that prevent oligomer formation, disaggregate preformed oligomers, or accelerate the conversion of oligomers to amyloid fibrils. Compounds with any of these activities may be useful for targeting amyloid oligomers *in vivo*.

It is widely believed that the soluble oligomeric forms of amyloids may represent the primary pathogenic species in many different amyloid-related degenerative diseases. The anti-oligomer antibody has been shown to block the toxicity of many different amyloid oligomers *in vitro* (Kayed *et al.*, 2003). If anti-oligomer antibodies are capable of either blocking the toxicity of oligomers or facilitating their clearance *in vivo*, they may be useful therapeutic agents as a result of either active vaccination or passive immunization. Vaccination against amyloid Aβ42 fibrils has produced promising results in transgenic animal models and human clinical trials, but the first-generation vaccines were associated with inflammatory side effects in human trials. Specifically targeting amyloid oligomers for vaccine development may offer some significant advantages over other vaccination strategies. Because the generic oligomer epitope is specific to misfolded protein structures that are specific to pathological states, the antibodies do not recognize the natively folded proteins or the mature amyloid fibrils that accumulate in disease. Because of this, the antibodies would not be expected to be associated with autoimmune inflammatory complications. Because the oligomer epitope is not represented in native protein structures, it may not be susceptible to tolerance mechanisms that suppress the immune response to native protein sequences. This may explain why a high-titer immune response against the oligomer mimic antigen may be obtained without the addition of adjuvants (C. G. Glabe and R. Kayed, unpublished observations). The amount of soluble amyloid oligomers is generally much lower than the amount of mature amyloid fibrils that accumulates in the disease; thus, lower concentrations of oligomer-specific antibodies may be effective in neutralizing or clearing them. Antibodies directed against amino terminal epitopes of Aβ that bind to fibrillar Aβ plaques are also associated with an elevated incidence of microhemorrhage in transgenic animals (Pfeifer *et al.*, 2002). Because anti-oligomer antibodies do not bind to mature plaque deposits, they may avoid this type of complication as well. Because many different types of oligomers display this generic oligomer epitope, a single vaccine may be an effective therapeutic agent for several different amyloid-related degenerative diseases. If this possibility turns out to be true, vaccination against amyloid oligomers may be broad-spectrum "silver bullet" therapy for many significant age-related human diseases.

Conclusions

Vaccination of animals with a novel molecular mimic of amyloid oligomers gives rise to unique antibodies that recognize a generic peptide backbone epitope common to amyloid oligomers of diverse sequences. These antibodies are capable of specifically targeting and neutralizing amyloid oligomers that are widely believed to represent the primary toxic species of amyloids. The anti-oligomer antibody has several potential uses: (1) to localize and quantify oligomers to determine their role in pathogenesis, (2) to examine the pathway of amyloid aggregation and test the role of oligomers in the fibril formation pathway and in pathogenic mechanisms, (3) as a diagnostic tool to evaluate the levels of amyloid oligomers in complex mixtures and biological fluids, (4) to screen for compounds that can prevent the formation or reverse the aggregation of these toxic intermediates, and (5) as a vaccine to produce therapeutic antibodies that neutralize the toxicity of amyloid oligomers or enhance their clearance. Thus, the antigen may be useful for the development of a specific vaccine, or the antibody may prove useful for passive immunization therapeutic approaches, especially because it specifically targets the toxic intermediates without any reactivity toward the natively folded proteins.

Acknowledgments

This work was supported by National Institute of Health grants AG00538 and NS31230 and a grant from the Larry L. Hillblom Foundation (to C.G. Glabe). The authors are grateful to C.G. Glabe's laboratory staff members for their help and suggestions.

References

Anguiano, M., Nowak, R. J., and Lansbury, P. T., Jr. (2002). Protofibrillar islet amyloid polypeptide permeabilizes synthetic vesicles by a pore-like mechanism that may be relevant to type II diabetes. *Biochemistry* **41**, 11338–11343.

Armen, R. S., DeMarco, M. L., Alonso, D. O., and Daggett, V. (2004). Pauling and Corey's alpha-pleated sheet structure may define the prefibrillar amyloidogenic intermediate in amyloid disease. *Proc. Natl. Acad. Sci. USA* **101**, 11622–11627.

Benzinger, T. L., Gregory, D. M., Burkoth, T. S., Miller-Auer, H., Lynn, D. G., Botto, R. E., and Meredith, S. C. (1998). Propagating structure of Alzheimer's beta-amyloid(10-35) is parallel beta-sheet with residues in exact register. *Proc. Natl. Acad. Sci. USA* **95**, 13407–13412.

Bocharova, O. V., Breydo, L., Parfenov, A. S., Salnikov, V. V., and Baskakov, I. V. (2005). *In vitro* conversion of full-length mammalian prion protein produces amyloid form with physical properties of PrP(Sc). *J. Mol. Biol.* **346**, 645–659.

Bucciantini, M., Giannoni, E., Chiti, F., Baroni, F., Formigli, L., Zurdo, J., Taddei, N., Ramponi, G., Dobson, C. M., and Stefani, M. (2002). Inherent toxicity of aggregates implies a common mechanism for protein misfolding diseases. *Nature* **416**, 507–511.

Burdick, D., Soreghan, B., Kwon, M., Kosmoski, J., Knauer, M., Henschen, A., Yates, J., Cotman, C., and Glabe, C. (1992). Assembly and aggregation properties of synthetic Alzheimer's A4/beta amyloid peptide analogs. *J. Biol. Chem.* **267**, 546–554.

Cardoso, I., Goldsbury, C. S., Muller, S. A., Olivieri, V., Wirtz, S., Damas, A. M., Aebi, U., and Saraiva, M. J. (2002). Transthyretin fibrillogenesis entails the assembly of monomers: A molecular model for *in vitro* assembled transthyretin amyloid-like fibrils. *J. Mol. Biol.* **317**, 683–695.

Chiti, F., Stefani, M., Taddei, N., Ramponi, G., and Dobson, C. M. (2003). Rationalization of the effects of mutations on peptide and protein aggregation rates. *Nature* **424**, 805–808.

Chiti, F., Webster, P., Taddei, N., Clark, A., Stefani, M., Ramponi, G., and Dobson, C. M. (1999). Designing conditions for *in vitro* formation of amyloid protofilaments and fibrils. *Proc. Natl. Acad. Sci. USA* **96**, 3590–3594.

Collins, S. R., Douglass, A., Vale, R. D., and Weissman, J. S. (2004). Mechanism of prion propagation: Amyloid growth occurs by monomer addition. *PLoS Biol.* **2**, 1582–1590.

Conway, K. A., Lee, S. J., Rochet, J. C., Ding, T. T., Williamson, R. E., and Lansbury, P. T. (2000). Acceleration of oligomerization, not fibrillization, is a shared property of both alpha-synuclein mutations linked to early-onset Parkinson's disease: Implications for pathogenesis and therapy. *Proc. Natl. Acad. Sci. USA* **97,** 571–576.

Crowther, D. C., Kinghorn, K. J., Miranda, E., Page, R., Curry, J. A., Duthie, F. A., Gubb, D. C., and Lomas, D. A. (2005). Intraneuronal Abeta, non-amyloid aggregates and neurodegeneration in a *Drosophila* model of Alzheimer's disease. *Neuroscience* **132,** 123–135.

Dahlgren, K. N., Manelli, A. M., Stine, W. B., Jr., Baker, L. K., Krafft, G. A., and LaDu, M. J. (2002). Oligomeric and fibrillar species of amyloid-beta peptides differentially affect neuronal viability. *J. Biol. Chem.* **277,** 32046–32053.

Dobson, C. M. (1999). Protein misfolding, evolution and disease. *Trends Biochem. Sci.* **24,** 329–332.

Dobson, C. M. (2001). The structural basis of protein folding and its links with human disease. *Philos. Trans. R. Soc. Lond. B Biol. Sci.* **356,** 133–145.

DuBay, K. F., Pawar, A. P., Chiti, F., Zurdo, J., Dobson, C. M., and Vendruscolo, M. (2004). Prediction of the absolute aggregation rates of amyloidogenic polypeptide chains. *J. Mol. Biol.* **341,** 1317–1326.

Fandrich, M., and Dobson, C. M. (2002). The behaviour of polyamino acids reveals an inverse side chain effect in amyloid structure formation. *EMBO J.* **21,** 5682–5690.

Fezoui, Y., Hartley, D. M., Harper, J. D., Khurana, R., Walsh, D. M., Condron, M. M., Selkoe, D. J., Lansbury, P. T., Jr., Fink, A. L., and Teplow, D. B. (2000). An improved method of preparing the amyloid beta-protein for fibrillogenesis and neurotoxicity experiments. *Amyloid* **7,** 166–178.

Garzon-Rodriguez, W., Vega, A., Sepulveda-Becerra, M., Milton, S., Johnson, D. A., Yatsimirsky, A. K., and Glabe, C. G. (2000). A conformation change in the carboxyl terminus of Alzheimer's Abeta (1-40) accompanies the transition from dimer to fibril as revealed by fluorescence quenching analysis. *J. Biol. Chem.* **275,** 22645–22649.

Harper, J. D., Wong, S. S., Lieber, C. M., and Lansbury, P. T. (1997). Observation of metastable Abeta amyloid protofibrils by atomic force microscopy. *Chem. Biol.* **4,** 119–125.

Helms, L. R., and Wetzel, R. (1996). Specificity of abnormal assembly in immunoglobulin light chain deposition disease and amyloidosis. *J. Mol. Biol.* **257,** 77–86.

Hrncic, R., Wall, J., Wolfenbarger, D. A., Murphy, C. L., Schell, M., Weiss, D. T., and Solomon, A. (2000). Antibody-mediated resolution of light chain-associated amyloid deposits. *Am. J. Pathol.* **157,** 1239–1246.

Ingenito, R., Bianchi, E., Fattori, D., and Pessi, A. (1999). Solid phase synthesis of peptide C-terminal thioesters by Fmoc/t-Bu chemistry. *J. Am. Chem. Soc.* **121,** 11369–11374.

Jarrett, J. T., Berger, E. P., and Lansbury, P. T., Jr. (1993). The carboxy terminus of the beta amyloid protein is critical for the seeding of amyloid formation: Implications for the pathogenesis of Alzheimer's disease. *Biochemistry* **32,** 4693–4697.

Kayed, R., Head, E., Thompson, J. L., McIntire, T. M., Milton, S. C., Cotman, C. W., and Glabe, C. G. (2003). Common structure of soluble amyloid oligomers implies common mechanism of pathogenesis. *Science* **300,** 486–489.

Kelly, J. W. (1996). Alternative conformations of amyloidogenic proteins govern their behavior. *Curr. Opin. Struct. Biol.* **6,** 11–17.

Khurana, R., Gillespie, J. R., Talapatra, A., Minert, L. J., Ionescu-Zanetti, C., Millett, I., and Fink, A. L. (2001). Partially folded intermediates as critical precursors of light chain amyloid fibrils and amorphous aggregates. *Biochemistry* **40,** 3525–3535.

Klein, W. L., Krafft, G. A., and Finch, C. E. (2001). Targeting small Abeta oligomers: The solution to an Alzheimer's disease conundrum? *Trends Neurosci.* **24,** 219–224.

Kokubo, H., Kayed, R., Glabe, C. G., and Yamaguchi, H. (2005). Soluble Abeta oligomers ultrastructurally localize to cell processes and might be related to synaptic dysfunction in Alzheimer's disease brain. *Brain. Res.* **1031,** 222–228.

Lambert, M. P., Barlow, A. K., Chromy, B. A., Edwards, C., Freed, R., Liosatos, M., Morgan, T. E., Rozovsky, I., Trommer, B., Viola, K. L., Wals, P., Zhang, C., Finch, C. E., Krafft, G. A., and Klein, W. L. (1998). Diffusible, nonfibrillar ligands derived from Abeta1-42 are potent central nervous system neurotoxins. *Proc. Natl. Acad. Sci. USA* **95,** 6448–6453.

Lashuel, H. A., Hartley, D., Petre, B. M., Walz, T., and Lansbury, P. T., Jr. (2002). Neurodegenerative disease: Amyloid pores from pathogenic mutations. *Nature* **418,** 291.

Lashuel, H. A., Hartley, D. M., Petre, B. M., Wall, J. S., Simon, M. N., Walz, T., and Lansbury, P. T., Jr. (2003). Mixtures of wild-type and a pathogenic (E22G) form of Abeta40 *in vitro* accumulate protofibrils, including amyloid pores. *J. Mol. Biol.* **332,** 795–808.

Lomakin, A., Teplow, D. B., Kirschner, D. A., and Benedek, G. B. (1997). Kinetic theory of fibrillogenesis of amyloid beta-protein. *Proc. Natl. Acad. Sci. USA* **94,** 7942–7947.

Luibl, V., Isas, M., Glabe, C., Kayed, R., Chen, J., and Langen, R. (2004). Similarities between AMD and amyloid diseases suggested by the presence of toxic, amyloidogenic oligomers in drusen. *Invest. Ophthalmol. Vis. Sci.* **45,** 1789.

MacPhee, C. E., and Dobson, C. M. (2000). Formation of mixed fibrils demonstrates the generic nature and potential utility of amyloid nanostructures. *J. Am. Chem. Soc.* **122,** 12707–12713.

Matsubara, K., Mizuguchi, M., Igarashi, K., Shinohara, Y., Takeuchi, M., Matsuura, A., Saitoh, T., Mori, Y., Shinoda, H., and Kawano, K. (2005). Dimeric transthyretin variant assembles into spherical neurotoxins. *Biochemistry* **44,** 3280–3288.

Niemeyer, C. M. (2001). Nanoparticles, proteins, and nucleic acids: Biotechnology meets materials science. *Angew. Chem. Int. Ed. Engl.* **40,** 4128–4158.

Nilsberth, C., Westlind-Danielsson, A., Eckman, C. B., Condron, M. M., Axelman, K., Forsell, C., Stenh, C., Luthman, J., Teplow, D. B., Younkin, S. G., Naslund, J., and Lannfelt, L. (2001). The 'Arctic' APP mutation (E693G) causes Alzheimer's disease by enhanced Abeta protofibril formation. *Nat. Neurosci.* **4,** 887–893.

Oddo, S., Caccamo, A., Shepherd, J. D., Murphy, M. P., Golde, T. E., Kayed, R., Metherate, R., Mattson, M. P., Akbari, Y., and LaFerla, F. M. (2003). Triple-transgenic model of Alzheimer's disease with plaques and tangles: Intracellular Abeta and synaptic dysfunction. *Neuron* **39,** 409–421.

O'Nuallain, B., and Wetzel, R. (2002). Conformational Abs recognizing a generic amyloid fibril epitope. *Proc. Natl. Acad. Sci. USA* **99,** 1485–1490.

Pepys, M. B., Hawkins, P. N., Booth, D. R., Vigushin, D. M., Tennent, G. A., Soutar, A. K., Totty, N., Nguyen, O., Blake, C. C. F., Terry, C. J., Feest, T. G., Zalin, A. M., and Hsuan, J. J. (1993). Human lysozyme gene mutations cause hereditary systemic amyloidosis. *Nature* **362,** 553–557.

Pfeifer, M., Boncristiano, S., Bondolfi, L., Stalder, A., Deller, T., Staufenbiel, M., Mathews, P. M., and Jucker, M. (2002). Cerebral hemorrhage after passive anti-Abeta immunotherapy. *Science* **298,** 1379.

Sanbe, A., Osinska, H., Saffitz, J. E., Glabe, C. G., Kayed, R., Maloyan, A., and Robbins, J. (2004). Desmin-related cardiomyopathy in transgenic mice: a cardiac amyloidosis. *Proc. Natl. Acad. Sci. USA* **101,** 10132–10136.

Shorter, J., and Lindquist, S. (2004). Hsp104 catalyzes formation and elimination of self-replicating Sup35 prion conformers. *Science* **304**, 1793–1797.

Soreghan, B., Kosmoski, J., and Glabe, C. (1994). Surfactant properties of Alzheimer's A beta peptides and the mechanism of amyloid aggregation. *J. Biol. Chem.* **269**, 28551–28554.

Tseng, B. P., Esler, W. P., Clish, C. B., Stimson, E. R., Ghilardi, J. R., Vinters, H. V., Mantyh, P. W., Lee, J. P., and Maggio, J. E. (1999). Deposition of monomeric, not oligomeric, Abeta mediates growth of Alzheimer's disease amyloid plaques in human brain preparations. *Biochemistry* **38**, 10424–10431.

Walsh, D. M., Lomakin, A., Benedek, G. B., Condron, M. M., and Teplow, D. B. (1997). Amyloid beta-protein fibrillogenesis. Detection of a protofibrillar intermediate. *J. Biol. Chem.* **272**, 22364–22372.

Williams, A. D., Sega, M., Chen, M., Kheterpal, I., Geva, M., Berthelier, V., Kaleta, D. T., Cook, K. D., and Wetzel, R. (2005). Structural properties of Abeta protofibrils stabilized by a small molecule. *Proc. Natl. Acad. Sci. USA* **102**, 7115–7120.

Yang, F., Lim, G. P., Begum, A. N., Ubeda, O. J., Simmons, M. R., Ambegaokar, S. S., Chen, P. P., Kayed, R., Glabe, C. G., Frautschy, S. A., and Cole, G. M. (2005). Curcumin inhibits formation of amyloid beta oligomers and fibrils, binds plaques, and reduces amyloid *in vivo*. *J. Biol. Chem.* **280**, 5892–5901.

Zagorski, M. G., Yang, J., Shao, H., Ma, K., Zeng, H., and Hong, A. (1999). Methodological and chemical factors affecting amyloid beta peptide amyloidogenicity. *Methods Enzymol.* **309**, 189–204.

Author Index

Subject Index

A

Acylphosphatase, native protein aggregation
   studies from *Sulfolobus solfataricus*
   aggregation condition selection, 87
   aggregation mechanism elucidation, 88–89
   apparent rate constant for aggregation
      determination, 88–89
   conformation analysis, 87–88
AFM, *see* Atomic force microscopy
Amyloid A, high hydrostatic pressure-
   induced dissolution of fibrils, 241
Amyloid-β
   aggregation, *see* Amyloid aggregates
   metastability, 22–23
   polydispersity, 22–23
   total internal reflection fluorescence
      microscopy of fibril growth, *see* Total
      internal reflection fluorescence
      microscopy
Amyloid aggregates
   amyloid cascade hypothesis, 218–219
   antibodies, conformation-dependent
      affinity purification, 337
      amyloid oligomer preparation, 329–331
      antibody generation, 334, 336
      applications, 337–341
      common targets in fibrillogenesis,
         326–329
      conformational specificity, 336
      nanoparticle molecular mimic of
         spherical amyloid oligomers
         antigen preparation and
            characterization, 333–334
         synthesis, 331–332
   assembly, 20
   diseases, 20, 218
   electron paramagnetic resonance, *see*
      Electron paramagnetic resonance
   elongation assay for modulator screening
      challenges, 314–315
      data analysis, 319
      fluorescence reading, 318–319

high throughput, 319
incubation conditions, 317–318
lead compounds, 320–322
plate preparation, 315–317
small molecule inhibitor discovery, 314
troubleshooting, 319–320
validation of results, 322–323
fibrillogenesis pathways, 274–275,
   325–326
high hydrostatic pressure studies,
   see High hydrostatic pressure
high-performance liquid chromatography-
   based sedimentation assay
   advantages, 70–71
   aggregate stock preparation
      aggregate weight concentration
         determination, 48
      amyloid-β(1-40), 46
      overview, 43–46
      polyglutamine aggregates, 47
   autosampler storage conditions, 52–53
   disaggregation of peptides and
      centrifugation
      amyloid-β peptides, 42–43
      overview, 37–41
      polyglutamine peptides, 41–42
   elongation kinetics analysis, 58–59,
      61–62
   elongation thermodynamics analysis,
      67–70
   nucleation kinetics analysis, 63, 65–67
   overview, 36–37
   standard curve construction, 49–52
   titration of fibril growing ends, 53–55,
      57–58
hydrogen/deuterium exchange studies,
   see Hydrogen/deuterium exchange
native proteins, *see* Native protein
   aggregation
nonfibrillar components, 199, 200
nuclear magnetic resonance of structure,
   see Nuclear magnetic resonance
nucleation, 92

369